125
130
135
140
35
30
25
20
120
PO-HAI
TUNG-HAI
Boundaries of Eastern Han
Boundaries of Western Han
Swamps
Major agricultural areas
Irrigation and transportation canals
NOTE: In Han times, rivers as parts of canal systems were considered canals.
(Adapted from Shih Nien-hai, *Shan-ho-chi*)

Han Dynasty China

VOLUME II: HAN AGRICULTURE

School of International Studies
University of Washington

HAN DYNASTY CHINA

Jack L. Dull, *Editor*

I. *Han Social Structure*, by T'ung-tsu Ch'ü
II. *Han Agriculture*, by Cho-yun Hsu

HAN AGRICULTURE

THE FORMATION OF EARLY CHINESE AGRARIAN ECONOMY (206 B.C.–A.D. 220)

By Cho-yun Hsu

EDITED BY JACK L. DULL

UNIVERSITY OF WASHINGTON PRESS
SEATTLE AND LONDON

Library of Congress Cataloging in Publication Data
Hsu, Cho-yun.
Han agriculture.

(Han Dynasty, China ; v. 2)
"Documents of Han agriculture": p.
Bibliography: p.
Includes index.
1. Agriculture—Economic aspects—China.
2. Agriculture—China—History—Sources. 3. China—History—Han dynasty, 202 B.C. - 220 A.D.—Sources.
I. Dull, Jack L. II. Title. III. Series.
DS748.H28 vol. 2 [HD2097] 931'.04s [338.1'0931]
ISBN 0-295-95676-3 79-4920

CONTENTS

SECTION III: The Farmer's Livelihood

SECTION IV: Farming Resources

ILLUSTRATIONS

Map

Plates

Figures

TABLES

EDITOR'S PREFACE

THIS volume surveys Han agriculture from essentially two perspectives, that of the government and that of the farmer; most attention is devoted to the latter. The work cannot be construed as economic history in the narrow sense of the term, for the materials from the Han period do not permit the kind of analysis that economic historians practice. We do not, for example, have a typical or average grain price for any of the four hundred years of the Han, to say nothing of local price fluctuations. But the volume does tell us much about the economy of the period and even more about the nature of agriculture during China's first sustained imperium.

The opening chapter introduces two aspects of the late Warring States period that provide bases for comparison throughout the volume: the urban economic activity—the commercialism—and the level of agricultural development. There is a persistent belief that the potential for economic development, particularly in the realm of commerce, that was such a noticeable feature of the Warring States period was forever stifled by the rise of the imperial bureaucratic system. There is some evidence for the validity of this belief, but the issue is highly complex and cannot be fully unfolded without looking at the entire economy of the Han period. The issue demands our attention, however, for it at least demonstrates a particular kind of imperial concern. The state of agricultural development in the Warring States period is crucial to an understanding of the great strides that were made during the Han. In the repertoire of crops, in technology (e.g., the use of iron implements), and in the development of irrigation, the Warring States period does no more than prefigure the first long-lasting dynastic era. Even though the pre-Han data are far less rich than the very limited data of the Han period, the evidence suffices to suggest a much simpler and more primitive level than that of the Han.

Chapter 1 introduces the Han government's attempts to deal with broad questions pertaining to agriculture. We begin with a survey of the demographic data, focusing on the Western Han period for which the data are comparatively rich. The evidence leads to the conclusion that the population growth rate was as high as 1.6 percent, which would have caused the total population to have doubled and perhaps redoubled during the Former Han. Even if our data have misled us somewhat, it is still likely that the population of the Former Han period at least doubled; severe demographic pressures in a few key areas would have created acute concern for population growth and concentration. Very early in the period, memorialists such as Chia I and Ch'ao Ts'o began to demand that the state take an active role in increasing agricultural production. The state, however, did not perceive the problem as a grave one until well over a century after these entreaties. Indeed, the response of the early emperors was little more than symbolic: admonitions to increase production, recognition of outstanding farmers, and so on. Farmers who left the countryside during the first few generations of the Han were probably attracted by the greater incomes possible in the cities. At this time land was still plentiful and although there were complaints that grain stores were inadequate, the movement of farmers to the cities suggests that grain prices were low. That is, farmers were not being driven from their land by engrossers who took it away from them. When the state did intervene by moving masses of people, it was at least as much interested in defense as in increasing production: farmers were moved to the frontier areas in the north and northwest, but not to the more fertile areas of the south. Finally, the government responded to the growing population pressures by opening up government parks and preserves. This supply of land was largely exhausted by the end of Western Han and throughout the Eastern Han period the government had to settle for much more modest relief measures. Attempts were made, most notably by Wang Mang, to redistribute land, but by that time the social forces had changed sufficiently that the government had to annul its land reform measures. In effect, the government admitted that it could not directly confront the rising power of landed elements.

The second chapter draws attention to a contradiction in imperial policies. On the one hand, beginning in the reign of Han Kao-tsu and culminating in the reign of Emperor Wu, the government legislated against entrepreneurs; thereafter, although much of the prejudicial legislation persisted throughout the remainder of the two Han dynasties, the government's maintenance of monopolies and its enforcement of anti-merchant laws was either weak or nonexistent. The state pursued these policies to prevent powerful economic interests from jeopardizing its own position. On the other hand, to the extent that the state was successful in dissuading people from investing in entrepreneurial activities it encouraged investment in land. Over the course of many generations the state inadvertently contributed to the development of landlordism. Any success it may have achieved in moving against merchants and manufacturers was probably offset in the long run by the creation of social forces over which it ultimately could not exercise adequate control.

Chapter 3 concentrates on farm households. Our translation of the *Ssu-min-yüeh-ling* provides, for the first time in English, a detailed description of the manifold array of activities of a well-to-do household. At the opposite end of the socioeconomic spectrum we have attempted to create a sketch of the major expenditures of a farm family. The picture that emerges, although lacking in detail, suggests that the renowned statements of such memorialists as Tung Chung-shu and Ch'ao Ts'o are so hyperbolic as to be atypical, at least for the first half of Former Han and probably for much of that period. For the level of tenancy was quite low, although undoubtedly on the rise throughout the two hundred years of Later Han. At both the top and the bottom of society, farmers regularly engaged in marketing activities, so long as disorder in the society did not prevent them from doing so. Thus in a society in which the state attempted to suppress entrepreneurial activities, the farmer was dependent upon the marketplace for a significant portion of his income.

Initially, the Han age was well endowed with land, relative to the population. As late as A.D. 2, there were 2.27 acres of land under cultivation per capita, whereas in North China in ca. 1400 the comparable figure is 1.72 acres.[1] But as population increased and as landlordism slowly grew, there were powerful incentives for the Han farmer to increase his productivity. Chapter 4 discusses the plants that the Han farmer cultivated, soil conditions and how he modified them, and governmental and private investment in irrigation. All three of these factors figured prominently in the intensification of farming.

1. These figures are computed from data in tables B.2 and B.5 (pp. 220 and 225) in Dwight H. Perkins, *Agricultural Development in China 1368-1968* (Chicago: Aldine Publishing Company, 1969).

Chapter 5 delves into the methods and techniques of Han agriculture. The government played an active role by encouraging new planting patterns and the use of the plow, but private initiative was also important in both wet and dry agriculture. The development of high quality wrought-iron tools also contributed to more efficient farming. These together made possible a general rise in the standard of living; although the comparison is gross, for we are forced to deal with two periods of two centuries each, the Eastern Han farmer was better off than his Western Han ancestor.

As noted earlier, the Han farmer was involved in a marketing network; he was not a simple, self-sufficient peasant. Chapter 6 surveys his nonagricultural economic activities. It goes beyond that in noting that when marketing patterns were disrupted or when the farmer was faced with major problems due to drought or flood, he had several options, none of which was taken lightly. He could migrate, either on a short-term basis out of the disaster area or move permanently to another part of the empire. There is a striking contrast between government-sponsored and voluntary migration. The state sought to populate the northern and northwestern frontiers and its aim was largely strategic. But when the Han people resettled themselves they generally moved to the south, to an area rich in land where during much of the period primitive slash-and-burn agriculture was practiced. As a last resort and in areas where the marketing system was less well developed, the farmer would instigate or participate in uprisings as a means to alter the socioeconomic forces that he otherwise could not control. The state, as one aspect of its attempts to restore order, availed itself of the *t'un-t'ien,* garrison farm, system that was originally designed to decrease the amount of grain that had to be shipped to the frontier by having farmer-soldiers produce their own. In periods when the government was restoring order and increasing agricultural production, as at the beginning of the Later Han and of the Three Kingdoms periods, these agriculture-military institutions were established inside China. Their efficacy is seen in the trend to extend their activities from grain production to restoration of normal marketing activities.

Professor Hsu highlights these major themes in the Epilogue. He reminds us again of the extent to which the Han farmer integrated his activities with market forces.

We have provided translations of the essential documents on which this study is based. Two comments must be made concerning the documents. First, in spite of the difficulty in understanding some of the texts and in spite of the consternation expressed by commentators regarding some of the passages, we have attempted to minimize the philological apparatus that accompanies the translations. The expert in the field will have no difficulty in consulting the original texts and we assume that he will be able to determine why our translation reads as it does. Second, we have made no attempt to mine the documents for all they are worth. For example, there are materials on religion and medicine that we have included because they are parts of our documents, but which we felt did not warrant analysis for our purposes. By making the materials available, we trust that we have made easier the task of analyzing those materials and we hope that others will find them useful.

We have not avoided speculation when shortage of data would have forced us otherwise not to mount an argument. Speculation at least forces one to address the issues and opens the way for alternative formulations. We have, however, adequately warned the reader when we are speculative. We would not be disappointed by other interpretations of greater validity.

Many people have helped me in my work on this volume; the list would be long were I to record all of them. Since they have been personally thanked already, I shall limit my public expression of appreciation to only one person: Margery Lang. She has labored long on the manuscript and she deserves immense credit for the final work. The Ford Foundation provided funding in the early stages of this project, and we are fully appreciative of that support when it was most needed to initiate the work.

JACK L. DULL, *Editor*
University of Washington
September 1979

PREFACE

IT is more than a decade since Professor Dull approached me to ask about contributing to the Han history project and now the volume on Han agriculture is finally complete. During these intervening years I have, not without some turbulence and excitement, passed several landmarks in my own life: marriage, birth of a son, change of jobs, moving to the United States—all these, as well as intermittent frustrations resulting from my deep involvement in political and academic affairs in Taiwan, from time to time have interrupted the progress of my research on the development of Han agriculture. Nevertheless, the main cause of delay has been the progress of my own intellectual grasp of the material, during the course of which a mass of facts and information was gradually transformed into a work with a definite theme interpreting a certain phenomenon in Chinese history.

Shifting my base of operations to Pittsburgh, I found the atmosphere in the Department of History extremely stimulating. Many of the topics touched on in those evening Peasant Studies Group meetings aroused questions relevant to the development of Chinese agrarian culture, not only in the Han period but also in the later stages of Chinese history. One of the central questions is the apparent persistence of intensive farming in small farmsteads. The term "intensive," of course, should always be regarded as relative to the conditions prevalent at a given time and place. Rich archaeological findings of recent years have provided information that was previously unavailable and also often presented me with compelling reasons for reworking certain parts of already written sections and thus further delaying completion of a final draft. At any rate, I now regard this work as not merely an account of farming and farmers' lives in Han China, but rather an attempt to answer questions dealing with population, land tenure, physiocratic ideology, and government intervention in commerce and manufacturing. Had the scope of this volume not been defined as part of a series of Han studies, I would have been tempted to cover even more ground.

I take this opportunity to acknowledge help from friends and colleagues at various stages in my research and writing. May I express gratitude to Jack L. Dull, editor of the series, who has put much time and effort, no less than mine, into editing the draft. His comments were frequently inspiring. I am grateful to Margery Lang, whose meticulous care on details was helpful in preventing errors, to artist Cynthia Louie for the maps and figures, and to the staff of the publications program of the School of International Studies at the University of Washington, all of whom participated in the production of the book.

Professor Kan Lao, my beloved teacher, has been my inspiration and guide ever since I was a sophomore in his class in Ch'in-Han history at National Taiwan University. This volume is honored by his elegant Chinese calligraphy on the book jacket. I have also benefited at various stages of research and writing from the advice, inspiration, and encouragement of many learned scholars, including Gary Allinson, Kang Chao, Mu Ch'ien, Han-sheng Ch'üan, H. G. Creel, Han-kwang Mao, Sadao Nishijima, Dwight Perkins, Evelyn Rawski, Thomas Rawski, Julius Rubin, T. H. Shen, Beatrice Spade, Richard Smethurst, Hsi-sheng T'ao, Martin Wilbur, C. K. Yang, Keng-wang Yen, and Ying-shih Yü, to mention only a few. I must thank at least three groups of scholars who have influenced the shaping of my thought: the Peasant Studies Group at the University of Pittsburgh, members of the Columbia University Seminar on Traditional China, and my colleagues at Academia Sinica.

I have used facilities and materials at the Fu Ssu-nien Library of the Institute of History and Philology, Academia Sinica, the Harvard-Yenching Library of Harvard University, Far Eastern Library of the University of Chicago, East Asian Library at the University of Pittsburgh, and the National Historical Museum at Taipei. I am grateful for the kind assistance of Dr. T. H. Tsien, Dr. Eugene Wu, Dr. Thomas Kuo, Mr. Hao-t'ien Ho, and their staffs at these institutions. Without support from administration and staff of the Department of History, Asian Studies Program, and the University Center of International Studies at the University of Pittsburgh, this volume could not have been finished. I thank them for their valuable services.

Finally my love to Man-li and Leo (Lo-p'ung) for their understanding and affection, from an absent-minded husband and father who during many an hour was preoccupied with the problems of a farmer some two thousand years ago.

This volume is sincerely dedicated to Chinese farmers in the past as well as the present day. It is these unsung heroes, toiling in the fields generation after generation, who created one of the most sophisticated farming systems in human history.

CHO-YUN HSU

Part One: Analysis of Han Agriculture

INTRODUCTION

THIS volume investigates the development of agriculture in the Han dynasty (206 B.C.-219 A.D.). As it was in the Han dynasty that the unification of China was consolidated after the brief rule of Ch'in, and in which the direction of Chinese civilization seemed to have been oriented, it is significant for us to pose the inquiry on one of the very essential phenomena that Chinese history has presented—a prolonged agrarian civilization. It appears that once the Chinese stepped upon the path of agriculture, they stayed on that course without deviation. Progress and change occurred from time to time, but agriculture remained paramount in the Chinese way of life. It is not a historical certainty, however, that China should have her development so oriented. In the years of turmoil from the fifth to the third century B.C., there was the strong possibility of developing a predominately urban-centered economic life rather than a rural-based agrarian economy. Large and prosperous market centers flourished and the urban mentality of profit making and of contractual reciprocity prevailed; both conditions would be fertile ground for the blossoming of commercialism.[1] Even within its first hundred years, the imperial power of the Han dynasty was struggling against influential urban leaders who, with their social strength, threatened political authority.[2]

1. Cho-yun Hsu, *Ancient China in Transition*, pp. 116-23, 134-38.

2. Cho-yun Hsu, "The Changing Relationship between Local Society and the Central Political Power in Former Han," *Comparative Studies in Society and History* 7, no. 4 (1965): 385 ff.

It is my intention to single out the key factors in the Han period that effected China's transformation to an agricultural economy. It is therefore relevant to look backward to the agriculture prior to Han, that of the Chan-kuo.

Significant in the development of agriculture in the Chan-kuo period is the popularization of iron implements. In recent years, excavations at ancient settlements and tombs have unearthed many iron tools, and molds for iron casting have been found at the site of blacksmith foundries. Shapes varied a little from one region to another, and various names were attributed, but they were, in broad categories, spades, hoes, sickles, plows, knives, axes, and chisels. Their shapes and sizes indicate their use at different stages of cultivation. The number of hoes and spades recorded in the archeological reports reveals that weeding and soil crumbling probably required most of the farmer's time between planting and harvest. In other words, the Chan-kuo iron implements are evidence of the existence of the beginning of intensive farming: the iron axes and chisels could be very useful in the felling of trees to open new land; the iron plowshare probably means the beginning of animal-hauled tilling, a big step forward from hand tilling with digging sticks. Although iron implements are sharper and sturdier than those of wood or stone or sea shells, the Chan-kuo iron implements were cast in molds and were relatively small and light. The iron plowshare that was discovered at Hui-hsien is a small V-shaped iron piece attached to a wooden body. With no ridge and with an obtuse angle, this plowshare, which served only as the cutting edge of a wooden plow, could not plow deeply. Therefore, even though Chan-kuo farming was intensive, and possibly used animal power, it reached only a relatively low level in comparison with farming in the Han period.[3]

Irrigation systems were built in the Chan-kuo period, some of them on quite a large scale. Yet they were small in comparison with the Ch'in-Han irrigation and water-control projects, and existed only in the valleys of the Wei, the Tsang, the Shansi rivers, and in a small portion of the Huai River valley. For instance, the earliest Chan-kuo irrigation system of much size in the eastern part of the Yellow River valley is the T'ien-chin-yen (built about the end of the fifth century, B.C.). It covers an area of no more than twenty *li* along the Tsang River. The Cheng Kuo-ch'ü project, conducted by a Cheng engineer in the state of Ch'in, is more than ten times the size of T'ien-chin-yen.

3. Huang Chan-yüeh, "Chin-nien ch'u-t'u-ti Chan-kuo Liang-Han t'ieh-ch'i," *K'ao-ku hsüeh-pao,* 1957, no. 3. Cf. Hsü Cho-yun, "Liang-Chou nung-tso yü chi-shu," *Bulletin of the Institute of History and Philology* 42, pt. 4 (1971): 809 ff.

The Han irrigation project Pai-ch'ü, however, flowing a length of some two hundred *li,* was built in 95 B.C. in the Ching-Wei valley.[4] Extensive construction of large-scale irrigation projects was not undertaken until the Ch'in-Han period, and especially in the Han dynasty. Those efforts in the Han period led to the expansion of arable land labeled by Kimura as the formation of "secondary farming land," a condition he related to the formation of the Ancient Empire.[5] Water sources for farming in pre-Han times, as archeological evidence shows, were more likely from small ponds and wells in the vicinity of the field than from large-scale irrigation systems.[6] It should be noted, nevertheless, that the Chan-kuo irrigation undertakings, although less in both scale and scope than those in the Han, did pave the way for the Han projects to follow.

The main pre-Han cereal foodstuffs were millet *(Setaria italica),* panicled millet *(Panicum miliaceum),* wheat, and rice. Both millet and panicled millet are indigenous to China and had been the major foodstuffs for the Chinese since Neolithic times. It was not until the Chan-kuo period that wheat and *Setaria italica* gained their positions as basic subsistence. Soybeans also became popular then, and were especially important as a crop for the mountainous areas. Rice was restricted to the south, particularly in the Yangtze River valley where the swampy terrain provided natural paddy fields with an ample supply of water.[7] Therefore all the major crops in North China, which was then China proper, were dry-farming plants. The types of crops explain why irrigation was not as crucial a need as it is in rice cultivation.[8]

Several principles that were to remain significant in Chinese intensive farming emerged in the Chan-kuo period. One of these is the stress on timing. The Confucian ideal government as presented by Mencius is a government that should never deprive people of the possibility of cultivating the land *on time.* Mencius once said that prompt action in the fields should count more than the possession of sharp, good implements. He repeatedly said that if the proper time were not missed in cultivating one hundred mu, a household of eight

4. Yu Yü, "Kuan-tzu Tu-ti-p'ien t'an-wei," *Nung-shih yen-chiu chi-k'an* 1: 1-3. An ancient *li* is equivalent to 1364.1 feet, Nancy Lee Swann, *Food and Money in Ancient China,* p. 362.

5. Kimura Masao, *Chugoku kodai teikoku no keisei,* pp. 146 ff.

6. Hsü, "Liang-Chou nung-tso yü chi-shu," pp. 815-16.

7. Ibid., pp. 803-8. Also, Amano Motonosuke, *Chūgoku nōgyōshi kenkyū,* pp. 3-100.

8. Ping-ti Ho, *The Cradle of the East,* pp. 45-60.

mouths need not worry about starvation, and that a country would have an abundance of grain, more than enough for consumption. Hsün-tzu, in many ways an ardent critic of Mencius, agreed with him on the importance of timing. He said that the five kinds of grain would follow each other when the proper times for cultivation in spring, weeding in summer, harvesting in fall, and storage in winter were not missed.[9] In North China, where frost comes early and moisture evaporates rapidly in the spring, one indeed has to seize the time and act quickly for the plant to make the most of its time.

The ancient people also stressed the necessity of deep plowing. It should be understood, of course, that *deep plowing* in ancient times could not be measured by modern standards. A more adequate translation for the Chinese phrase *shen-keng* probably is *deep tilling.* In the ancient documents deep tilling is normally followed by a phrase that means "thorough preparation of the soil by crumbling earth clods and weeding the soil free of grasses." In North China the dry loess soil would easily lose its moisture if the topsoil, once turned over, were not crumbled to fine-grained particles that would cover the surface of the tilled field. Meanwhile, the seed planted deep in the soil would be benefited by the moisture siphoned from underground water and be nourished by mineral and organic materials.[10] Ground dressing, prolonged through the whole growing period, consists of weeding, turning, piling soil around the plant, and deepening the shallow drainages in furrows. It demands the full-time attention of the farmer, and Chinese farmers spent more labor and time on this sort of work than on the pre-planting tillage and the harvest put together. The preponderance of spades and hoes among the excavated Chan-kuo implements strongly suggests that ground dressing with the use of these tools was the most important aspect of their agriculture.

The utilization of fertilizers was common in the Chan-kuo period. Possibly as early as the Shang dynasty, the most usual fertilizers were animal and human night soil. The character denoting fertilizer was precisely the same character for dung. Mencius said that one hundred mu of land, with fertilizer, and tilled by a first-rate farmer, would

9. P'an Hung-sheng and Yang Ch'ao-po, "Chan-kuo shih-tai-ti liu-kuo nung-yeh sheng-ch'an," *Nung-shih yen-chiu chi-k'an* 2:58. The citation of Mencius and Hsün-tzu are from James Legge, trans., *The Chinese Classics* 2: 130-31; 135-36, 183, and *Hsün-tzu,* SPPY ed., 5:8 (hereafter cited as HT).

10. P'an and Yang, "Chan-kuo shih-tai-ti liu-kuo nung-yeh sheng-ch'an," pp. 58-59. Cf. Ping-ti Ho, "The Loess and the Origin of Chinese Agriculture," *American Historical Review* 75, no. 1 (1969): 1 ff.

be good enough to feed nine mouths, but that in years of natural calamity, even fertilization would not make sufficient crops. Hsün-tzu and Han Fei interpreted one phrase in the *Lao-tzu* that mentioned horse dung in a discussion on using fertilizer and irrigation, both of which would increase wealth through agriculture.[11]

As early as the time of the *Book of Odes,* people had prayed to deities to get rid of locusts and other harmful insects. Fire, however, is also mentioned in the poem of prayer so that it is possible that people gathered and burned insects or that the farmers burned the fields before tilling the soil.[12] For centuries the Chinese farmer relied entirely upon his own hands when he was doing the ground-dressing work to free the soil of insects.

Farming in the Chan-kuo period is well summarized in three chapters included in the *Lü-shih ch'un-ch'iu,* an encyclopaedic work compiled by scholars sponsored by the Ch'in state chancellor Lü Pu-wei. Three chapters, "Jen-ti," "Pien-t'u," and "Shen-shih" were probably written by scholars who perhaps lacked firsthand experience in the fields for the information given is somewhat fragmented. Nevertheless, such knowledge in all likelihood represented the most established techniques so that even the theoretician would lay them down as basic farming principles. The text is so brief and some parts are so corrupted that it has required much effort by classical scholars and agricultural historians to restore its meaning.[13]

The text of "Jen-ti" ("On Land") starts with ten questions allegedly posed by Hou-chi, the legendary cultural hero of agriculture. Freely translated, "Can you transform the wet lowland to high? Can you transform the dry land to moist? Can you level the field so that it will be evenly covered by water? Can you make the plant grow well with adequate moisture? Can you stop the spread of undersized grasses and weeds? Can you give plants in your field good ventilation? Can you make the plant well-branched and strong-stemmed? Can you

11. P'an and Yang, "Chan-kuo shih-tai-ti liu-kuo nung-yeh sheng-ch'an," pp. 59-60. The citations are from Legge, *Chinese Classics* 2:376, 241; *HT* 6:2; *Han Fei-tzu*, SPPY ed., 6:7 (hereafter cited as *HFT*).

12. *Mao-shih cheng-i*, SPPY ed., 14A:9b-10a (hereafter cited as MSCI).

13. The most eminent scholars involved in such tasks include Yin Chung-jung, Wang Shu-min, Wan Kuo-ting, Hsia Wei-ying, Ōshima, and others. Cf. Yin Chung-jung, *Lü-shih ch'un-ch'iu chiao-shih;* Wang Shu-min, *Lü-shih ch'un-ch'iu chiao-pu;* Wan Kuo-ting, "Lü-shih ch'un-ch'iu-chung-ti keng-tso yüan-li," in *Chung-kuo nung-hsüeh-shih,* chap. 4, and his "Lü-shih ch'ung-ch'iu-ti hsing-chih chi ch'i tsai nung-hsüeh-shih-shang-ti chia-chih," *Nung-shih yen-chiu chi-k'an* 2: 175-84. Ōshima Riichi, "Ryo-shi shunjū nōgyō shihen ni mieru nōgyō gijutsu ni tsuite," *Shirin* 49, no. 1 (1966): 57 ff.

make the spikes large, strong, and even? Can you make the grain round, and the husks thin? And can you make the cereal tasty and nourishing as food?"[14] These questions reveal the main concerns of a farmer at the end of the Chan-kuo period: problems ranging from field preparation, ground dressing, plantation, and moisture preservation, to the water supply to the plant. Yet the optimum conditions still relied upon prompt action. Hence, the whole chapter of the "Shen-shih" ("On Timing") is devoted to the discussion of morphological distinctions between the crop that grows "on time" and that which does not. The crops mentioned were millet, rice, panicled millet, hemp, beans, and wheat. In summary, it is pointed out in the "Shen-shih" that more is harvested from the crop growing at the proper time than from either the prematurely or the tardily planted one. The timely farmer produces more grain that weighs heavier and tastes better and is more nutritious than those that missed the proper time. The text gives a generally concise description of variations in six kinds of crops, indicating that the crop that was planted too early tended to grow stems and leaves instead of grain while the one planted too late permitted no time for the grain to grow fully.[15]

The chapters of "Jen-ti" and "Pien-t'u" ("On Soil") supplement each other in many ways. The general problems discussed are preparation of the field and modification of the soil. Immediately following the ten questions, "Jen-ti" continues with the statement that the farmer could actually change the condition of the field to provide plants with better ground. "The general principles: the hard soil will be made tender, the loose will be solidified, the exhausted will be restored, the rested will be put in use, the lean will be enriched, the overrich will be reduced, the hard will be softened, the soft will be hardened, the dry will be moistened, the wet will be dried."[16] These principles, of course, involve mainly soil modification and as fertilizer is one of the means to modify the soil, it is made explicit in the "Jen-ti" chapter that the land can be fertilized by adding manure together with moisture, but that this alone will not create wholly favorable conditions: useful ground dressing also has certain effects, and frequent plowing is relevant to making the soil fertile and soft. Plowing five times and weeding and dressing five times, carefully done, can make the plant take deep root so that it reaches the moist soil beneath and

14. Yin, *Lü-shih ch'un-ch'iu chiao-shih,* p. 161. The translation, however, is largely based on the interpretation of Wan Kuo-ting, "Lü-shih ch'un-ch'iu-chung-ti keng-tso yüan-li," p. 89.

15. Yin, *Lü-shih ch'un-ch'iu,* pp. 164-65.

16. Ibid., p. 161.

avoids the problem of long weeds and insects, since the deep soil normally contains neither.[17] One must first till the hard, claylike soil because the moisture contained there is difficult to preserve.[18] Much labor indeed is required in such farming even before the seed is sown.

The author of the "Pien-t'u" chapter warned that an otherwise good crop could be damaged in three ways: (1) by lacking order in arrangement, causing the plants to cluster together and to deprive each other of the chance to grow; (2) by wasting land by allowing more space for furrows than for plants, and (3), by allowing weeds to rob the field of fertility. It is then suggested that plants should be aligned along parallel lines with a width of one *ch'ih* ("foot") between rows. If the field was on high and dry land the crop should grow in furrows so to enjoy the protection of mounds on both sides and the moisture at the bottom of furrows. If the land was low and wet the crop should be planted on the top of the mound between the furrows, to avoid too much water. The advantages were good ventilation and ample growing space for the crop, in addition to adapting the planting of the field to the location. The text included another tip, that a good crop depended upon the proper density of plants. When more than three plants grew in a cluster, the smaller one should be uprooted to give the stronger more space and nutrients.[19]

It is difficult to say, of course, if such advanced farming was in practice everywhere in the Chan-kuo period. Nevertheless, the scholars who wrote such passages abstracted their theory from real practice.

All of these texts on farming techniques indicate that Chan-kuo agriculture was moving toward intensive farming, which the Chinese farmer has continued to develop and refine to this day. Such intensive farming, even at the relatively rudimentary level of Chan-kuo agriculture, requires tremendous labor. Meanwhile, such ancient documents as the *Mencius* and the *Hsün-tzu* were both written with the assumption that the standard landholding of one farming household was one hundred mu (4.764 acres).[20] This household was presumed to include the aged parents, the couple, and four children—eight mouths in one house. Moreover, it is important to understand that the household would grow all its own vegetables, raise its own poultry and pork for meat, and weave its own silk (and, possibly more important, hemp). In the ideal well-field system advocated by Mencius, he

17. Ibid.

18. The text (ibid., p. 162) is hopelessly corrupted. The interpretation here is based on Wan, "Lü-shih ch'un-ch'iu," p. 95.

19. Yin, *Lü-shih ch'un-ch'iu*, pp. 162-63.

20. Swann, *Food and Money*, p. 361.

visualized a basic farming commune consisting of eight households, each of which would receive from the lord one hundred mu of farming land and five mu of land for residence and for growing mulberry trees and rearing household animals. The eight households would cooperate to grow grain in one hundred mu of the lord's reserved share as corvée.[21] This idealized land system partly represents the Mencian utopia of leveled wealth, and partly reflects a basic principle of ancient Chinese feudalism—that land was a granted privilege and labor service was a form of repayment.[22] What most deserves our attention is the obvious implication of a small farm operation within one small household that does not go much beyond the stem family. The institution appears quite distinctive from the serfs described in the Western Chou classic, the *Book of Odes,* who, though living with their own families, received provisions, including clothes and possibly also foodstuffs, from the lord, and who cultivated as teams together in hundreds and thousands in the field of the lord supervised by the lord's agent. The crops grown went to the lord's granary.[23] If this were the case in the Western Chou, the Chou feudal lord was still perhaps playing his role as military colonizer and conqueror who utilizes the local population as serfs and who collected crops as his income, part of which, of course, would be turned over to the granary of his own lord master.[24] By the Chan-kuo period, the feudal states had been replaced by territorial monarchical states; the independent farmers had appeared when the feudal society had disintegrated and no feudal lord could hold a manor and serfs. The emergence of independent farmers can be attributed to several factors, among which are the opening of nonarable land by the use of iron implements, and the fall of old feudal aristocrats through intra- and interstate power struggles. The monarch became the only lord above all his subjects, whether they were officials or farmers.[25]

Although free of feudal aristocrats, independent farmers in the Chan-kuo period were not free of state control. When Mencius suggested

21. Legge, *Chinese Classics* 2: 244-45.

22. Hsu, *Ancient China in Transition,* pp. 195-96.

23. Ibid., pp. 8-14. Shirakawa Shizuka, "Shikyō ni mieru nōjishi (jo)," *Ritsumeikan bungakū* 1, no. 138 (1956): 18-45.

24. Wolfram Eberhard, *Conquerors and Rulers,* pp. 33 ff. However, I disagree with Prof. Eberhard on the theme of ethnic difference between the Chou conquerors and the conquered. I believe that the theory presented by Prof. Eberhard is still valid with regard to the Chou being military conquerors, superimposed over the conquered, but that both the conqueror and the conquered belonged to the same ethnic stock.

25. Hsu, *Ancient China in Transition,* pp. 78 ff.

his ideal well-field program, he seemed to take for granted that the state had legitimate authority and the power to alter the size of landholdings and make them into equal lots and with the authority that a Chan-kuo monarch had, almost anything could be accomplished by coercion. If the T'eng ruler, to whom Mencius tried to sell his idea, adopted the suggestion, it is very likely that he could have done it, for the Chan-kuo ruler did have certain land at his direct disposal. The *Mencius* includes mention of some agriculturists, Hsü Hsing and his disciples, who came to the same court when Mencius was there and pleaded with the ruler to be granted a piece of land and to be accepted as citizens.[26]

In recent years scholars have hypothesized that the Chan-kuo rulers did have full control of some land that was labeled "public land" *(kung-t'ien).*[27] According to Masubuchi Tatsuo, the patriarch who led a clan to establish a fortified city state might reserve one certain area outside of the city for his exclusive use. Although the large uncultivated land areas in the vicinity should belong to the collective body of the city (or, more precisely, the whole ruling group of the city), designated areas might not be very "exclusive" at all at the beginning. For instance, the *Mencius* contained a dialogue between Mencius and King Hsüan of Ch'i in which Mencius mentioned that the park of King Wen, the first king of the Chou dynasty, was open to the public for gathering firewood and small game, while the park of King Hsüan of Ch'i was completely closed. This distinction indicates the gradually increasing monopoly of public land by the monarch as his power grew.[28] The ancient sources give the names of many such "parks," some of which were tremendously extensive. The Yün-meng swamp of Ch'u, for instance, covered the whole lake area in the middle section of the Yangtse River. These so-called parks perhaps included not only the reserved game park, but also whatever was not cultivated because of unfavorable terrain–mountains, woods, swamps, and lowlands. It would be difficult to open such waste land with implements made of stone, wood, or shell, but when iron became economically available, felling huge trees to clear woods or digging trenches to drain excessive water from a swamp would be feasible. By the time of Mencius, farmers such as Hsü Hsing and his followers traveled with their own

26. Legge, *Chinese Classics* 2: 246.

27. The most convincing argument was presented by Masubuchi Tatsuo; see his "Sen-Shin jidai no sanrin sōtaku to Shin no kōden," *Chūgoku kodai no shakai to kokka,* pp. 279 ff. Ho Ch'ang-ch'ün, "Ch'in-Han-chien ke-t'i-hsiao-nung-ti hsing-ch'eng ho fa-chan" in *Han-T'ang-chien feng-chien t'u-ti-so-yu-chih hsing-shih yen-chiu.*

28. Masubuchi, "Sen-Shin jidai no sanrin sōtaku to Shin no kōden," pp. 295 ff.

implements.[29] On the other hand, King Hui of Liang, a contemporary of Mencius, granted land in his Feng-chi swamp to the people, according to a fragment of an annal.[30]

It is interesting that the same person, King Hui of Liang, once expressed his concern that his population was not increasing.[31] Putting all these bits of scattered information together, it is logical to surmise that there was plenty of uncultivated land and some shortage of labor. The pacifist thinker Mo-tzu confirmed this when he pointed out that any major state, a state of ten thousand vehicles, had too much land for its own population.[32] The *Chan-kuo-ts'e,* however, contained information conflicting with these two reports. In the state of Wei (Liang), it is said that the land had been so full of houses and cultivated fields that there was no space for pastures and that the traffic of people and vehicles looked much like the movement of large bodies of troops.[33] Yu Yü tried to explain such a contradiction by suggesting that the population in the Chan-kuo states, especially the six eastern ones including Wei, was too great on the already cultivated land, yet still relatively small for the large uncultivated areas.[34] If people had to live in crowded areas before they were allowed to migrate to where they could be granted new land to start a new life, a farmer's only solution would be the maximum utility by whatever means of the small patch of land at his disposal. The result is intensive farming. Therefore, the Chan-kuo period marked the inception of the development, at least in the well-developed areas, of Chinese agriculture toward intensive farming. It was a natural consequence of institutional restriction and technological limitation, as well as the result of the pressure of a large population on a limited amount of land.

Allowing land to lie fallow, as had been the custom, became impossible as the pressure to farm intensively increased, and people turned to the cultivation of leguminous crops to restore field fertility. As a consequence, legumes became one of the most common foodstuffs—if Fan Sheng-chih's information reflects the Chan-kuo situation, 40 percent of the productive acreage would have been in leguminous crops.[35]

29. Legge, *Chinese Classics* 2: 246-47.

30. *Han-shu pu-chu,* I-wen reprint ed., (hereafter cited as *HS*) 28A: 73.

31. Legge, *Chinese Classics* 2: 129.

32. *Mo-tzu,* SPPY ed. 5: 3 (hereafter cited as *MT*).

33. *Chan-kuo-ts'e,* SPPY ed. 22: 3 (hereafter cited as *CKT*).

34. Yu Yü, "Yu Hsi-Chou tao Ch'ien-Han-ti keng-tso chih-tu yen-ke," *Nung-shih yen-chiu chi-k'an* 2: 7-8. He estimated a population of some twenty million in all of Chan-kuo China.

35. On the significance of legumes, see for instance, *MT* 2:4, and Legge,

Multicrop rotation was also adopted in lieu of fallowing. The *Lü-shih ch'un-ch'iu* mentioned that after tilling five times, weeding five times, and after deep plowing, one could expect a good millet crop one year, and a good wheat crop in the next.[36] By rotating crops one after another, the land would not be exhausted, even though it was given no fallow period.

Parallel to the technological change in agriculture in the Chan-kuo period were the socioeconomic changes. Political reforms experimented with in several states in one way or another also affected agriculture.[37] The reform with the most far-reaching consequences, of course, was carried out by Lord Shang, or Shang Yang, in the state of Ch'in. He, by rearranging the local administrative units, delivered people from the feudal structure, and a society under the control of a strong territorial state monarchy emerged.[38] The official recognition of private land ownership, in fact, signaled the significance of the free, independent small farmer.[39]

The reward of rank for military merits lifted the Ch'in farmer, who was also a soldier in military service, to a better social status, a level between the real aristocrats and the old-day serfs who belonged to the feudal lord.[40] Meanwhile, Ch'in was the only state that made full and systematic use of its public land to attract foreign laborers to increase national growth of agricultural produce. This was the main concept of Lord Shang's strategy, to free Ch'in's own labor for military duties outside of Ch'in while foreign laborers replaced them for production.[41] The new immigrants to Ch'in, although inferior socially to the Ch'in citizens, who could advance their social status through military merits, still constituted a new kind of farmer, one whose newly cultivated land was free from the bondage of the feudal manor. Such free small farmers owned the land that they were tilling.[42]

It seems that when Ch'in Shih-huang-ti, the first emperor of Ch'in, had a stele inscribed with the comment that the free farmers who were no longer feudal serfs were able to make themselves wealthy

---

Chinese Classics 2: 463. For the estimation of acreage for legumes, see Yu Yü, "Yu Hsi-Chou tao Ch'ien-Han-ti keng-tso chih-tu yen-ke," p. 9.

36. Yin, *Lü-shih ch'un-ch'iu,* p. 161.

37. Hsu, *Ancient China in Transition,* pp. 92 ff.

38. Ibid., pp. 93-94, 112, 117.

39. Ho, "Ch'in-Han-chien," pp. 88 ff.

40. *Shang-chün-shu,* SPPY ed., 5: 1a-2b.

41. Ibid., 2:1-4. Miao Ch'i-yü et al., "Shang-chün-shu shih-tai-ti she-hui pien-ke ho nung-yeh pien-ke," in *Chung-kuo nung-hsüeh-shih,* chap. 3.

42. Hiranaka Reiji, "Shindai no jujitsuden ni tsuite," *Ritsumeikan bungaku* 180 (1960), pp. 325 ff.

by enjoying agriculture, it was a fair statement.[43] Ch'in established the first imperial dynasty in Chinese history and, as Ho Ch'ang-ch'ün ably argued, the dynasty signified a social revolution that created free and independent small farmers, and in turn it was the new productive force of the small farmers that supported the new regime.[44] Han was to inherit from Ch'in the same group of small farmers as pillars of the imperial power, at least for some time.

43. *Shih-chi hui-chu k'ao-cheng,* I-wen reprint (hereafter cited as *SC*) 6:36, 46; Ho, "Ch'in-Han-chien," pp. 94-96.

44. Ibid., pp. 93-97.

I

GOVERNMENT RESPONSES TO POPULATION AND LAND TENURE PROBLEMS

Agriculture and the People

After seven years of civil war, Liu Pang finally defeated his principal contender, Hsiang Yü, and proclaimed himself Emperor of Han. In 202 B.C., immediately following the proclamation, a general demobilization was announced and all people were urged to return home. It was decreed that those who returned to their home prefectures be given back their own land and houses, and that the meritorious ones be granted land and houses (document I, 1). During the next half century, agricultural production gradually recovered, largely as a consequence of light taxes and small government expenditures (*HS* 24A: 9b-10a). Meanwhile, the population of China also grew rapidly and population registration was improved. In the following pages we discuss demographic growth and its relationship to agricultural development.

By the time of Emperor Wen (179-157 B.C.), agricultural production of the empire was still too low and there was little grain stored, so Chia I, one of the most important statesmen of the Han dynasty, submitted a memorial in which he called the emperor's attention to the increasing flight of farmers to nonproductive secondary professions—mainly commercial businesses. Chia I argued that production should be spurred by urging people to return to farming in order to assure food reserves (I, 2). Chia's contemporary, Ch'ao Ts'o, also emphasized the significance of agriculture by pointing out that the security of the empire, as well as the livelihood of the people, depended upon a flourishing and prosperous farming population. In one memorial he suggested encouraging agricultural productivity and upgrading the social status of farmers by having the government take grain as payment for rank sale and crime redemption. The life of the farmer, according to the memorial of Ch'ao Ts'o, was far from a rewarding and comfortable one. The household of a free farmer who tilled one hundred mu of his own land could not make both ends meet even in a year of normal crops. After toiling in the field all year round, the farmer had to borrow money at high interest in order to pay his taxes (I, 3). The situation supposedly provoked a series of reductions in the land-tax rate. By 155 B.C. a farmer needed only to pay one-thirtieth of the crop, possibly the lowest rate a farmer ever had to pay in Chinese history (*HS* 24B: 15a). Nevertheless, only one century after the establishment of the dynasty, and no more than two generations after Ch'ao Ts'o, the poor living conditions of the independent farmer are again depicted. Tung Chung-shu presented a gloomy picture in which he mentioned some farmers who had to sell their land to the wealthy and become tenants, paying one-half of the crop for rent, in addition to the heavy corvée obligations that could not be evaded (I, 4). Though Tung Chung-shu did not give a percentage of farmers who had to sell their land, the mere emergence of a tenant who lost his property to the wealthy marked a great deviation from the farmer in the Ch'ao Ts'o memorial who, though poor, still remained his own master. The life of a tenant, however, was to be the fate of quite a large proportion of Han farmers.

Meanwhile, the population of the Han Empire, after one full generation of relative peace, grew considerably. Ssu-ma Ch'ien noted that at the very beginning of the dynasty, only 20 to 30 percent of the registered population stayed in their home areas. After less than one century, by Ssu-ma Ch'ien's own time, the population in various localities had doubled, tripled, and even more (*SC* 18:3-4).

These rough estimates give very little help in the drawing of a demographic profile of the first century of the Han dynasty. Fortunately, a small group of samples are available. In the *Han-shu,* populations of nineteen marquisates are given at their creation in about 200 B.C. and at the final stage when the marquisates were terminated by decree for various reasons. The statistics and the mode, the medium, and the means are all given in table 1. The value and reliability of these figures depends upon at least two variables: First, the number of households in 200 B.C. may have been underestimated, and second, the increase in the marquisates may be the consequence of migration and boundary changes as well as of natural growth. Nevertheless, the first becomes less likely as the growth rate is reflected in data of the same source.

The second pitfall also appears less threatening since these marquisates are distributed in virtually all major regions of Han China, which was already revealed by the wide range in population density. What impresses us is the consistency of growing two or three times within two or three cohorts. For the areas with a very high rate of growth, regional differentiation should then be considered. Three other independent cases of later date were also helpful in confirming the consistency of the possible growth rates. They are given in table 2.

In table 1, there are nineteen cases of population growth with the periods of increase ranging from 46 to 110 years. Table 2, meanwhile, shows the growth in the first century B.C. The geometric mean of all the twenty-two cases is an annual growth rate of 1.60 percent. In table 1, the median time span is fifty-four years, while the annual median growth rate is 1.6 percent. The table shows that if the time span is longer than the median (Period B), seven of the nine cases have annual growth rates less than or equal to the median, 1.6 percent. On the other hand, if the time span is shorter than fifty-four years (Period A), six of the nine cases have annual growth rates greater than 1.6 percent. The high growth rate of the early years perhaps should be attributed not only to the natural growth but also to the normalization of registration under a stable government. The rates in the longer periods show the gradual leveling after the high peak and table 2 probably reflects the growth in a peaceful period. If the population grew at the rate of a moderate 1 percent per annum, a base population of 100 will be 128 after twenty-five years, 164 after fifty years, and 270 after one hundred years. Thus, during one century of Western Han, the population probably doubled, and the increase could easily have been even more than that.

Vagrancy, the phenomenon of large groups of persons roaming the countryside, is one of the indications of demographic crisis. In the Western Han period, the imperial annals registered only nine cases (119, 83, 71, 70, 67, 52, 25, 23, and 17 B.C.) in a period of more than two centuries, while individual biographies and treatises gave fourteen cases, some of which, however, were regional or actually referring to the same ones mentioned in the annals. With very few exceptions, which mentioned simply the phenomenon of wandering populations, the references to vagrants in the Western Han period were repeatedly related to some major famine caused by natural disasters.[1] On the other hand, among the twenty-three mentions of

1. Wang Chung-lo, "Kuan-yü Chung-kuo nu-li she-hui-ti wa-chieh chi feng-chien kuan-hsi-ti hsing-ch'eng wen-t'i," *Chung-kuo ku-tai-shih fen-ch'i wen-t'i t'ao-lun-chi,* pp. 450-51.

TABLE 1

POPULATION GROWTH IN NINETEEN MARQUISATES DURING SECOND CENTURY B.C.

Marquisate	Commandery	Number of Households Beginning	Number of Households End	Years B.C.	Span (Years)	Annual Growth Rate	Population Density (Persons/km^2)
Yang-tu	Cheng-yang	7,800	17,000	201-155	46	1.7%	6.1
Wei-ch'i	Lang-ya	1,000	3,000	201-154	47	2.4	45.9
Wu-ling	Chi-nan	740	1,500	201-153	48	1.5	87.5
Tsan	Pei	8,000	26,000	201-152	49	2.4	5.5
Cheng	Cho	2,500	5,600	201-151	50	1.6	48.8
Tung-wu	Lang-ya	2,000	10,100	201-151	50	3.3	45.9
Nan-an	Chien-wei	900	2,100	201-149	52	1.6	3.9
Chun-cheng	Tung-lai	4,000	9,300	201-149	52	1.7	46.2
Chu-chou	Kwang-p'ing	4,000	18,000	201-148	53	2.8	177.4
Tu-yen	Nan-yang	1,700	3,400	200-146	54	1.3	48.0
P'ing	Ho-nan	1,300	2,300	201-145	56	1.5	155.0
Liu-chiu	Po-hai	1,000	3,000	201-143	58	1.9	35.4
Kao-yüan	Ch'ien-ch'eng	1,600	3,200	201-138	63	1.2	80.9
Ying-yin	Ying-ch'uan	5,000	8,400	201-134	67	0.78	207.0
Ch'u-ni	Chung-shan	5,000	16,000	201-130	71	1.6	72.3
Ch'ao-yang	Chi-nan	1,000	5,000	200-127	73	2.2	87.5
Ch'i	Nan	1,000	4,000	200-116	84	1.6	9.5
Lo-ch'eng	Ho-chien	1,300	2,300	201-113	88	1.0	61.1
P'ing-yang	Ho-tung	10,000	23,000	201-91	110	0.75	28.0

	Mode	Median	Arithmetic Mean	Geometric Mean
All marquisates	1.6	1.6	1.70	1.62

SOURCE: *Han-shu, chüan* 16. Li Chien-nung, *Hsien-Ch'in Liang-Han ching-chi-shih kao,* pp. 236-37. The dates of the cases are often in error in Li's table. They are herewith corrected. The population density is taken from Lao Kan, "Liang-Han chün-kuo mien-chi-chih ku-chi chi k'ou-shu tseng-chien-chih t'ui-ts'e," *Bulletin of the Institute of History and Philology* 5, no. 2: 215 ff. The data Lao tabulated are those of A.D. 2, given in the *Han-shu, chüan* 28. The annual growth rate is determined according to the interpolation method; cf. T. L. Smith and P.E. Zopf, *Demography: Principles and Methods,* pp. 552-53.

vagrants in the Eastern Han, nineteen were listed as major events in the imperial annals (A.D. 76, 76, 93, 94, 100, 100, 102, 103, 106, 108, 110, 115, 127, 131, 139, 146, and 153). Combining incidents occurring in the same year, there are still sixteen occurrences in a period of less than two centuries (A.D. 25-220), and only four are attributed to famine or natural disasters. In other words, having a sizable population roaming around became routine and commonplace in the Eastern Han period.[2] If population growth began to gather momentum after two cohorts from the starting point, it is not sheer coincidence that, in both the Western and the Eastern Han dynasties, the first mention of vagrancy did not appear until at least half a century after the founding of the dynasty.

The population growth already placed a critical strain on the inadequate food supply. The man/land ratio in table 3 shows a very small per capita landholding. The problem was further complicated because of the uneven distribution of population. There were densely populated areas with as many as several hundred persons per square kilometer and others with less than one person per square kilometer. The high population density was in the vicinity of the capital and in a few commanderies located in the well-developed central plains along the middle and lower reaches of the Yellow River.[3] The areas where the population was concentrated and agriculture was well developed actually constitute a fairly small proportion of all of China.

The price of fertile land near the capital could be as expensive as one piece of gold bar for one mu, which makes it one hundred times the price of a mu of land in a frontier commandery.[4] The valuable land in the crowded areas became desirable investments for the rich, but difficult for ordinary farmers to possess. Land seizure by the wealthy and the powerful is to be discussed in the next chapter. The following, however, is a discussion of the government attitude toward inequality of landholding.

2. Ibid., pp. 451-52. The last years of the Eastern Han brought the great chaos that persisted into the period of the Three Kingdoms. From A.D. 168 on, there was not a single year when some part of the population did not have to leave home. Cf. Lien-sheng Yang, "Notes on the Economic History of the Chin Dynasty," in his *Studies in Chinese Institutional History*, pp. 126 ff.

3. Lao Kan, "Liang-Han chün-kuo mien-chi-chih ku-chi chi k'ou-shu tseng-chien-chih t'ui-ts'e," *Bulletin of the Institute of History and Philology*, vol. 5, no. 2, table 1. Nishijima Sadao, *Shin Kan teikoku*, in *Chūgoku no rekishi*, vol. 2, maps on front pages.

4. *HS* 65:7a. One piece of gold was valued at ten thousand cash. *HS* 2: 2a, the comment by Cheng Hsüan. The price of land in the frontier commandery Chü-yen is recorded on a Han tablet as one hundred cash for one mu (III, 5, 6).

TABLE 2

POPULATION GROWTH IN THREE LOCATIONS DURING THE FIRST CENTURY B.C.

Name	Commandery	Households		Duration	Annual Growth
Yin-p'ing	Chi-nan	1,279	2,944	73-10 B.C.	1.3%
Fu-yang	Pei	711	1,420	72 B.C.-A.D. 3	0.93
Shan-yang	Shan-yang	93,000	172,800	65 B.C.-A.D. 2	2.5
		Median		Arithmetic mean	Geometric mean
Annual Growth Rate		1.3		1.58	1.44

SOURCE: *HS* 18:13 and 14; 76:14; Lao Kan, "Liang-Han chün-kuo . . . ," pp. 180-216, and Smith and Zopf, *Demography*, pp. 552-53. The case of Shan-yang is not included in Li Chien-nung, *Hsien-Ch'in Liang Han*. . . . The date 65 B.C. is established by estimating the tenure of Chang Ch'ang as the grand administrator of the Shan-yang commandery between 66 and 64 B.C.

TABLE 3

AVERAGE MAN/LAND RATIO

A	B	C	D	E	F	G
2	827,053,600	12,235,062	58,594,978	4.87	67.61	13.88
105	732,017,080	9,237,112	53,256,229	5.76	79.25	13.76
122	694,289,323	9,647,838	48,690,789	5.04	71.96	14.28
144	689,627,150	9,946,919	49,720,550	4.99	69.33	13.89
145	695,767,820	9,937,680	49,524,181	4.99	70.01	14.03
146	693,012,338	9,248,227	47,566,772	5.08	74.13	14.59

A = year (A.D.); B = arable land acreage (in Han mu); C = households; D = population; E = average persons for each household; F = average landholding per household (in Han mu); G = average per capita landholding (Ma Ch'eng-feng, *Chung-kuo ching-chi-shih,* p. 352).

Restriction of Landholdings

It is not coincidence that people in the reigns of Emperors Ai and P'ing began to diagnose the agricultural problem as a symptom of uneven distribution of land that was correlated to the polarization of the rich and the poor. The demand for restriction of landholdings gradually became commonplace. Confucian scholars, especially, dreamed of an ideal well-field system of even land distribution according to which every adult would receive one hundred mu while each eight households would constitute a community unit jointly tilling another one hundred mu of land, the produce from which was for the state (or more correctly, for the feudal lord).[5] From the time of Emperor Yüan, the Han emperors received a thorough Confucian education and Confucian scholars gradually dominated the court.[6] During the second half of the first century B.C., the members of the Confucian school made their first efforts to put their ideology—at least in part—into practical application. In 7 B.C., when Emperor Ai had just come to the throne, Shih Tan, renowned as a learned Confucian scholar, made the first suggestion that the ancient well-field system should serve as a model to redistribute the land. He pointed out that after generations of prosperity through free development, the wealthy had aggrandized their landholdings at the expense of the poor and that the uneven distribution of land had polarized society. In order to meet the crisis, he urged that certain limitations on landholdings be established. The proposal was then handed to the court for discussion. K'ung Kuang, the chancellor, recommended a limitation with a high ceiling of three thousand mu of land and certain limitations on the number of slaves owned by an individual. The emperor was, however, a weak personality. His cousins and maternal uncles, as well as a favorite, often received tremendous amounts of largesse, which sometimes included large pieces of land in addition to other properties acquired by various means. Since a limitation on landholding would mean the end of their vested interest, the imperial cousins of the Ting and Fu families, aided by Tung Hsien, the favored imperial companion, all exercised their influence upon the emperor, who then shelved the proposal, nullifying the first attempt to remedy the land concentration crisis[7] (I, 5).

Wang Mang usurped the throne in A.D. 9 after he had been

5. James Legge, trans., *The Chinese Classics* 2:245. Cf. our introduction.

6. *HS* 9:1a-b, 10:1b; also H. H. Dubs, trans., *The History of the Former Han Dynasty* 2:299, 301, and 373 (hereafter cited as *HFHD*).

7. For the biographies of Shih Tan and K'ung Kuang, and their being Confucian scholars, see *HS* 81 and 86.

regent.[8] Among his many policy changes, the one of greatest importance for this study was his effort to nationalize all the land and to revive the ancient well-field system. No sooner had he ascended the throne than he issued a decree in which Han taxation and labor corvée systems were condemned. He might actually have located the real problem of Han agricultural development by pointing out that the farmer bore a much heavier burden than the seemingly light land tax. A farmer had to pay a corvée tax and even a conscript tax that was to substitute for the actual labor corvée for the court, the local government, and garrison duty. Also, a farmer often lost his own land, for various reasons, to rich persons, who might use heavy usury, political pressure, or other means, to seize the land and compel the tiller to pay the self-claimed landlord the "rent" that normally amounted to one-half of the crop. Wang Mang's accusation was quite justified. The farmer was at the mercy of powerful personalities, who in some cases simply subjugated a whole region. Land bestowed upon the poor only provided more ready prey for the voracious and powerful.[9]

Wang Mang thought that the only thoroughgoing solution was an even distribution of land that could be tilled by everyone yet owned by none. His reform was a reaction to the moderate yet inept proposal of setting an upper ceiling on land ownership. With the intention of combining the Confucian ideas of equal distribution of land and harmonious relationships among family members, Wang Mang instructed that the family whose land exceeded the average allowance of nine hundred mu per eight adult males should divide the surplus among their relatives and neighbors. The result was far from successful. After three years of confusion and chaos, the law was reversed to permit land transactions once again. Private ownership was again recognized by law and no restriction was imposed on the size of landholding (I, 6).

Wang Mang, a Confucian himself, was perhaps sincerely trying to revive the ancient equal landholding system. In fact, he explicitly stated that the trouble of land concentration was the consequence of the collapse of the well-field, which he attributed to the time of Yao and Shun, the ancient sage kings (I, 7). Wang Mang perhaps considered his policy of stopping all land transactions an effective means of preventing people from selling their own land, for a poor farmer often had to sell in time of emergency the land that could even have been recently received from the government as a famine relief meas-

8. H. H. Dubs, "Wang Mang and His Economic Reforms," *T'oung Pao* 35: 219-65. Dubs, *HFHD*, vol. 3.

9. A detailed discussion of land engrossment is included in chap. 3 of this work.

ure.[10] What Wang Mang failed to appreciate was the reason that the farmer had to sell land. Kung Yü, an advocate of natural economy, made a statement on the livelihood of a Han farmer, saying that as well as the regular taxes there were also miscellaneous taxes to pay and unofficial exploitation by authorities at the district and village level, adding that the possibility of handsome profit in trade also attracted the farmer to leave the land (I, 8). In other words, a Han farmer was unable to make both ends meet, largely because of the heavy burden of nonagricultural expenses. Constant exploitation drained farming resources until the final resource—land—was transferred into money to meet the demand. As long as Wang Mang was not in a position to stop exploitation at the local level, no institutional reform could work. More detailed discussion of the exploitation that contributed greatly to the creating of the rural proletarian, the landless peasant, will be in chapter 2.

Noneconomic Measures

The early emperors had been aware of the problems in agriculture though they may have missed the nature of the problem. When Emperor Wen revived the ritual of imperial plowing in 178 B.C., he declared that agriculture was the foundation of the empire and that his personal participation should be appreciated as leading in the cultivation of land to supply provisions for use at the ancestral temple (I, 9). Thus started a tradition of semireligious and semipolitical ritual in the Han imperial house. Emperors from time to time would symbolically push the plow three times in the spring, in the ancient rite specifically advocated by Confucian scholars[11] (for illustrations, see I, 10-16).

Meanwhile, several edicts were issued that expressed imperial concern for agriculture—which showed symptoms of being unable to supply sufficient food. The edict of 168 B.C. was an admonition

10. Cf. Ho Ch'ang-ch'ün, "Ch'in-Han-chien ke-t'i-hsiao-nung-ti hsing-ch'eng ho fa-chan," in *Han-T'ang-chien feng-chien t'u-ti-so-yu-chih hsing-shih yen-chiu.*

11. The rite of cultivating the ceremonial field can be traced as far back as the Western Chou dynasty or even to the Shang period. The early Chou bronze inscriptions bear the term *chi-t'ien.* Cf. Kuo Mo-jo, *Liang-Chou chin-wen-ts'u ta-hsi,* pp. 30b, 150b. The Han ritual included the ceremony in the field as well as the visit of the emperor to the temple of the agricultural deity *(Hsien-nung)* where he prayed for a good harvest. Cf. *Hou-Han-shu chi-chieh* 4:5b-6a (hereafter cited as *HHS*) and *Li-chi chu-shu* 14:12a (hereafter cited as *LCCS*). For further discussion, cf. Dubs, *HFHD* 1:281-83; Nancy Lee Swann, *Food and Money in Ancient China,* p. 157 and n. 161.

to officials to exert themselves to encourage people to grow food, since the lack of expansion of arable land had led to a shortage of food reserves. The same edict contained an order to reduce the land tax by one-half and another edict in 167 B.C. ordered a universal waiving of the land tax as a special grace (I, 17, 18).

In 163 B.C. an edict was a serious inquiry into the reasons for poor agricultural development in general and for the shortage of food in particular. The hypothesis posed to explain the food shortage seemed to have missed many points. Wine-making and husbandry had never been so developed in China that grain consumed for wine or animal feed would affect the normal food supply. The assumption that too many people engaged in commercial activities and thereby caused a shortage of labor in farming, however, was a very handy accusation against the relatively prosperous urban professions. It, in fact, remained one of the most common recurring themes in the argument for government promotion of agriculture[12] (I, 19, 20, 21).

The government promotion of agriculture during the reigns of the early emperors of Han amounted to more than oratory. Other than the grace of tax exemption and reduction, the throne attempted to provide people with a model by selecting the best farmers in each area. The *li-t'ien,* or "diligent cultivator," was for the first time selected in 191 B.C. and was granted exemption from taxes and labor service, an indulgence obvious enough to induce emulation by other farmers (I, 22). In 187 B.C. a decree was issued to each commandery to recommend one such *li-t'ien* to the court, possibly as a candidate for a government position (*HS* 3:2a). In 168 B.C. *li-t'ien* became one of the three categories of local recommendees, together with *san-lao* ("thrice venerables") and *hsiao-ti* (the "filial and brotherly loving"). They were to be recommended in numbers proportionate to the local population (I, 23). From that time, the *li-t'ien* continued to receive imperial gifts along with the other recommendees, though they, perhaps, remained the lowest in rank (*HS* 6:13b). No less than thirty mentions of *li-t'ien* are in the imperial annals covering the whole span of the two Han dynasties, so it remained institutionalized throughout the whole period. Meanwhile, routinization of the *li-t'ien* recommendation also, ironically, reduced the impact that the invention of the title was intended to create (for illustration, see I, 24).

Related to such measures is the grant of the honor of rank to the people. The earliest mentions of the grant of titles were when Liu Pang first entered the Ch'in capital and permitted people to cul-

12. The accusation against merchants and the so-called secondary occupations is based upon the logic that grain only is the real and valuable object, namely food, while trade does not produce food (I, 20, 21; cf. I, 4, 14, 17).

tivate the formerly restricted imperial reserved land and gave males steps of noble rank (I, 25, 26). Similar graces, no less than two hundred, were granted from time to time throughout the two Han dynasties.

Such indiscriminately bestowed largesse would have sounded impractical if the rank were an extraordinary honor. It, however, does make sense when the origins of rank-bestowing are examined. When the state of Ch'in initiated the system of honorific ranks, the feudalistic social stratification was still a recent memory and entering the rank system was probably regarded as a steppingstone uplifting a commoner to the privileged stratum. According to the Ch'in system, a rank holder indeed could expect some material benefit, such as house, land, and service of a subordinate, being granted as part of the largesse.[13] In the decree issued by Liu Pang in 202 B.C. there is a clear implication that a rank holder at the beginning of Han was entitled to demand some material benefit (for the high ranks) and exemption from labor service (for the lower).

Nishijima suggests that the universal bestowing of titles on the people helped to forge direct links between the emperor and the people.[14] Since the majority of the Han populace were farmers or their dependents, the institution of rank-granting should have had an impact on the farming population.

The value of the rank-title was more than honor. It could be sold for cash and entitled the holder to reduced punishment in case he faced such a verdict (*HS* 24B: 9a; I, 2). Considering the numerous categories of forced labor for a convict, who had to pass from one level of inferior status to another through many years of lost freedom, one can easily understand the value of ransom (*HS* 23: 14a-b). Titles of rank would be protective cushions to absorb the danger that sometimes might press a free man into a low social status. The rank-grant system, however, assumed undesirable aspects when the government began to sell ranks for either cash or grain. This measure was suggested by Ch'ao Ts'o in 178 B.C. to elevate the value of grain and therefore to elevate the status of the farmers (I, 3). The sale of ranks was mentioned repeatedly, and probably persisted throughout the two Han

13. Scholars have engaged in much debate on the nature of the Ch'in military rank system. Some agreed that the material reward was an integral part of the rank-title. Some thought that these two were separated. For the former, see Hiranaka Reiji, "Shindai tochi seido no ichi kosatsu: myōden-taku ni tsuite," *Ritsumeikan bungaku* 70 (1961). For the latter, see Moriya Mitsuo, *Chūgoku kodai no kazoku to kokka*, chap. 1.

14. Nishijima Sadao, *Chūgoku kodai teikoku no keisei to kōzō*, pp. 126-27.

dynasties.[15] The consequence was the creation of a special class in society that could use wealth to purchase honor, office, tax and corvée exemption, and crime redemption. The privileges, therefore, only made social differentiation further pronounced.[16]

Summing up the various efforts of the early Han emperors to promote agriculture, one finds a common underlying presupposition, that farmers gave up farming because they lacked sufficient initiative and that profit from business and other "secondary" professions lured people away from the "fundamental" profession. The tax privilege, the honor of rank, and the creation of the *li-t'ien* title, all were designed to provide motivation. Even the ceremony of imperial plowing was intended to demonstrate governmental encouragement of agriculture. The inquiry made by Emperor Wen in 168 B.C. and in 163 B.C. into the causes that led to the shortage of food reflects the ignorance of the court of the essential problems existing in Han agriculture (I, 17, 19).

Migration and Removal of the Populace

The uneven distribution of population was dealt with by the mass moving of people to the less populated areas. The first such government-initiated migration, however, was taken for national defense. The northern frontier for years had been easy prey for the Hsiung-nu nomads who could penetrate the sparsely populated northern front in many places. In 178 B.C., Ch'ao Ts'o recommended that permanent settlements of farmers with dependents should be established on the frontier. He stressed the bestowing of social status for the migrants by granting freedom to former slaves and ex-convicts. Titles of rank and exemption from labor corvée were also given to the migrants in order to induce volunteers, and the government even paid ransom to buy back the freedom of those who had lost it as captives (I, 27).

In 165 B.C. Ch'ao Ts'o once again urged the government to settle migrants in the northern border regions. He suggested that they should be provided with easy conveniences and facilities in the new communities, including towns at ideal locations, cultivable land with good resources, and be furnished residences with two bedrooms for

15. It is interesting to see that the price went downwards from 2,000 cash per rank in 194 B.C. to 1,000 cash per rank in 18 B.C. Cf. *HS* 2:3b and 10:10a. The price of 2,000 cash, nevertheless, was cited by the Eastern Han commentator Ying Shao as the standard price (*HS* 2:4a).

16. Li Chien-nung, *Hsien-Ch'in Liang-Han ching-chi-shih kao,* p. 261.

each household, in addition to good medical care and opportunities for social and religious activities (I, 28).

Although Ho Ch'ang-ch'ün may have been mistaken in assuming that these documents on frontier conditions were evidence that the Han Empire took care of all the farmers in the manner described above, it is still quite right to say that the Han court wanted to see the farmers able to maintain a good, independent livelihood and that the size of the residence indicates that the independent farming households should be small ones that the two-bedroom house could accommodate.[17] In other words, the policy of Han frontier defense was linked with the policy of redistributing the population in order to make more farmers independent and self-sufficient. The frontier migration policy remained active throughout almost the whole Han period.

It was not until 156 B.C., however, that the Han court faced the problem of uneven landholding by permitting people to move from crowded regions into regions with more arable land. This particular decree first mentioned the famine that caused starvation and death, and then pointed out that the land in some regions was infertile and insufficient while some other regions were fertile yet not in use, and finally granted permission for people to migrate from the former to the latter (I, 29). From the time of the Ch'in, people had not been allowed to migrate freely. In an early Han record in 205 B.C., there had been another decree granting people the right to move from one region to another, but there was a distinction between these two imperial decrees. The one in 205 B.C. actually granted nothing other than permission for people in famine-affected regions to move into food-abundant regions to receive better care, no more than a temporary relief measure to avoid shipping large amounts of grain to the famine-stricken areas (I, 30). The decree of 156 B.C., on the other hand, obviously viewed as desirable a permanent migration to reduce population pressure on the land by moving the excess population from one region to another.

Another large-scale operation was carried out in 119 B.C. during the reign of Emperor Wu, when a flood caused a major famine in the eastern commanderies of the empire. The flood was so severe and the eastern region was so widely affected that the government granaries were exhausted and private resources were called upon to make loans to the poor. As a last resort, the government moved some seven hundred and twenty-five thousand poor people from the famine-stricken commanderies into four northwestern frontier commanderies that were only recently free from the danger of nomadic invasion,

17. Ho Ch'ang-ch'ün, "Ch'in-Han-chien . . . ," pp. 100-101.

following the Hsiung-nu surrender in 121 B.C. The migration movement was immense. The government provided the migrants with food and clothing for several years and real estate was granted to migrants, who were organized into divisions supervised by government agents. The national treasury was almost emptied as a consequence[18] (I, 31). This massive movement not only relieved the flood-stricken people of the east, but also helped to populate the frontier regions.[19]

The most effective migration was the continuous flow toward the southern commanderies where the soil was fertile and the weather was mild. Though the government never took as active a role to arrange the southbound movement as it did for migration toward the northern borders, the growth of the southern population in the Eastern Han dynasty was remarkable. The areas that increased in population during the Han included the Yangtse Valley, the Szechwan Basin and the northeastern region. More discussion of this topic can be found in chapter 4; see especially table 15.

Population pressure in the most crowded regions was greatly reduced by emigration. According to Lao Kan's tabulation of population densities in the commanderies during the Western and the Eastern Han dynasties, however, ten of the thirteen most densely populated commanderies in the former period were still among the seventeen top commanderies in the latter.[20] Migration did help to absorb the excessive population relative to land but did not solve the problem completely.

Opening Up of Government Land

The Han emperors, the masters of the whole empire, were entitled to possess directly land over which no private ownership had been established and registered. Thus the Han imperial household held large areas, mostly mountain regions, swamps, forests, and ponds, some of which had been reserved for imperial recreation parks for hunting and fishing, and others reserved for imperial pastures. Nevertheless, the unclaimed land around the whole empire, even though not designated for any imperial use, was still under the control of the imperial household.[21] In the Western Han government there were always two

18. Other than the four northwestern commanderies, a southern commandery, K'uai-chi, was also to take in migrants, according to the Annals (*HS* 6:15a-15b).

19. In 127 B.C., one hundred thousand people had been recruited to settle in the northern frontier (ibid., p. 10b).

20. Lao Kan, "Liang-Han chün-kuo mien-chi . . . ," pp. 216 ff.

21. The best discussion on this topic is Masubuchi Tatsuo, *Chūgoku kodai no shakai to kokka*, pp. 265 ff.

separate organs to handle government and imperial household financial affairs. The Privy Treasury *(shao-fu)* handled the revenues from mountains and hills, swamps and marshes, seas, rivers, ponds and lakes, produce from government land, parks and gardens, and tribute from feudal lords. The payments included goods and money for palace expenses, imperial ancestor temple expenses, and gifts and largesse. On the other hand, the Grand Ministry of Agriculture *(ta ssu-nung)* managed the land tax and property tax to cover the state expenditures, national defense costs, public works, and salaries for government officials[22] (I, 32, 33, 34). The land of the imperial parks was so much regarded as exclusively for imperial use that Kao-tsu angrily jailed Hsiao Ho, his most meritorious lieutenant and the first chancellor of the Han dynasty, when Hsiao suggested that the wasteland in the Shang-lin Park be made available for people to cultivate (*HS* 39: 5b-6a). Included in the government land was the marginal land in the marshes and riverbeds after construction of irrigation networks. During the reign of Emperor Wu, for instance, land in a deserted canal bed in the capital area was granted to the Yüeh people, a southern ethnic group who were moved to the capital district. Rent was paid to the Privy Treasury (V, 14).

The Han court also directly possessed extensive confiscated estates. Some of the land thus obtained by the imperial house was under the management of various agencies. Presumably the government would retain the original tenants to till the land and would merely collect rent. The largest amount of such land under imperial control was seized during the reign of Emperor Wu during the last quarter of the second century B.C. The imperial house, even in other periods, still held quite sizable areas of land obtained from convicted citizens and officials (I, 35).

It seems that various government offices, either an organ of the central government or the local administration, might possess land, although it is not clear how such ownership came into existence. In one case, the grand administrator of a commandery ordered that the land possessed by his office be granted to the people free, presumably instead of renting it to them. The case suggests that otherwise the office of the grand administrator, or even the official as a person, could actually enjoy the rent as a source of revenue (I, 36).

22. Katō Shigeshi, *Shina keizai-shi kōshō*, vol. 1. Yoshida Torao, *Ryōkan sozei no kenkyū*, pp. 234-41. Masubuchi, *Chūgoku kodai*, pp. 267-69. Li Chien-nung, *Hsien-Chin Liang-Han*, pp. 230-44. Ying-shih Yü, *Trade and Expansion in Han China*, pp. 60-64. Yü-ch'üan Wang, "An Outline of the Central Government of the Former Han Dynasty," *Harvard Journal of Asiatic Studies* 12 (1949): 134-87.

In one very unusual circumstance, government offices were instructed by decree to collect land tax by themselves, presumably from the people who tilled the land assigned to certain government agencies. This incident, dated A.D. 190-95, occurred when the court was newly re-established in Ch'ang-an after a powerful general moved the new emperor there. During the emergency, as a regular fiscal system was not in operation, various agencies might have the discretion of finding ways to finance the operation of their own offices. Yet it is hardly plausible that government organs could command the means of seizing land to exact tax or rent. More likely, the land referred to in the document was nominally assigned to these government offices, and the concerned officials were given the responsibility of collecting taxes thereupon. Though a practice in a time of emergency should not be assumed to be a regular pattern of assigning land to government offices, the possibility nevertheless should not be ruled out that offices, instead of individuals, also could be given land by the court as a source of revenue (I, 37).

In short, the Han government did have direct control of a sizable amount of land that could be at the disposal of the court. By the time of Emperor Wu, the polarization between the rich and the landless farmers seems to have been so severe a problem that the government had to take measures to curtail the continuous deterioration of the society's structure. The Han court began at the other end of the problem by granting land to those who had lost their own. From the first mention of government land granted in 140 B.C. to the last action taken in A.D. 2, the annals of the *Han-shu* register eleven grants[23] (I, 38-49). It is interesting to note that cases occurring from 69 B.C. to 43 B.C., which is more than half of the total, mentioned the term *land lent.* In other words, with the exception of two early cases and the last three, government land was often lent rather than given to the people. Meanwhile, two of the last three cases, the events of 6 B.C. and A.D. 2, relate to private properties that were yielded to the government for redistribution (I, 46, 47). After repeatedly giving away land of the parks, as well as other reserves, and land belonging to local governments, the Han court seems to have exhausted its land resources and had to ask imperial relatives and ranking officials to give up their landholdings in order to resettle the landless. The very last effort in A.D. 2 indicates, perhaps, the end of land grants from imperial holdings, for on that occasion a part of An-ting, a frontier commandery in the northwest, was used to relocate the hungry farmers from

23. Amano Motonosuke has made a chart of these events in his "Kandai gōzoku no daitochi keiei shiron," p. 8.

the eastern regions, instead of land in the vicinity of the famine-stricken area.

The request made to grant private land to the poor implies that uncultivated land was no longer available to the farmers who needed land. Since the private land normally should have been under cultivation, the grant of such land to the poor means nothing other than ejection of the original tiller who perhaps was a tenant. Indeed, there was no gain in such a practice. The plea to the rich and the powerful to grant land to the people was a basic change from creating more cultivated acreage toward a better distribution of landholding. Therefore, the cases of the reigns of Emperors Ai and P'ing in 6 B.C. and A.D. 2 indicate a change of policy from lending the land to redistributing some of the private land. One paragraph in the *Discourses on Salt and Iron* gives us a clue to the real situation of land lending. The significant passage says that local elites who were summoned to court to participate in a discussion on the advantages and disadvantages of salt and iron monopolies challenged the justification of the land lease program. From the answers of the responsible officials it is clear that the state or the government actually was the biggest landlord in the empire, and collected heavy rent from the tiller. Meanwhile, the poor who had received land from the government often had to sell that land to the rich and powerful families when faced with a major hardship. Thus the government land actually was transferred to the rich and the poor farmers once again fell into the lowly status of tenants. In either case, whether the farmer became a tenant of the government or of the powerful personalities who seized land from the farmers, the farmer was the one who suffered (I, 49).

The Eastern Han court copied the measures taken in the preceding period: in A.D.66, public lands everywhere in the commanderies and kingdoms were ordered to be given to the poor; in A.D. 70, the lowland along a canal leading out of the Yellow River was given; in A.D. 76 the bird preserve in the imperial park at Shang-lin was distributed (I, 50, 51, 52); in A.D. 84, a decree once again instructed all the local administrations to grant government land to the poor, as well as to provide them with loans for seed, provisions, and rented implements. Taxes were also exempted for a few years. This decree reveals that the man/land ratio in the empire was by no means even and that some farmers had neither land nor capital with which to cultivate the land (I, 53). Yet in A.D. 86, the emperor, traveling extensively through the eastern part of the empire, noticed that there was still considerable fertile land lying unused in several commanderies and kingdoms and decreed that it be given to the poor. It seems either that the previous decrees releasing the land were not adequately executed or that the local administrations continued to acquire land (I, 54; cf. I, 50). By the end of the first century, the grace of the court to the poor no longer took the form of land grants. The imperial decrees to open up government lands, such as palace parks, imperial reserves, and so on, involved no more than permission for the people to gather firewood and fodder, to fish, and to collect things in these otherwise restricted areas (I, 55, 56, 57). It deserves attention that after the decree of A.D. 109 (I, 58), no more cases of government land being made available to the poor were recorded in the annals of the *Hou-Han-shu*. Instead, distribution of grain became the major means of providing help in time of famine.[24]

There seemed to be a gradual diminishment of benevolence in the Eastern Han aid-to-the-poor programs. At the beginning land was granted to the people from the imperial directly controlled estates. Then the emperors had to trust the local administrators to execute the order. Thereafter imperial grace was reduced to granting permission to collect food and other items in the imperial parks. And finally the government could extend relief to the poor only by distributing food. The court of the Eastern Han did not stay as long in the position of the biggest landlord as the Western Han emperors had. Of course, the lack of as much uncultivated waste land as the Western Han court had possessed was also significant. The Eastern Han decrees normally instructed that land be *given* to the poor, instead of being lent, thus constituting a nonreversible action. With the domination of Eastern Han politics by local powerful families, one may speculate that the powerful families would welcome the granting of land to the poor as a convenient means for the powerful ultimately to seize the land for themselves.[25]

Both Han governments also made efforts at the local level to increase the acreage of cultivated land and the production per unit by improving irrigation or by other actions. In the Western Han period, the emperors repeatedly issued edicts urging local officials to promote agriculture and sericulture. Punishments and rewards were promised to assure that officials would function with the utmost effect (I, 59, 60, 61, 62). Commissioners were even dispatched from the central government to supervise the promotion of agriculture (I, 63, 64).

24. For instance, in A.D. 113, tax grain from five southern commanderies was shipped to seven famine-stricken commanderies for distribution (*HHS* 5:9b-10a).

25. The term *loan (chia)* in the Western Han documents implied the lease of land; in the Eastern Han documents, it is used only in the cases of making certain government property available for people to collect subsistence materials without paying use tax.

Yet the cases of accomplishment in the *Han-shu* are not many. The most controversial person, of course, was Wang Ch'eng, who in 67 B.C. was ennobled to the rank of marquis as a reward for his merits in resettling eighty thousand vagrants in the region under his charge. (*HS* 8: 8a-b). After his death, however, it was discovered that Wang Ch'eng actually had made a false report (*HS* 89: 36).

One of the most successful local officials was Shao Hsin-ch'en, a grand administrator of Nan-yang, around 35 B.C. He spent a great deal of time visiting the villages in an effort to promote agriculture. In his tenure, several tens of water-control works that irrigated an area finally reaching thirty thousand *ch'ing* were constructed and the local population doubled as a consequence of the new prosperity (IV, 23). His contemporary, Ch'en Li, also was rewarded for his merits in promoting agriculture (I, 65).

The Eastern Han emperors issued instructions very similar to those issued in the Western Han period. Officials were told to refrain from disturbing the people in the busy season, to exterminate insects, and to be diligent in encouraging agriculture (I, 66, 67). The model officials in the Eastern Han period, as cited in the *Hou-Han-shu,* were more numerous than their counterparts in the previous period. Among the outstanding local officials of Eastern Han, most achieved either an increase in cultivated land, or an increase in registered population, or, more often, both. The translations give six cases, representing practically the whole span of the dynasty. Examination of their records shows that these cases involve not only the frontier where open land was readily available, but also the heartland of China where the civil war had made waste out of cultivated fields (I, 68, 69, 70). In other words, what was opened was often actually deserted land reclaimed once again for cultivation, or land improved by using more effective irrigation.

Summary

The governments of both Han dynasties were determined to maintain the security and prosperity of the farming population, which was regarded as the foundation of the empire. The principal means of production in agriculture is land. Yet the growth of population, especially its uneven regional distribution, made it difficult for the government to assure the farmer land. In the most crowded regions, elevation of social status and reducing the tax burden could not provide real help in easing the pressure of too many persons on too little land. Migration, of course, was a way to release some of the pressure in the most crowded areas, but migration toward the south, which was the most significant long-term demographic trend in Chinese history, was not actively sponsored by the government. The government-sponsored population movements were largely related to the defense of the northern frontier and northern emigration was never comparable to the southbound movement in either effect or scale. The granting of government land to the people should be viewed as a measure to open up uncultivated land and therefore to augment arable acreage. The speed of increase of cultivated acreage, nevertheless, could never match the growth rate of the population. The frequent references to the vagrant population are indications of the overflow of population relative to available land. The problem of an inadequate man/land ratio is further complicated by the problem of uneven landholding distribution between the wealthy and the poor. The latter, having once lost the status of independent farmer and having become a tenant or even a serf of the former, had to face a very narrow margin between subsistence and starvation.

II

LAND AS WEALTH

The Development of Enterprises

It has been shown in the preceding chapter that the growth of population after the unification of China led to a small per capita arable land ratio, leaving farming households with small farms. If agriculture were not labor-intensive, excess labor must either emigrate to new land or enter nonfarming professions, such as manufacturing or trading.

Indeed, once in early Chinese history it seemed as if manufacturing and trading would have an opportunity to expand. As early as the Chan-kuo period, in conjunction with the frequent interregional contacts, the development of monetary systems, and the growth of urban centers, there was commercial prosperity for the first time in Chinese history and those who succeeded in making fortunes also won status in society.[1]

In the early Han, business activities flourished, continuing commercial prosperity and the spectacular urbanization of the Chan-kuo period. The Grand Historian, Ssu-ma Ch'ien, in the chapter on entrepreneurs in the *Shih-chi,* listed a number of important Chinese urban centers of the second century B.C. Ssu-ma Ch'ien named one key center in each region. It had either sizable hinterland to

1. Cho-yun Hsu, *Ancient China in Transition,* pp. 138-39.

provide basic goods or foreign countries to provide imported materials, and had highways or waterways that permitted easy transportation.[2] Many of these same cities were mentioned as commercial centers in the *Discourses on Salt and Iron*, the record of a court debate upon economic policy held in 81 B.C.[3]

A comparison of the Ch'un-ch'iu and the Chan-kuo periods indicates that the active trade in the latter was owing to its better communication system, which was facilitated by highways constructed for war or other interstate contacts, and the free moving about within each of the major states, which were much larger than the feudal domains of the Ch'un-ch'iu period.[4] One would expect, then, more vigorous commercial activities in the Han period, as people enjoyed an even better road system in a unified China with fewer of the barriers and restrictions that were erected by the contending states along the borders during the Chan-kuo period. The best road system, of course, was the one constructed by Ch'in Shih-huang-ti and inherited by the Han Empire. It is said the highway, five paces in breadth and with pounded earth as its base, was reinforced by metal poles along the foundation and was shaded with lines of pines. The network connected the eastern coastal region and the Yangtse valley to the capital (*HS* 51: 2a-b). Ssu-ma Ch'ien pointed out that wealthy merchants and businessmen traveled around China and the commodities flew from one region to another as the empire was unified and that there was neither tariff nor restrictions on any natural resource (see document II, 1). It is apparent that the interregional dependence that had been felt in the Chan-kuo period led to active interregional trade when the compartmentalization imposed by the contending states of the Chan-kuo period was lifted under the unified empire[5] (II, 2).

Regional specialization meant the possibility of large-scale production of one particular local staple. Ssu-ma Ch'ien listed a number of enterprises that could reward the owner with an income of no less than the equivalent of an enfeoffed noble lord. These included raising animals and fruit and nut trees, and producing forest and garden

2. *SC* 129: 17-26. English translation in Nancy Lee Swann, *Food and Money in Ancient China*, pp. 437 ff. Cf. Shih Nien-hai, "Shih-chi Huo-chih lieh-chuan so-shou-ti T'ao-wei-t'ien-hsia-chih-chung. Chien-lun Chan-kuo shih-tai-ti ching-chi tu-hui," *Jen-wen tsa-chih* 2 (1958): 77 ff.

3. *Yen-t'ieh-lun*, Ssu-pu-pei-yao ed., 1: 6b-7b, hereafter cited as *YTL*. Also translated in E. M. Gale, *Discourses on Salt and Iron*, p. 18.

4. Hsu, *Ancient China in Transition*, pp. 116-19.

5. Kageyama Takeshi, "Kandai no keizaikan ni tsuite," *Wada hakushi koki kinen Tōyōshi ronsō*, pp. 291-93.

products (II, 3). In another passage, Ssu-ma Ch'ien extended the list to include the production of grain, furs and hides, vehicles, timber, housewares, textiles, dried and preserved food products, and various kinds of brokerages and credit-loan businesses as the fields in which one could expect a handsome profit if done on a large scale.[6]

In spite of the unfavorable laws that Kao-tsu legislated to impede the growth of merchants, there were a number of successful enterprises in the early stages of the Han. Ssu-ma Ch'ien gave cases of the successful money-makers to illustrate the worthiness of such professions. Table 4 lists the businesses in which these Ch'in and Han persons made their fortunes.

TABLE 4

BUSINESSES OF WEALTHY PERSONS

Names	Professions and Activities
The Chos	Iron smelting; foreign trade
Ch'eng Cheng	Iron smelting; foreign trade
The K'ungs	Iron smelting; owning ponds and fields
The Pings	Iron smelting; trade; credit loans
Tiao Hsien	Peddling fish and salt
Shih Shih	Nation-wide peddling
Jen	Grain storage and dealing; farming and pasturing
Ch'iao T'ao	Pasturing horses, cattle, and sheep
Wu-yen	Moneylending
The T'iens	Trade
The Lis	Trade
The Tus	Trade

SOURCES: *SC* 129: 39-42; Watson, *Grand Historian* 2: 495-98; Swann, *Food and Money*, pp. 452-56.

It seems that smelting and interregional as well as foreign trade were the most common ways of making great wealth. And these were also fields where much labor would be in demand.

CURTAILMENT OF ENTREPRENEURIAL ACTIVITIES

Prosperous commercial groups suffered political checks almost as early as the founding of the dynasty, for the Han government was

6. *SC* 129: 34-36; *HS* 91: 7a-8b; for translation see Burton Watson, *Records of the Grand Historian of China* 2: 494-95, and Swann, *Food and Money*, pp. 434-37.

far from friendly to merchants. The Han's first emperor forbade merchants to wear silk and to ride in carriages and the trade tax was increased to hamper and humiliate them. During the reigns of successive rulers, neither merchants nor their children and grandchildren were allowed to serve in the government (II, 4).

The motivation for such measures is not apparent in the early Han documents. It is possible that the Han court simply inherited the Ch'in law that made soldiers and farmers the pillars of the state (*SC* 69:8-9). On the other hand, the Ch'in court did have such eminent merchants as Lü Pu-wei, who was chancellor for a long period, and wealthy personages such as Lady Ch'ing, a widow of Pa commandery, did enjoy honor bestowed by Ch'in Shih-huang-ti (*SC* 85:1, 9; 129: 15-16). What is speculated, therefore, does not seem a satisfactory explanation. It is likely that the profitable activities of the merchants had aroused much envy and antagonism among other professions. Kao-tsu, who was the son of a farmer, probably took punitive measures that reflected a jealousy shared by many in the noncommercial sectors. A fundamental concern of the Han monarchy, of course, was the stability of the empire, which conforms to what Eisenstadt terms the internal condition in the political sphere to develop a centralized empire. In such nonpolitical fields as economic activities, resources should be made "free-floating" instead of being imbedded within any particular group.[7] Wealth created by commercial activities, independent from the political institution, would be one of the resources that the ruler would rather not have committed to and embedded within a group of merchants outside of political control.

The Han Empire of the mid-second century B.C., after more than one generation of relative peace, felt the pressure of the growing population, which had to be fed adequately if the empire were to remain secure. Therefore, at the court of Emperor Wen (179-167 B.C.), Chia I, who first recognized the threat of a food shortage, concluded that there were too few producers for the large number of consumers, and urged the idling and parasitic population to return to farming (I, 2). Chia's contemporary, Ch'ao Ts'o, went further, asserting that the easy profit of commercial activities was responsible for luring people away from the productive profession, farming (I, 3). So the looming food shortage that was in all likelihood related to the increase in population was misunderstood in the mid-second century B.C. and attributed to the outflow of the farming population to commerce. This lack of understanding paved the way for the destruction of private trade during the reign of Emperor Wu.

7. S. N. Eisenstadt, *The Political Systems of Empires*, pp. 26-27.

As Eisenstadt observes, the ruler of a centralized empire often tends to implement the control of resources by one of two means: either directly accumulating resources or encouraging various groups in the society to develop them. The frequent foreign wars, large-scale destruction, extravagant luxuries, and the ritualistic visit to the sacred mountain T'ai-shan during the reign of Emperor Wu all cost tremendously,[8] and the ensuing financial crisis compelled his court to mobilize more resources. The first step taken by the government was the sale of ranks and titles and pardon privileges, that is, the use of fines as substitutes for imprisonment or other punishment.[9] Floods and the heavy burden of maintaining the frontier defense then emptied the imperial coffers. Meanwhile, the monetary crisis that was precipitated by the existence of counterfeit coins was worsened by the devaluation of legal currencies. Inflation, although severely damaging to the government and the general public, did not hamper the commercial sector, but instead, usury and hoarding added to the opportunity of earning windfall profits.[10] The government monopoly of salt and iron, both of which had been extremely profitable businesses, probably was the first effective measure taken by the government upon entering the sphere of private trade (II, 5). The motivation for such government monopolies was twofold: the obvious one was the need for money to solve financial difficulties; less obvious was the need to reassert the authority of Han imperial power. The increasing influence of the wealthy merchants and others who "belabored and exploited the poor people" was a threat that could not be ignored. A passage in the *Discourses on Salt and Iron* made it explicit that curtailing the growing influence of the wealthy was one of the main goals of the state monopolies, for merchants were able to gather a large horde of followers (II, 6, 7, 13).

The government, in addition to the salt and iron monopolies, developed other measures that affected the merchants and traders. In 129 B.C., it was decreed that merchants' wagons should be taxed (*HS* 6: 7b), an act that may be related to the earlier prohibition on the wearing of silk and riding in carriages by merchants. Meanwhile, some business taxes were exacted (II, 4), and in 120 B.C. a more elaborate taxation system was imposed upon merchants and artisans. The tax on transportation facilities remained paramount, but property owned by them was also taxed at the rate of 6 percent of its value

8. *HS* 24B: 6b-10b; Swann, *Food and Money*, pp. 241-65; Yoshida Torao, *Ryōkan sozei no kenkyū*, pp. 180-94.

9. *HS* 24B: 8a-9b; Swann, *Food and Money*, pp. 245-46; 252-54.

10. *HS* 24B: 10b-12a; Swann, *Food and Money*, pp. 264-68.

(II, 8). The most far-reaching effect of this particular resolution, however, was the award granted to anyone who was able to accuse another of violating the regulation. Thence, three years later, in 117 B.C., a certain Yang K'o was put in charge of such accusations. Prosecutions resulted in the complete ruination of "well-to-do and above" whose reports of their own property evaluation were found to be incomplete and false. Their landholdings and slaves as well as other properties were all confiscated by imperial order. And such wealthy persons included the "traders and merchants" in the text (II, 9).

It has been suggested that this particular property tax would drive the merchant to invest his capital in land, since "land did not come under the new property taxes."[11] Note, however, that in this decree (II, 8), merchants were forbidden to possess land and that the land of law violators was confiscated. The implication seems to be that there had been acquisition of land by merchants before 120 B.C. and the land owned by them amounted to so much that its possession was seen as detrimental to the economy of the state. Furthermore, the property registration on the Chü-yen tablets includes land as well as residents, slaves, animals, and wagons (III, 5, 6). Thus, land was included in property registration and the property tax was meant to take land away from the merchants and put it at the direct disposal of the government (II, 9, 10, 16).

The merchants suffered greatly from the property tax. Yet, after all, the merchants, who probably made up the majority of the destroyed wealthy group, probably saw their bad luck shared by other wealthy persons as well. A government measure that made the merchants its only victims was the government's entrance into trade. Not only did the government monopolize salt and iron, but it competed with private enterprise at large by buying and selling other products at profit (II, 11).

Government enterprises continued throughout the Western Han period and in the reign of Wang Mang. In edicts announced in A.D. 10 and in A.D. 17, the government imposed more controls upon commerce, including compulsory standardization of prices. Nevertheless, the complicated procedure as well as bureaucratic flaws made the system more of a nuisance than a help to the people (II, 11, 12, 13). In comparison, the market tax in the A.D. 10 edict, which was one-eleventh, or 9 percent of the profit, was much less than the property tax imposed on the merchants, which was 6 percent of the capital, a common interest rate suggested by Han contemporaries.[12] The exis-

11. John A. Harrison, *The Chinese Empire*, p. 144.
12. Swann, *Food and Money*, pp. 436-37.

tence of the market tax indicates that the property tax probably did not continue in Wang Mang's reign.[13] In the Eastern Han period, no property tax was mentioned; probably the tax on commercial profit remained at one-eleventh. This probably was a tax burden much higher than the one on agriculture, which normally was one-thirtieth of the yield. We do not know whether the merchants were still forbidden to own land after the prohibition decreed in 120 B.C. However, at a very early date in the Eastern Han period, a memorial presented in about A.D. 26-30 mentioned the large amounts of land possessed by rich businessmen that was leased to tenants to till (II, 14). It is likely that the prohibition was only enforced in the reign of Emperor Wu. And more likely, as the law pertaining to accusation of false tax declarations ended, so also did the prohibition.[14] In other words, the merchants in the days before 120 B.C. had invested in land, a fact that can be established by the 120 B.C. decree and the land confiscation in 117 B.C., and the disparity of the tax burden between one-eleventh for businessmen and one-thirtieth for agriculture was large enough to encourage the merchants to invest their earnings in land as soon as the prohibition against merchants owning land was not enforced.[15]

Therefore, Ssu-ma Ch'ien correctly points out that the most reliable form of wealth was still land. "They gained their wealth in the secondary occupations, and held on to it by investing in agriculture; they seized it in times of crisis and maintained it in times of stability."[16]

The making of land the most preferred valuable reflects, perhaps, not only the generally held truth that land would last but reflects also the need for expediency. The principle on which Ssu-ma Ch'ien presented a detailed discussion as a general theory for engaging in trade is mainly on the relationship between supply and demand of grain, which was regarded as the basic commodity (*SC* 129: 8-9, 13-14). It

13. There is the possibility that the property tax actually was terminated in 113 B.C. (*SC* 30:37, especially the comment). Yet in the *HS*, it is recorded that the monopoly of wine and liquor was terminated in 81 B.C., while the other economic measures survived the reigns of five emperors and until the period of Emperor P'ing (A.D. 1-5) (*HS* 24B: 20b).

14. The date of 113 B.C., which has been regarded as the termination of the property tax, perhaps actually was the date to end the encouragement of accusations for false tax declaration; the text has *kao-min* (report of tax-cheating), not *chan-min* (tax declaration).

15. John Harrison would be correct if he means the shifting of capital to land by the businessmen in this context. Cf. Harrison, *Chinese Empire*, p. 148.

16. Watson, *Grand Historian* 2:498-99; Swann, *Food and Money*, p. 462.

is very likely that these businessmen felt the necessity to acquire land to assure a supply of grain. Ssu-ma Ch'ien declared that the money earned in trade was guarded by investment in the "basic profession," which meant farming (*SC* 129: 43). And among the successful money-makers of the Han period, at least two were persons who made their fortunes in trade and had acquired some landholdings (*SC* 129: 38, 41; 117: 3).

The pressure imposed upon entrepreneurial development, therefore, not only reduced the capacity of nonfarming activities to absorb surplus labor, but also drove capital to land, the primary resource for agricultural development.

The State Moves against the Powerful

No sooner had private landownership been established, as feudalism gradually deteriorated, than land became desirable for the powerful and the wealthy to seize and to accumulate. Chao Kuo, a general of Chao, spent his time and money looking for good land to acquire. It was possible through purchase (II, 15), and it was also possible by other means, such as favor granted by the ruler (II, 16), or by using one's influence to provide ordinary people with tax shelter and thus acquiring control of their property (II, 17).

The fall of the feudal states after the Ch'in-Han unification did not terminate completely the influence of the formerly powerful households that had their connections with the feudal courts of earlier times. As noted by Masubuchi Tatsuo, some of the local powerful families during the first century after the founding of the Han dynasty could trace their ancestry back to the feudal aristocrats of the same localities. Being wealthy, they led extravagant lives and could command the services of followers and hired swordsmen to exercise their control. They competed with each other and even intimidated the local administrators—for instance, the descendants of the aristocrats of the former Ch'in state remained so powerful that the government had to assign its toughest administrators to govern that region.[17]

After the founding of the Han, new groups of powerful persons began to use political influence to enrich themselves. These were the meritorious followers of the founding emperors and their descendants. Kuan Fu, one of the lesser meritorious military men, for example, made a fortune and acquired good land. Ssu-ma Ch'ien implied in his

17. Masubuchi Tatsuo, "Kandai gunkensei no chiikibetsuteki kōsatsu ni tsuite," in *Chūgoku kodaishi kenkyū*, pp. 291-305.

account that the wealth of Kuan Fu was built by using his political and social connections (II, 18).

During the first three generations of the dynasty, the meritorious deputies and generals and their descendants dominated the empire's political scene. The chancellorship, for instance, was assigned to a marquis only—and the marquises in the early Han were those who were ennobled because of merit in founding the dynasty, or their children, or very close relatives of the imperial consort families.[18] The incomplete information on other lesser positions in the court demonstrates a comparable situation (*HS* 19B: 1a-20b). Although information on the local administrators is rare, the meritorious group seems often to have had their members occupying the positions of grand administrators of commanderies or the state chancellors of the kingdoms, as suggested by the career of Kuan Fu, who had been both a local administrator and a ranking official in the court (*SC* 107: 13-14).

The meritorious members of the group ennobled during the reign of Emperor Kao-tsu amounted to 143 marquises; 12 were ennobled during the reign of Empress Lü and 10 designated in the reign of Emperor Wen. The total is 165 (*HS* 16: 1a-69b). These families, together with their former retainers and relatives (see, for example, the relationship between Kuan Fu's father and the Kuan family) constituted the most influential households in the empire. They monopolized the political offices of the central government although the local leaders remained powerful at local levels. Both had the potential to acquire large landholdings for their own benefit.[19]

Although little hard evidence survives for the historian to substantiate the foregoing speculation, Han contemporaries did record the loss of titles by the poor to land they cultivated, while the wealthy possessed extensive land. Even though Tung Chung-shu did not specify which social groups acquired title to the land of the farmers, the likelihood is that it was the meritorious deputies and their descendants (I, 4).

Expanding the model proposed by G. A. Almond to apply to the interrelationship between economic performance and bureaucratic structure, Bert F. Hoselitz noted that the historical course of a society could be viewed as passing through several phases: first,

18. *HS* 19B: 1a-20b. Chao Chou was the last one who was advanced to the chancellorship on the basis of being a descendant of a meritorious general. 112 B.C. was the end of Chao's tenure.

19. For the relationship between these two, see Cho-yun Hsu, "The Changing Relationship between Local Society and the Central Political Power in Former Han: 206 B.C. - 8 A.D.," *Comparative Studies in Society and History* 7, no. 4 (1965): 358-61.

confrontation with the problems of solidarity or integration; second, confrontation with those of systemic goal attainment; and, finally, those of adaptation.[20]

Han China in the second century B.C. not only faced the problem of reintegrating the nation but also attacked the profound problem of defining its goals. During the first half-century of the Han dynasty, there were at least five civil wars. Internal warring did not cease in China proper until the reign of Emperor Wu. In the first five imperial reigns, the most ambitious contenders for the throne, including the powerful warlords and the imperial princes who were enfeoffed with extensive domains, had fallen after military confrontations with the throne. Yet the long arm of imperial authority did not effectively reach the local level until the reign of Emperor Wu, when rich men, ranking officials, and "local elites" were forcibly removed from their local areas, a measure aimed mainly at uprooting the survivors of the Chan-kuo period[21] (II, 19).

Furthermore, the first example of a grand administrator using extremely harsh punishment to eliminate local powerful households also occurred during the reign of Emperor Wu.[22] It should be noted here that such local leaders and elite were often the ones who had "encroached upon the small people," or had been "engrossers."[23]

Ironically, the powerful families in a region could consist of both the meritorious group who were the pillars of the Han regime and the old local leaders whom the throne viewed with suspicion. In fact, descendants of the members of the meritorious group, who had shared

20. Bert F. Hoselitz, "Levels of Economic Performance and Bureaucratic Structures," in *Bureaucracy and Political Development,* ed. Joseph La Palombara, pp. 168 ff., especially p. 188. Hoselitz here recapitulated in historical terms the Almond functional model on a social system of four sectors (latency, integrative, goal gratification, and adaptive). Cf. Gabriel A. Almond, "A Functional Approach to Comparative Politics," in *The Politics of Developing Areas,* ed. Gabriel A. Almond and James S. Coleman, pp. 3-64. Also see B. F. Hoselitz, "Economic Policy and Economic Development" in *The State and Economic Growth,* ed. Hugh G. J. Aitken, pp. 333 ff.

21. Hsu, "Changing Relationship Between Local Society and the Central Political Power in Former Han," pp. 361-63, especially n. 21, and T'ung-tsu Ch'ü, *Han Social Structure,* pp. 196-98. Note that although the descendants of the ruling houses of the Chan-kuo states had been removed from their homes to settle near Ch'ang-an, large-scale migration affecting the whole empire was ordered in 127 B.C. *HS* 6: 10b and *HS* 64: 19b; Ch'ü translations III, 18, 19 (pp. 419-20).

22. Ch'ü, *Han Social Structure,* p. 200; Hsu, "Changing Relationship . . . ," p. 362.

23. Ch'ü, *Han Social Structure,* translations II, 19 (pp. 419-20), and II, 37 (pp. 356-57).

the political power with the throne, witnessed the end of their period of power in the reign of Emperor Wu, at least two generations after the founding of the dynasty. The formerly fast ties between the throne and the meritorious deputies of the founding emperor or between the emperor and his distant cousins were much weakened. As the structure of informal relationship was no longer functioning, the successor, Emperor Wu, had to introduce a new structure with formal procedures to assure smooth functioning of the government. This was the time when the Han government had to recruit other social groups to man the staff and when the old-time constituents had to go.[24]

Nishijima has pointed out that the Han imperial system had to secure direct control of the individual subject and that the government would not tolerate other social forces standing between the ruler and the ruled.[25] Thus almost all the meritorious marquises were stripped of their domains and titles by the last two years of Emperor Wu's reign (88-87 B.C.). Verdicts were often based on extremely trivial accusations. The old group was gone so completely that Emperor Hsüan (r. 73-49 B.C.) found that the descendants of the most celebrated meritorious generals were working as hired hands and in other servile statuses (*HS* 16: 2b-3b).

So it was that the two types of powerful families who built local influence, the descendants of the old-day local leaders and the meritorious deputies of the Han founding emperors, probably saw their land, if any, taken away by that most powerful authority, the hand of Emperor Wu.

Imperial Favorites and Powerful Persons

Ironically, the mighty Han court of Emperor Wu, which suspiciously and jealously curtailed the growth of both the wealthy and the powerful who could have challenged imperial authority, nourished its own influential groups who also would encroach on the lands of the commoners. First were the ranking courtiers, all the political figures who directly exercised or used imperial authority, such as

24. Peter Blau cited an interesting case in which a new plant manager, in the absence of informal channels of communication that an old manager had built up to keep him informed about the work situation, replaced the old workers by outsiders who were readily inclined to obey a disciplinarian approach; hence he developed a formal bureaucratic method of administration. Peter Blau, "The Implication of Weber's Construct," in *Bureaucracy in Historical Perspective,* ed. Michael T. Dalby and Michael S. Werthman, pp. 20-21.

25. Nishijima Sadao, *Chūgoku kodai teikoku no keisei to kōzō.*

consort families, eunuchs, favorite courtiers, high officials in the court, imperial princes, princesses, and their relatives and descendants. It was a group whose flourishing depended upon the personal favor of the emperor, or the authority derived therefrom. It was, in a way, the incarnation of the Han monarchy.

The consort families and the eunuchs were especially significant because both were often the major members of the Inner Court, which was, in fact, the private secretariat of the emperor. The government, headed by the chancellor, was the Outer Court. The Inner Court was more powerful since it spoke in the name of the emperor.[26] Indeed, in the Western Han period, whatever title the leading figure of the consort family might bear, he often dominated the throne or at least was a right-hand man of the emperor. From the early days of the dynasty, when the Lü family rose to power with the assumption of imperial authority by Empress Lü after the death of her son, Emperor Hui, a preceding reign's consort family fell from the zenith of power and a new consort family, whose daughter had married the reigning emperor came to power. There were no less than ten such families in the Western Han period (Lü, Tou, T'ien, Ho, Wei, Hsü, Shih, Fu, Ting, and Wang), and eight in the Eastern Han (Kuo, Ma, Yin, Teng, Tou, Yen, Liang, and Ho). Several of them, such as the Wangs and the Liangs, repeatedly returned to the center of power as marital ties were arranged by the family in control of the Inner Court, thereby strengthening the connection between the consort family and the imperial house.[27]

Powerful persons often acquired tremendous amounts of land, either government land granted by the throne or private land seized from its owners (II, 20). The consort families were notorious for such practices and very likely the grant of public land to these families was so common that it rarely entered the historical record; the forceful seizure of land, however, was often mentioned. One of the early cases was the conflict between Tou Ying and T'ien Fen. Both were related

26. Ch'ü, *Han Social Structure*, pp. 170-71, 216 ff.; cf. Lao Kan, "Lun Han-tai-ti nei-ch'ao yü wai-ch'ao," *Bulletin of the Institute of History and Philology* 13 (1948): 231-32, and Yü-ch'üan Wang, "An Outline of the Central Government of the Former Han Dynasty," *Harvard Journal of Asiatic Studies* 12 (1949): 166 ff.

27. Ho Kuang's wife did her utmost to make her daughter empress in 71 B.C. Her conspiracy included the murder of the reigning empress (*HS* 97A: 22b-24b). The Wangs, who controlled the court for decades toward the end of the Western Han period, repeatedly established their daughters as empresses (ibid., 99A:9). In the Eastern Han period, the Liangs also arranged an imperial marriage when the family was in power (*HHS* 10B 5b-6a).

to the throne by marriage and both had been appointed great generals and chancellors. T'ien, at the zenith of power when Tou was living in semiretirement, asked Tou for some fields in a fertile district. Tou refused and the offended T'ien brought false charges against Tou, resulting in Tou's execution (II, 21). In another case, a member of the Wang family falsely claimed government land that had been rented and already cultivated by others. He then sold this land to the government and was paid an inflated price for it (II, 22).

In the Eastern Han period, imperial consort families usually had position and wealth already. Most of them, in fact, had been wealthy and influential before Emperor Kuang-wu unified the empire. Empress Kuo, the first wife of Emperor Kuang-wu, was a daughter of Kuo Ch'ang, who had houses and land valued at several million cash (II, 23). Yet there was still forceful acquisition of property by consort families in the Eastern Han. A very well-known case was Tou Hsien's seizure when he was in power of land belonging to a princess (II, 24). Allegedly they were the most conscientious members of a consort family, yet Ma Fang and his brother still managed to acquire large amounts of fertile land in the capital district (II, 25).

Liang Chi, an imperial consort family member who exercised the regential power for almost two decades (A.D. 141-159), was notorious for owning tremendous areas of land in the capital district. According to historical records, his extensive lands were used as a game reserve; Liang Chi must have taken much of the arable land away from the original owners, since the capital district was renowned for its fertile soil. However, as soon as Liang Chi was deprived of his power in 159, his property was confiscated by the government. The parks and gardens were then distributed to the landless poor to provide them with livelihood (II, 26).

Eunuchs, the palace servants, were very closely associated with the emperor. Although they were in a servile status and their castration evoked scorn and contempt, their proximity made them often the only persons with whom the emperor could forge an alliance to regain power lost to other powerful elements, such as the consort families. The services of eunuchs to the emperor as his private secretaries in the palace also provided them with influence in government routines.[28] In the Eastern Han period they acquired an even stronger position after the coup d'état of Cheng Chung, who in A.D. 93 helped Emperor Ho recapture power from the hands of Tou Hsien, an imperial

28. Ch'ü, *Han Social Structure*, pp. 97-101, and Ho Ch'ang-ch'ün, "Ch'in-Han-chien ke-t'i-hsiao-nung-ti hsing-ch'eng ho fa-chan," in *Han-T'ang-chien feng-chien t'u-ti-so-yu-chih hsing-shih yen-chiu*, pp. 175 ff.

maternal uncle (*HHS* 4:5b; 78:3b-4a). At least five times thereafter the eunuchs held supreme power in the court by overthrowing the de facto regents of the consort families (*HHS* 78: 3b-4a). Although a eunuch was without progeny, his extended family and even foster children constituted a considerable group at the core of power. In the biographies of powerful eunuchs, there are several examples of greedy behavior on the part of eunuchs and their relatives. Most detailed is the biography of Hou Lan, who held power throughout the reign of Emperor Huan (A.D. 147-167) and down to 172. From bribes and other illegal income, Hou Lan acquired large landholdings. In A.D. 169, a local official of Hou's home prefecture reported that Hou Lan had seized from others no less than 381 residences and 11,800 mu of land. Hou was probably the largest landlord in his home prefecture when he was in power (II, 27).

Others who availed themselves of imperial power were imperial favorites and courtiers. Although they did not reach the political significance of consorts and eunuchs, they were still able to obtain sizable fortunes owing to imperial favor. The most notorious was Tung Hsien, a courtier who enjoyed the love and trust of Emperor Ai (r. 6-1 B.C.), and who possessed such large landholdings that he opposed the limitation on land held by individuals (I, 5). The lesser imperial favorites included those who either were granted land by the emperor or could seize it through their influence in the palace and in the court. Among such persons were the wet nurse, very remote imperial relatives, and others, in addition to favored officials (II, 28-31).

This acquisition of land by consort families, eunuchs, and favorites seems to reflect a common idea that those who enjoyed imperial power, or power derived therefrom, normally would expect their influence to be transformed into durable wealth in land. They therefore pursued this end either by requesting the throne to bestow government land upon them, or by seizing the property of others. In later reigns of Han the "government land," over which the emperor had direct control, became less and less; the likelihood of seizing private possessions thus would have increased.

This group of persons had the power to acquire large landholdings because they were the favorites of a particular emperor; the power usurper of one reign, however, usually could not maintain his position throughout the next reign because of power struggles in the court. Nevertheless, there were always certain persons in favor in each reign who had the opportunity to seize sizable amounts of property by various means. Regardless of the quick turnover in reigns, the presence of such groups was constant.

The Powerful at the Local Level

The power holders at the local level were those who could utilize sociopolitical advantage to build up economic strength. This category, the local establishment, consisted of members of the hereditary nobility and the local elites who reached influential positions through government service. The former were the enfeoffed imperial children who bore the titles of princes and marquises; the latter were the members of the officialdom, imperial as well as local. It was not unusual for these two groups to be associated. Since the establishment of the Han dynasty, kingdoms and marquisates had been scattered amidst the regular administrative units, commanderies, and prefectures. The gradual concentration of power in the hands of the throne took away from the feudal lords their administrative power over the enfeoffed domains, which were then governed by a chancellor appointed by the imperial court. The princes, who were of the Liu imperial house, and the marquises, who were either meritorious officials or imperial relatives, still took residence at the places where they were enfeoffed, living upon the tax revenue.[29] Despite having no actual administrative authority, the status of these aristocrats nevertheless gave them considerable influence in their particular localities. They were in a position to acquire privately owned land from which they could draw rent, as well as receiving income from their domains. One such case entered the historical record. Prince K'ang, one of the sons of Emperor Kuang-wu, was enfeoffed in the kingdom of Chi-an. In A.D. 83, the reigning Emperor Chang (r. 76-88) returned to him domains that had been previously taken from the kingdom. The prince then began to build up his personal fortune, including houses, slaves, and real estate of 800 *ch'ing* of cultivated land (II, 32).

Private land might be acquired by several means. One, of course, was by purchase, another was by reclamation. Utsunomiya made an interesting observation about the financial situation of the house of Marquis Ch'ung-ling, to which Emperor Kuang-wu (Liu Hsiu) belonged. The marquisate of Ch'ung-ling was located in Nan-yang, present-day southern Honan, and was no more than a subdistrict, below the prefectural level. The first marquis was a son of the prince of Ch'ang-sha. For four generations, the ancestors of Liu Hsiu had left the marquisate to serve as civil servants. It is obvious that Liu Hsiu and his three generations of forefathers did not inherit the title of marquis, although they were still members of the marquisate house-

29. *Kuan-nei-hou,* however, are in a separate category. Lacking a definite domain, they stayed in the capital and were provided by the government treasury with a fixed amount of revenue.

hold. Liu Hsiu, in his youth, represented the marquisate in a tax appeal case (II, 33). It seems that the Liu household of about twenty male members remained as a kinship group, sharing the property together.[30] The tax alleged to be unpaid included both grain and hay money. Utsunomiya, basing his conclusion upon the mention of hay money, surmised that the Liu household property included some newly reclaimed cultivated fields, from which no regular tax was levied other than that on hay. He then figured that the unpaid tax in kind, which amounted to twenty-five thousand *hu* of grain, should be equal to the tax of ten years, beginning with the change of government from the Han to the Wang Mang regime. He concluded that this amount represented the tax exacted from 250 to 313 *ch'ing* of land, which he hypothesized as private property of the Liu's, in addition to the originally enfeoffed domain, which was about 200 *ch'ing* tilled by 467 households of Ch'ung-ling subjects.[31] Although the Ch'ung-ling marquisate household was not the only feudal lord of the Ch'ung-ling subdistrict, the land reclaimed and possessed by the Ch'ung-ling marquisate was about one-third of the arable land in that area, so the marquisate was the largest landlord in Ch'ung-ling.

What happened to the Ch'ung-ling marquisate was recorded in history because one of the sons, Liu Hsiu, later became the founding emperor of the Eastern Han dynasty. With the tremendous amount of revenue from these domains, the Han enfeoffed nobles, including the marquis of Ch'ung-ling, whose marquisate was at the level of a subdistrict and therefore a modest one, certainly would be able to build up great fortunes that included large landholdings. Since there were tens of kingdoms and hundreds of marquisates scattered throughout China, the existence of such transitional types of landlordship, which had an aristocratic background, was probably common in Han society. It would not be unusual if some of them simply remained rich landlords after they lost either their aristocratic status because of a dynastic change or their political influence because of remoteness of kinship ties to the ruling house. After all, the rent collected from each tenant would be one-half or more of the tenant's annual yield. The income of a landlord was indeed better than that of an aristocrat who could only exact 200 cash annually from each subject's household.[32]

30. The genealogy of Ch'ung-ling marquisate is reconstructed by Utsunomiya Kiyoyoshi in his *Kandai shakai keizaishi kenkyū*, p. 376.

31. Ibid., pp. 386-89. The text given in *HHS* has the tax figure 26,000 *hu* of grain, instead of 25,000 *hu*, as given by Utsunomiya. The period from A.D. 10, when Wang Mang abolished many of the Han nobles' titles, to the end of A.D. 20, should be eleven years instead of ten years as Utsunomiya suggested. These minor mistakes, however, affect his estimation very little. Cf. II, 33.

32. Li Ya-nung, *Hsin-jan-chai shih-lun-chi*, pp. 960-63.

The local leaders, after the full consolidation of imperial power in the reign of Emperor Wu, were gradually crystallized into a group of local elites from among whom the imperial authority recruited the candidates for governmental services. From the reign of Emperor Wen the selection of members of the Han bureaucracy was gradually shaped into an established institution. The reigns of Emperors Wen, Ching, and Wu witnessed the various stages in the development of this institution.[33] On the other hand, families of powerful officials normally could expect to monopolize the recommendations by exercising their influence upon the local officials to recommend their own children. The result was the creation of local powerful families.[34] The distribution of local recommendees helped the spread of local powerful families throughout the empire, since every region had its share in the officialdom.

During the Han, salaries for government service at middle and high levels were quite generous. According to the tabulations made by Utsunomiya, the salary scale for Han officials ranged from that of the minister, who received 180 *hu* of grain monthly, down to the clerk, who received 8 *hu* of grain monthly. In A.D. 106 the official exchange rate from grain to cash was 71.42 cash per *hu*.[35] The annual income of a ranking official, therefore, was more than 120,000 cash, while the income of the middle rank should be one-third of this figure. In terms of actual production rate, a mu of land yielded about 3 *hu* of grain annually; the highest yield from the most fertile land was 6.4 *hu* per mu. The salary of a ranking official, therefore, was equivalent to the production of about 720 mu of ordinary land. The midrank had an income equivalent to the produce of 240 mu of ordinary land. The range of actual productivity, of course, could go above or below the 3 *hu*/mu figure considerably.[36]

In addition to the regular salary the official enjoyed food provisions and an official residence, as well as various gifts and largesse, from the throne. His personal income, therefore, was sufficiently ample for him to accumulate sizable savings. The case of Kung Yü serves as a good example. A respected Confucian scholar, he was an ordinary farmer

33. Yen Keng-wang, *Chung-kuo ti-fang hsing-cheng chih-tu-shih,* pt. 1, 1: 81-85.

34. Ch'ü, *Han Social Structure,* pp. 180 ff., 205 ff.

35. Utsunomiya, *Kandai shakai keizaishi kenkyū,* pp. 203 ff., especially p. 213. This estimation is somewhat different from the figure of 120 cash per bushel given by Ch'ü. Ch'ü, *Han Social Structure,* p. 90. Ch'ü's estimate is based on the documents in II, 34.

36. For the production rate of 1.5 and 6.4 *hu,* see Hiranaka Reiji, *Chūgoku kodai no densei to zeihō,* pp. 147-56. For the rate of 3 *hu* per mu, see III, 13.

who tilled 130 mu of land to support his wife and children. His property was registered as being worth less than 10,000 cash. When he was summoned by the throne, he had to sell 100 mu of land to make the trip to the capital, where he was appointed to a middle-ranking office with a monthly salary of 9,200 cash, eventually reaching the upper rank with a monthly salary of 12,000 cash (II, 34). In one month Kung Yü was able to earn the equivalent of the value of his whole property. Imagine a powerful family in a locality that produced several generations of recommendees to government offices and suppose each was able to reach a good standing in government—how much wealth would be accumulated if only the regular salary were counted! In some cases the high-ranking official was ennobled and enfeoffed with a domain. The income therefrom would enable the family to acquire more wealth. K'uang Heng, a learned man, rose from a hired swineherd to a high position with the title of marquis. He collected tremendous wealth simply from the tax on his domain (II, 35).[37] At a time when land was the most reliable investment available, the establishment of a successful local family that produced a few generations of government officials was likely to mean the emergence of a landlord family that continuously poured their surplus cash, saved from salaries and/or other revenues, into the acquisition of land. Besides buying land, the bureaucrats could utilize their connections and influence to have public land transferred to them, or could even seize the property of less influential persons. Favoritism, especially in the Eastern Han, helped the perpetuation of these families since they simply recruited candidates from their own peers.[38]

Development of Landlordism

In the period before the reign of Emperor Wu, landlords were the locally powerful, the court favorites, the meritorious generals, and the wealthy merchants. After the power consolidation of Emperor Wu, the imperial favorites would stay, but the landlords, who were locally based powerful groups, were integrated into the Han bureaucratic structure as they were recruited to take government jobs through the recommendation system.[39] The merchants' properties, at least for once, were forcibly seized by the imperial power. In order to survive, the merchants often had to establish ties with the bureaucrats or even with the court. For instance, one of the very rich persons

37. Ch'ü, *Han Social Structure,* translation II, 37 (pp. 356-57), and p. 92.
38. Ibid., pp. 93, 174-75.
39. Hsu, "Changing Relationship . . . ," pp. 364-70.

listed in the "Biography of Money-makers" made his fortune by establishing alliances with influential figures in the court who provided him with protection (II, 36). Conversely, bureaucrats also engaged in business. For example, Ning Ch'eng, an official dismissed from office, looked upon investment in land as a profitable enterprise so he borrowed money to acquire land that was tilled by tenants (II, 37). Brothers in the same household might divide their jobs, as did the family of Ho Wu. He and his four brothers all had been on the staff of the local administration and at least three of them had been ranking officials in the government, yet one of them was still registered as a merchant in their home-town market (II, 38).

The *Ordinances for the Four Peoples (Ssu-min-yüeh-ling)* is testimony that the bureaucrat, the landlord, and the trader-manufacturer could be combined in one household and that their activities were regarded as the normal pursuits of the upper class (III, 1). The intermarriage of the aristocratic landlord family of Liu Hsiu with the merchant-landlord families Li, Fan, and Yin reflects another kind of coalition of wealth and political power.[40]

Meanwhile, the aristocrats and the bureaucrats could overlap, since many noble children also could serve in the government. Liu Hsiu's great-grandfather, grandfather, and father all had been local administrators while they were still members of the household of the Marquis of Ch'ung-ling (*HHS* 1: 1b). And of course the imperial favorites, albeit of various origins, always held certain government appointments, even though they might not be career bureaucrats. The merging of these various social groups created a single layer of that trinity of the wealthy, the educated, and the powerful, by the time of the second half of the Western Han.

If the powerful people invested in land, it was not only because the commercial sector was relatively lacking in security and vulnerable to political pressures (which had been well demonstrated in the reign of Emperor Wu), but also because of the profitable return on land. The Han land tax was normally one-thirtieth of the annual production, while the rent collected from tenants was one-half or even more of the crop (I, 4, 6). The difference between the tax and the rent was pocketed by the landlords. When the population grew, the price of land would rise in response to the demand for it. The profits would be invested in acquiring more land. The snowballing process of land acquisition would take time to gain momentum, yet once under way, acceleration would be fast. This seems to be the main reason that

40. Ch'ü, *Han Social Structure*, pp. 243-44; Utsunomiya, *Kandai shakai keizaishi kenkyū*, pp. 275 ff.

land concentration did not create a social crisis until the reign of Emperor Ai (6 B.C.-A.D. 1), more than a century after the effort of Emperor Wu to take land away from the wealthy. Nevertheless, the crisis reached such proportions less than two decades later that Wang Mang had to try the drastic measure of nationalization of all the land (I, 6).

In the Eastern Han period, the problem of extensive landownership by the politically powerful was a thorny issue as early as the days of the founding emperor, who noticed that his relatives and his courtiers were owners of much land. His attempt to conduct a land survey throughout the empire ran into all kinds of obstacles, and numerous local officials were punished for their failure to fulfill the assignment (II, 39).

The shortage of land for independent farmers in the Eastern Han period probably was not solely the consequence of population growth but also reflected the increasing demand by landlords who searched for suitable land to acquire. Thus in the Eastern Han a good local administrator would devote much of his energy and attention to reclaiming arable land (II, 40, 41, 42, 43). Although these model officials probably did sincerely attempt to open up new land for those who would till it themselves, the urge for new land still could be a combination of demographic pressure and the stimulus of the market.

It is recognized that the establishment of the Eastern Han was the result of the joint efforts of intellectuals of landlord background, among whom Liu Hsiu and his associates were leaders.[41] Therefore, powerful families enjoyed a favorable climate during that period, perpetuating themselves and developing into a significant social force that was characteristic of the Eastern Han.[42] Their economic interests were protected by their political predominance.[43]

Throughout the Eastern Han period no attempt was made to curtail the uneven distribution of land; even the kind of effort made by Shih Tan and Wang Mang in the Western Han was not attempted. The situation became critical, as noted by Chung-ch'ang T'ung, who lamented the unchecked accumulation of wealth in the hands of

41. Yü Ying-shih, "Tung-Han cheng-ch'üan-chih chien-li yü shih-tsu ta-hsing-chih kuan-hsi," *Hsin-ya hsüeh-pao*, vol. 2 (1956); Hans Bielenstein, "The Restoration of the Han Dynasty," *Bulletin of the Museum of Far Eastern Antiquities*, no. 26 (1954).

42. Yang Lien-sheng, "Tung-Han-ti hao-tsu," *Tsing-hua hsüeh-pao*, vol. 11, no. 4 (1936); Utsunomiya, *Kandai shakai keizaishi kenkyū*, pp. 393-96, and idem, "Kandai gōzoku ron," *Tōhōgakuhō*, no. 23 (1962).

43. Ch'ü, *Han Social Structure*, pp. 203-4.

a few (II, 44). Chung-ch'ang T'ung, however, also wished that he himself might have fertile land tilled by tenants so he could devote his time to scholarly pursuits. His ideal, indeed, is a reflection of a life regarded as upper class[44] (II, 45).

By the end of the Han dynasty, large landholdings had become so common that Hsün Yüeh, who perhaps was the last of the Han officials to propose some restriction on landholdings, felt caught in the dilemma of deciding between the impracticality of restoring the ancient well-field system and the undesirability of allowing land acquisition without restriction (II, 46).

The final phase of landlordism was the polarization of the social differentiation between landowners and tenants. Tenants were totally subjugated to the control of the landlord, who in theory should enjoy only economic superiority by contract. What is described by Chung-ch'ang T'ung (II, 44) inevitably led to the development that occurred at the end of the Han period when the tenants became private retainers of the landlord as imperial control of individual subjects collapsed. On the eve of the Three Kingdoms, local leaders could command hundreds or thousands of retainers.[45] This kind of tie, of course, went beyond the landlord-tenant economic relationship and ushered in another social structure marked by the stratified society of the Wei-Chin period. Until this stage was reached, however, some tenants would remain directly and legally subjects of the state that levied taxes on them and provided protection for them, a situation that differed significantly from that of serfs on seignorial manors of feudalism.

Conclusion

The failure of further development of the entrepreneurial sector of Han society, owing to imperial pressure to destroy competition, meant the limited development of business in early Chinese economy and the relative favoring of agricultural development. In chapter 6,

44. Cf. Etienne Balazs, *Chinese Civilization and Bureaucracy*, trans. H. M. Wright, pp. 214 ff.

45. Such retainers were known as "guests" *(pin-k'e)* or "personal troops" *(pu-ch'ü);* for instance, Li Ch'ien had several thousand households of *pin-k'e* whom he led to join the army of Ts'ao Ts'ao. Li Ch'ien's nephew, Li Tien, inherited the command of this armed band after Li Ch'ien's death. Li Tien and his *pu-ch'ü* supplied Ts'ao Ts'ao with grain and clothes during a battle fought near their home prefecture. They obviously remained a group to engage in agriculture, yet were completely subordinated to the Li family. *San-kuo-chih,* I-wen ed., 18: 1a-2b (hereafter cited as *SKC*).

nonagricultural activities undertaken on the farm will be discussed further. The development of landlordism benefiting the most powerful groups in the central government as well as those at the local level meant that agriculture was to be regarded as a way—the principal way—to produce and to make profit. Small amounts of arable land per capita would lead to the intensification of agriculture to fulfill the hope of adequate production and profit. The large population provided abundant labor; the desire of the landlord to profit and of the farmer (independent as well as tenant) to survive with limited land provided strong motivation to develop such labor-intensive agriculture.

III

THE FARMER'S LIVELIHOOD

THE LANDLORDS

THE text of *Ssu-min-yüeh-ling (Ordinances for the Four Peoples)* by Ts'ui Shih in the second century provides interesting details of the life of a landlord.[1] Analyzing the existent parts of the text, Yang Lien-sheng and Shih Sheng-han reached similar conclusions: the *Ordinances* prescribed the methods and timetable to be followed by a gentleman farmer who supported a good-sized household by agriculture and who supplemented his income by the marketing of his agricultural surplus and by sale of handicraft items.[2] The contents of the *Ordinances* are divided into eight categories: (1) education, religious rites, and social activities; (2) seasonal demands of farming; (3) general sericulture, tailoring, laundering, mending, and so on; (4) food preparation, brewing, and food preservation; (5) house repairing and irrigation construction; (6) gathering of wild plants, especially medicinal herbs; (7) maintenance of household furnishings; (8) sale and purchase of miscellaneous items, such as hygienic supplies.

1. Ts'ui Shih's life is included in the biography of his grandfather, Ts'ui Yin. No precise dates of his birth and death are given (*HHS* 52: 14a-18b). Shih Sheng-han estimates that Ts'ui Shih was possibly born in 103, flourished 145-167, and died in 171 (see Shih, *Ssu-min-yüeh-ling chiao-chu*, pp. 79-88). The *Ordinances* were probably written shortly before A.D. 150 (ibid., pp. 84, 92).

2. Yang Lien-sheng, "Ts'ung Ssu-min-yüeh-ling so-chien-tao-ti Han-tai chia-tsu-ti sheng-ch'an," *Shih-huo* 1, no. 6: 8 ff. Shih, *Ssu-min-yüeh-ling chiao-chu*, p. 89.

The author, Ts'ui Shih, was a member of a distinguished bureaucratic clan, whose genealogy follows:

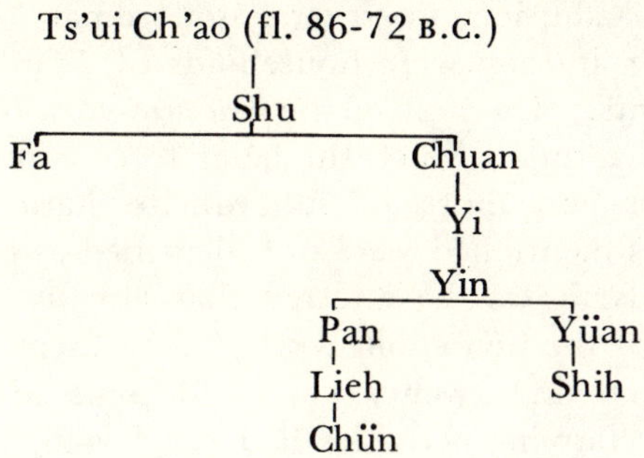

Ts'ui Ch'ao started his career as a staff member of the local administration and later served in the central government as an attendant imperial secretary. For generations, the Ts'ui family entered government as local staff members and eventually reached high positions in the central government. (Ts'ui Yi was an exception; because of illness he never worked in the government.) Ts'ui Shih was first recommended to the central court because of his filial piety; upon his arrival, he was appointed court attendant. For a brief period he served as a major on the staff of Liang Chi, the powerful grand general who dominated the government. Later he was appointed administrator of a frontier commandery and then returned to the capital because of illness. The last years of his life were spent in his home after his mother's death. Although Ts'ui Lieh, a cousin of Ts'ui Shih, was forced to contribute five million cash to obtain the position of grand minister, Ts'ui Shih was not rich, because his father had spent lavishly in entertaining friends and Ts'ui Shih had provided an expensive funeral for his father. A respected family name and the wealth of kinsmen helped little to save Ts'ui Shih from financial difficulties. When Ts'ui Shih died, he was too poor to have a coffin, and his friends purchased one for him and paid for his funeral (*HHS* 52: 1a-19b).

Ts'ui Shih's financial situation made it necessary for him to supplement his income by marketing wine and preserved food. The detailed information on all kinds of food preservation mentioned in the *Ssu-min-yüeh-ling* shows that Ts'ui Shih paid close attention to such trade. The text seems to be an actual record rather than a theoretical discussion, as some of the later agricultural encyclopedias appear to be. It records, for example, the appropriate time for purchasing and selling grain and silk for maximum profit, as well as Ts'ui's concern with the marketing of preserved food[3] (*HHS* 52: 18a; see document III, 1: B.8, J.5).

The *Ssu-min-yüeh-ling* farm might have consisted of a master household in addition to several subordinate households, though it is not clear if they were households of Ts'ui's grown children or those of tenants.[4] The text of the *Ssu-min-yüeh-ling* gives the seasonal activities of a farm in which the labor force consisted of both household members and those of subordinate status. Although the few working hands mentioned were not identified as being slaves, servants, or hired hands, it can be inferred that directly and indirectly wives and children of the household worked with them. Since male children between fifteen and twenty years old went to school only during winter when "farming activities had not begun," it appears that they were actively engaged in farming and were referred to as the children of the master. Perhaps some activities were made the responsibility of one of the participating units, albeit under certain supervision. For the third month, the busy season of growing silkworms, the text reads, "When the silkworms are all hatched, one joins together wives and children to diligently carry out the work. Do not allow them to devote themselves to other tasks and thereby bring disorder to the primary occupation. Those who disobey such an order should be punished without hesitation." The lines show that there were working units receiving orders from the master yet forming a household that was responsible for the production of silk.

The master of the house, as depicted in the *Ssu-min-yüeh-ling,* was an educated gentleman. He sent seasonal greetings to local officials, teachers, and officials under whom he had served. He observed the rituals of clan gatherings, ancestor worship, initiatory capping ceremony, and so on, which fit his social standing. His children were well educated. Furthermore, he was a local leader, taking care of the poor, especially those who were his clan members, as well as organizing local armed men to guard against bandits and thieves.

The clan organization appeared strong. Descendants of the same great-grandfather gathered on New Year's Day and other festivals to celebrate and offer sacrifice to their ancestor. The clan members were given priority for relief in time of need. On festive occasions

3. Yang, "Ts'ung Ssu-min-yüeh-ling . . . ," p. 10. Also, Nishijima Sadao, *Chūgoku keizaishi kenkyū,* pp. 54-56.

4. Yang Lien-sheng interprets the various kinds of workers as slaves or female slaves ("Ts'ung Ssu-min-yüeh-ling . . . ," p. 8). The terms *tailor, cook, silkworm girl,* and so forth, however, do not render any connotation other than descriptions of their tasks.

the activities were handled by the master of the household, assisted by clan members on special assignments. The individual members, nevertheless, could have different financial situations: some were rich and were able to provide aid to others, others were poor and received such aid.[5] *The Ordinances* clearly reveal the potential influence a kinship group could have exercised over the population in the general area because of political connections, economic strength, and the solidarity of the group. The gentleman farmer in the *Ssu-min-yüeh-ling* perhaps represented thousands of such local elite families scattered in every prefecture of Han China. They were neither extremely wealthy landlords who might be detached from actual farming activities nor ordinary farmers, since their lives were much more comfortable and more complex than those of simple peasants. Their social status and political role reinforced each other and they maintained a privileged position in the local community.

Fan Chung, the grandfather of Emperor Kuang-wu's mother, and Ning Ch'eng, a high official who acquired a large area of land after being dismissed from government service, embodied two variations of these landlords. Fan Chung differed from the Ts'ui family in that he had large landholdings and numerous servants and slaves. It is not clear whether or not he had tenants. The historians note that Fan was able to keep the entire household so self-sufficient that he grew catalpa and lacquer trees for making furniture at home (III, 2). Although Fan Chung was wealthier than Ts'ui Shih, he also engaged in some trade.

Ning Ch'eng was also an entrepreneur; he borrowed money and invested in land. However, Ning Ch'eng leased his land to thousands of tenants, whereas Fan Chung managed his himself (II, 39). Even the Fans, however, were in a position to extend their control beyond their own plantation. During an emergency, more than one thousand households in the vicinity went to live within the fortified compound constructed and defended by the Fans and their kinsmen.[6]

These examples illustrate the characteristics of leadership in the Han countryside. Moriya Mitsuo points out that the *Ordinances for the Four Peoples* actually listed activities that in the extant earlier almanacs were rituals and works to be supervised or observed by the

5. The historical reality of the Ts'ui case verifies the existence of differentiation among members of the same surname group. Note, however, that Ts'ui Lieh, although rich, did not extend such assistance to Ts'ui Shih, whose funeral was held only when a few friends provided help.

6. *HHS* 32:2a. The *Hsü-Han-shu* quoted in the commentary probably refers to this emergency when it records maintenance of weaponry. See III, 2.

state ruler.[7] Apparently the local elites saw themselves as practicing the Confucian ideology of the ancient teachings. Nishijima extrapolated from this observation to say that there had been in the Han period a popularization of the aristocratic rituals that had extended to the level of the common people.[8]

Local leadership, in the cases of the Fan and Ts'ui families, is characterized by its relationship with its kinsmen and relatives. The leader extended help and assistance to those who were in need, while his kinsmen rallied to his side when it was necessary to form an armed group to defend their homes. Religious ceremonies, such as worshipping common ancestors, and social interaction, such as feasts, provided activities that strengthened the solidarity of the kinship group (III, 1: A.1; III, 2). These strong and large kinship organizations, with the financial advantage of group security that was reinforced by mutual assistance, facilitated their own perpetuation.[9]

By the time of Wang Mang, the bonds of kinship formed the local armed bands who occupied fortified villages and who established virtually self-governing bodies. Those who responded to the uprising against the Wang Mang regime were often kinship groups numbering in the thousands. During social disturbances in the Eastern Han period, the large kinship groups again and again demonstrated their strength, especially during the last part of the Han when the empire was shattered by anarchy and civil wars. Local leaders mobilized their kinsmen to join a particular camp and to defend their area against intruders.[10]

The landlord, who had the collective strength of a kinship group at his disposal, certainly had a considerable financial potential to expand his farming operations. This was especially relevant to the opening up of new land, which would demand large amounts of capital as well as labor. Capital was essential to carry over the three- or four-year unproductive period until once wooded and brushy land could

7. Moriya Mitsuo, *Chūgoku kosaijiki no kenkyū*, pp. 16-18, especially p. 18. Moriya makes a very interesting comparison of the lives of the people in the *Ssu-min-yüeh-ling* and their counterparts in the Classics, such as the "Yüeh-ling" in the *Li-chi*, and the "Hsia-Hsiao-cheng" in *Ta-tai li-chi*. The parts he could not compare for want of relevant classical material relate to commercial activities and daily routines.

8. Nishijima Sadao, *Chūgoku keizaishi kenkyū*, p. 51.

9. T'ung-tsu Ch'ü, *Han Social Structure*, pp. 26 ff.

10. Yü Ying-shih, "Tung-Han cheng-ch'üan-chih chien-li yü shih-tsu ta-hsing-chih kuan-hsi," *Hsin-ya hsüeh-pao*, vol. 1, no. 2 (1956). Hsü Fu-kuan, *Chou Ch'in Han cheng-chih she-hui chieh-kou-chih yen-chiu*, pp. 327-39; Ch'ü, *Han Social Structure*, pp. 31-32, 202-9.

be used as grain fields. During the period of reclamation, not only did labor have to be hired, but also rather costly large implements and draft animals had to be maintained.[11] With capital, one could also acquire land directly by purchase or indirectly by foreclosing on mortgages. Not all landlords pursued the advantage that having money to lend gave them, of course. Fan Chung ordered his record of debts, which amounted to millions of cash, burned (III, 2). This was unusual, for most landlords probably would have taken possession of the land of debtors.

Interest on money lent was another source of income to the landlord. The interest rate in Han times was normally 20 percent annually, although government loans under Wang Mang were as high as 36 percent.[12] Huan T'an, a first-century scholar, noted the common practice of usury when he stated that many rich persons invest in land by means of extending loans and that the middle-class youth, who were guarantors of the loans, served the rich as servants in order to share some of the profit (II, 14).

Tenants and Other Farm Laborers

Laborers working in the fields of landlords were of several types. Hired hands, long-term as well as short-term, were perhaps the most common working force. There were also slaves. Hired hands were probably preferred to the slaves since the purchase of slaves meant capital was paid out in one payment at the time of transaction, while wages to hired hands were paid over a period of time.[13]

The cost of maintaining slaves may seem to involve only grain

11. Amano Motonosuke, "Kandai gōzoku no daitochi keiei shiron," in *Takikawa hakushi kanreki kinen ronbun shū* 1:13-14. Although the source material quoted by Amano is from the *Ch'i-min-yao-shu*, a work compiled in the sixth century, the technological level in Han times was about the same as that contained in the above book. Cf. Shih Sheng-han, *Ch'i-min-yao-shu chin-shih*, p. 3, secs. 1.2.1-1.2.3.

12. Dubs, *HFHD* 3: 300, 497, 533-34.

13. The price of a slave in the Han period varied from 5,000 to 20,000 cash. Utsunomiya Kiyoyoshi, *Kandai shakai keizaishi kenkyū*, pp. 254-55, n. 10. Amano Motonosuke and Makino Tatsumi debated the reasons for tenancy rather than large-scale use of slaves. Makino argued that the increase of population, the withering of the economy, the high price of slaves, and plowing were factors favoring the development of tenancy. Amano rejected the hypothesis. Cf. Makino, "Kandai ni okeru raikohō shinpo no imisurumono," *Tōzai gakujutsu kenkyūjō ronsō*, vol. 10 (1953), and Amano, "Chūgoku kodaishika no shosetsu o hyōsu," *Rekishigaku kenkyū*, no. 180, p. 36. I would recognize, however, that the high price of slaves would deter the development of large-scale slavery.

consumption, for there was no wage paid.[14] It even can be speculated that in large-scale land reclamation, which required extensive use of labor for felling trees or for irrigation construction and costly investment for maintaining large implements, the most economical means would be to use unpaid labor (slavery) or possibly hired hands. Once the field had been prepared for regular crops, the intensive use of labor required in Chinese farming then would make tenancy preferred.[15]

In such arguments, a few very essential variables against using slaves are missed. The landlord must pay the initial purchase price, as well as the slave's maintenance—housing, clothing, and so on. Furthermore, the landlord must bear all risks of production as well as the cost of tools, seeds, and animals. All these could amount to a sizable outlay of money in addition to the assumption of a great deal of responsibility. The favor then seems to tip toward the use of tenants. As for the arguments for the advantages of slaves in large-scale reclamation, it appears more likely that reclamation was a continuous ongoing project, one that added a bit of arable land each year and was probably pursued during agricultural off-seasons. Such tasks would not demand mobilization of sizable labor forces.

Furthermore, abundant supplies of cheap labor might discourage large-scale use of slaves for productive purposes. One of the advantages of using freemen for labor is that employment covered only the time needed for the job; the slave might become a nonproductive investment, from the owners' viewpoint, if he were sick or when there was insufficient employment.[16]

14. Utsunomiya estimated the cost of food necessary to maintain twenty slave laborers on a farm of 1,000 mu of land. He figured 530 *hu* of grain consumed annually. On the other hand, the value of the grain kept by tenants of ten households as their share is estimated by Utsunomiya as follows: After deducting two-thirds for rent, the average is 1320 *hu;* after deducting one-half for rent, the average is 1980 *hu.* Thus Utsunomiya concludes that the ratio of advantage is 60 percent in favor of the unpaid labor (slavery) against the two-thirds rent tenancy or 73 percent against the one-half rent tenancy. See *Kandai shakai keizaishi kenkyū,* pp. 301-9. He then acknowledges that in spite of such statistically determined disadvantages, tenancy would still be regarded as favorable because there was no need of supervision and no need for large capital paid out at one time (ibid., pp. 308 ff.).

15. Utsunomiya agrees that the reclamation of new land should demand quick mobilization of a large labor force, which would be more readily available by tenancy. See ibid., pp. 309-15. This argument is exactly contradictory to the one held by Amano in his "Kandai gōzoku no daitochi keiei shiron," pp. 5-13.

16. C. Martin Wilbur, *Slavery in China during the Former Han Dynasty,* pp. 203-4. Wilbur also estimated that slaves perhaps constituted less than 1 percent of the total population of Han China (ibid., pp. 176-77).

Although slaves were used in farming, the connection between landlordism and large-scale slavery is far from obvious. As long as the landlord acquired land by accumulating many small farms, the scattered holdings could be made profitable only through prompt renting to tenants.[17]

The landlord, therefore, had two alternatives: to keep the land as a large farm on which slaves or hired hands worked under the supervision of the master, or to divide the land into small farms to be leased to individual tenants. Both systems probably were used, though there is not sufficient material to reach a definite ratio between these two types. It is possible to deduce from contemporary observations that tenancy probably exceeded slavery as a common practice. One of the observations made by Chung-ch'ang T'ung was that a rich person had thousands of retainers and tens of thousands of "attached persons" who were probably tenants (III, 3). Ts'ui Shih observed in his essay on contemporary politics that there was a subordinate group that retained only limited independence by yielding themselves to a powerful person, yet lacked the protection that a slave could expect from a master (III, 4). The two references to subordinated and dominated populations include tenants as the principal constituents. Unless the tenants formed a good percentage of the rural population, it would be odd for both Tung Chung-shu and Wang Mang to cite the hardship imposed on farmers by heavy rent, one-half or even two-thirds of their produce (II, 4, 6).

Only a rough estimate of the Han tenant population can be given. In Emperor Wu's reign, the property tax was levied. Failure to give an honest report by the owner could bring a charge of defying the law, the penalty for which was the confiscation of one-half of the property in question with the other half going to the informant as a reward. The result, according to the record, was *the destruction of virtually the majority of the well-to-do.* In a large prefecture, confiscated land that then belonged to the government reached several ten thousand mu, while in a smaller prefecture, more than ten thousand mu (II, 9). If double the confiscated land represented less than one-half the land of the wealthy, then the total property of the wealthy in a large prefecture could be at least one hundred thousand mu, and in a small prefecture, several ten thousand mu. This would be land that was tilled by tenants or hired hands and slaves. If all this land were leased to the tenants, and each tenant household tilled fifty mu, the acreage needed for subsistence living, there then would be less than one thousand to several thousand tenant households in a

17. Ibid., pp. 212-15.

prefecture. However, the population in a prefecture was generally over ten thousand households, with forty to eighty thousand households for the most populated prefectures and three to five thousand for those least populated. The tenant population, therefore, would be hardly over 20 percent of the whole population.[18] The remaining land probably was tilled by the independent free farmers who operated small farms with labor drawn from the household. The speculation attempted above is, to be sure, very crude; it must be regarded as a suggestive estimate. A broad margin for errors and local deviations should be allowed.

Free Farmers and Small Farms

The number of free farmers who owned the greater part of the land, and the size of the landholdings, of course, still could vary a great deal. In the literary sources, there is the case of Kung Yü, who owned one hundred thirty mu of land, with a value of less than ten thousand cash. He regarded himself as a poor farmer (II, 34). Of course, the criteria of being rich varied according to the value of land in a given area. The Chü-yen tablets contain at least two property registers. One records the property of Li Chung—five hundred mu of land, one residence, three slaves, draft animals, and vehicles—which was valued at one hundred and fifty thousand cash (III, 5). The other records the property of Hsü Tsung, who possessed fifty mu of land, one residence, and two oxen, and which was valued at thirteen thousand cash (III, 6). It should be noted that Li Chung could have used the three slaves (two male and one female) and draft animals for cultivating the five hundred mu of land. Hsü Tsung, on the other hand, had four members of the household who were not of the nuclear family. Since Hsü Tsung was a low-ranking garrison officer and relatively poor, the possibility was small that these additional members resided

18. Suppose we take three hypothetical prefectures to calculate the number of tenants. *Case A* (a large prefecture): Total population, 50,000 households; acreage for tenancy, 100,000 mu (about three times more than II, 9 suggests as the maximum); tenant population, 100,000 divided by 50 equals 2,000 households. *Case B* (a small prefecture): Total population, 3,000 households; acreage for tenancy, 20,000 mu ("more than 10,000 mu"); tenant population, 20,000 divided by 50 equals 400 households. *Case C* (a small to average prefecture): Total population, 10,000 households; acreage for tenancy, 100,000 mu (about 5 times more than II, 9 suggests as maximum); tenant population, 100,000 divided by 50 equals 2,000 households. This rough estimate of the proportion of tenants to total population is 4 percent, 12 percent, and 20 percent, respectively. It should be noted that these figures are designed to maximize the extent of tenancy.

with him simply as his dependents; they probably lived with Hsü as working hands. The households of Li and Hsü probably drew some salary as well as official rations as frontier garrisons usually did. Their farming activities, carried out either by slaves or by non-nuclear-family working hands, still could have been a significant source of income.

Work on a farm was indeed varied. On a Han farm it could involve not merely agriculture but also animal husbandry, sericulture, gathering, hunting, and even some trading. The farmer had to manufacture and repair practically everything on the farm and in the house. He and his family were carpenters, masons, weavers, tailors, peddlers, and so on, in addition to being farmers. The verse jocularly composed by Wang Pao in the mid-first century B.C., using the form of a contract between a master and a slave, covered a wide range of activities on a farm. The whole literary piece is exaggerated, for it is hardly plausible that any single farm, not to say any single slave, could have actually taken up all of the tasks. However, it does provide us with a panoramic view of the many activities related to farming. The smaller the farm, of course, the fewer activities other than actual farming (III, 7).

The busy schedule on a Han farm, as reflected in the *Ordinances* and the mock contract, showed a very heavy working load on a farming household. Ch'ao Ts'o, in his famous memorial, told how hard a farmer and his family had to toil in order to make a living (I, 3). And it was expected that a farmer of medium means should be able to support seven persons [19] (III, 8).

Expenditures of a Farming Family

A large labor force on a farm means that a substantial portion of the farm's production is consumed by the labor, with a relatively small margin left to meet other expenses. An account of the income and expenditures of a small farm in the pre-Ch'in (Chan-kuo) period cited in the *Han-shu* gives a deficit of 10 percent of the annual income, presumably in a year of mediocre crops (III, 9). In the time of Ch'ao Ts'o the situation remained very much the same (I, 3).

19. Utsunomiya estimates that to work a 100-mu farm, the labor required should be 4.8 working persons without animals or two working persons with one animal (*Kandai shakai keizaishi kenkyū*, p. 303). Some scholars doubt that the three slaves (of III, 5) could be engaged in agricultural work as a labor force. The argument that the female slave was not strong enough to till the land, and the two male slaves were too young to be useful seems unconvincing because women and minors did work in Han times as they do in China today. Cf. Huang Lieh, "Shih Han-chien-chung yu-kuan Han-tai she-hui hsing-chih chu-li," *Li-shih yen-chiu*, no. 6 (1957), pp. 67-68.

Let us establish a hypothetical farming household and estimate its expenditures and income. Assume the household possesses a farm of seventy mu of cultivated land and is independent.[20] The household would provide its own tools, seeds, labor, and so forth, and would consist of five persons:[21] the master, forty years old, his wife, thirty-six, one parent, fifty-six, and two children, a son of twenty and a daughter eighteen years old. The expenditures would be in three categories: (1) household consumption, including food and clothing; (2) taxes, including land and poll taxes, and so forth, and (3) social activities, including religious worship and family gatherings.[22]

Estimates of food consumption in Han times vary from one source to another. Li K'uei, a Chan-kuo scholar, is attributed with an estimate of 1.5 *shih* of unhusked grain per person per month (III, 9). If the grain were unhusked millet, the capacity measurement would be 2.5 *hu*.[23] Chao Ch'ung-kuo, who first suggested in 61 B.C. the establishment of military colonies, reported that a troop of 10,281 persons consumed 27,363 *hu* of grain each month, or 2.66 *hu* per person per month[24] (III, 10). Another literary source, an essay attributed

20. The most commonly assumed farm size was 100 mu (see statements by Ch'ao Ts'o, Tung Chung-shu, etc.). This ideal derived from the alleged well-field system in the ancient days. The actual sizes of landholdings, of course, varied drastically.

21. This hypothetical case is based upon the average household size given by Ma Ch'eng-feng in table 3 of chapter 1. It is also consistent with the statement made by Ch'ao Ts'o (I, 3) in which he said that a household of five persons could include no less than two adult laborers. In our case, either the grandmother or the grandson should be considered as adult laborers. The actual household sizes, of course, could vary greatly; see Utsunomiya, *Kandai shakai keizaishi kenkyū*, p. 405 ff., and Hsü Cho-yun, "Han-tai chia-t'ing-ti ta-hsiao," *Ch'ing-chu Li Chi hsien-sheng ch'i-shih-sui lun-wen-chi*, pp. 789 ff.

22. Amano Motonosuke made a similar effort to estimate the income and expenditure balance of a farming household. He did not, however, go into detail. Cf. Amano, "Chūgoku kodaishika no shosetsu o hyōsu," p. 35.

23. Nancy Lee Swann, *Food and Money in Ancient China*, p. 365. Different grains of the same weight would measure at different capacities (ibid., pp. 138-39, n. 105; p. 140, n. 108; p. 364). The ratio of one *shih* of millet making 1.65 *hu* in capacity is calculated, according to Han lexical data, as one *shih* by weight equaled 20 *tou* of unhusked rice or 16.5 *tou* of unhusked millet, while 16.5 *tou* of unhusked millet produced 10 *tou* of husked *mi*, the capacity of one Han *hu*; see ibid., p. 365.

24. Michael Loewe cites two other examples of military provisions. One is an estimate made by Yen Yu of the cost of a proposed expedition against the Hsiung-nu that was eighteen *hu* of *pei* (dried cooked rice) as three-hundred day's provision per person (*HS* 94B:19b). Michael Loewe, (*Records of Han Administration*, p. 70. Yen Yu's estimate converts to a monthly ration of 1.8 *hu* and Li K'uei's to 1.5 *hu* of husked millet. Both figures are less than 2.66

to Ts'ui Shih, gives 6 *hu* of millet as food for two male adults in a month; the average provision was 3 *hu* of millet per person a month (III, 11).

The Han records of frontier garrisons give direct information on food provision. For a soldier, the monthly provision was 3.33 *shih* of grain. The measurement, however, is calculated on a "small scale," which is three-fifths of the "large scale." Converted to the measurement of capacity, it should be 4 *hu* of unhusked grain, a little more than the amount given by Ts'ui Shih.[25] There seems to be some distinction between the regular servicemen and the convicts, as well as between soldiers permanently living on the frontier and transient soldiers. The former received more rations than the latter, whose ration was discounted.[26] Dependents of soldiers, however, received fewer provisions. An adult (fifteen years and older) received 2.16 *shih* of unhusked grain; a child (seven to fourteen years), 1.66 *shih*; and an infant (under six years), 1.16 *shih*. In one case, a one-year old girl was given 0.8 *shih* of grain.[27] In terms of capacity, the dependents were given 2.1 *hu*, 1.2 *hu*, and 1.1 *hu* of grain for each adult, child,

---

hu. It should be noted, however, that both dried cooked food and husked grain are smaller in volume than unhusked grain. Taking 10:16.5 as the ratio for husked to unhusked millet, the capacity should be 3.13 and 2.48 *hu* respectively.

25. Loewe, *Records of Han Administration*, MD 8, pp. 75-81. Conversion to capacity measurements is calculated with a ratio of 1.65 x 3/5 x 3.33. For the ratio between small and large measurements, see ibid., pp. 321-33, and p. 330, n. 15. Lao Kan, "Ta-shih yü hsiao-shih," *Ta-lu tsa-chih*, vol. 1, no. 11. It has been suggested that the differences in *shih* weight was that the weight of the husked grain was called *ta-shih*, while that of the unhusked grain was called *hsiao-shih*; see Ch'en Kung-jou and Hsü P'ing-fang, "Kuan-yü Chü-yen Han-chien-ti fa-hsien ho yen-chiu," *K'ao-ku*, no. 1 (1960). Since there is a conversion rate of a *hsiao-shih* being 60 percent of a *ta-shih* given in the Han tablets, this interpretation seems unlikely.

26. Yang Lien-sheng questions what has been suggested by Wang Kuo-wei that a Han soldier received a flat rate of six *sheng* of grain each day. Yang suggests that there were two different rates. One was 6 *sheng* of husked grain per day (i.e., 1.8 *shih* of husked grain or 3 *shih* of unhusked grain in a full month of 30 days). He suggests that the higher rate went to officers and soldiers regularly guarding the watchtowers, whereas the lower one applied to convicts, soldiers working on agricultural colonies, and officers and soldiers who served on the frontier for a short period. See Lien-sheng Yang, "Numbers and Units in Chinese Economic History," in his *Studies in Chinese Institutional History*, p. 83, n. 31.

27. Loewe, *Records of Han Administration*, MD 10, pp. 86-91. The figures in MD 10 are given under each individual separately. MD 9, however, gives the provisions in total. The total disagrees with the sum of the individual rates. Cf. ibid., pp. 82-85.

and infant, respectively. These figures are compatible with the 3-*hu* standard derived from literary sources.[28]

Combining the information from the literary sources and the Chü-yen Han tablets, it is somewhat arbitrarily assumed that the monthly consumption of grain for an adult man was 3 *hu*; an adult dependent, 2.1; and a child, 1.2 *hu*. Thus the hypothetical household of five persons should consume 11.4 *hu* of grain each month, or approximately 140 *hu* of grain each year.

The cost for clothing is difficult to estimate. Li K'uei maintained that each individual needed to spend 300 cash for clothing. Since Li K'uei calculated the price of grain as 30 cash per *shih*, it can be deduced that the clothing for each individual in the household was valued at 10 *shih*.[29] The prices of various items quoted in the garrison records appear rather irregular. Regional variations, long- and short-term price fluctuations, and so on, all are factors over which we have no control; in general the prices appear to be around 300-400 cash per roll for hempen fabrics or less expensive silk, although fine silk could be as expensive as about 1,000 cash per roll.[30] If each person used two rolls of hempen fabric for clothing, five persons would spend approximately 3,000 to 4,000 cash on clothing each year.

28. These are not the only records among the Han wood tablets on food provisions. Michael Loewe collects others into two more groups of strips. UD 4 gives a rate of two *shih* of barley (or wheat) per month for a soldier. The information given in W 2, however, contains the total sum issued to a *t'ing* unit. The average seems quite irregular (ibid., pp. 192, 317 ff.).

29. Swann, *Food and Money*, p. 141.

30. Ch'en Chih, *Liang-Han ching-chi shih-liao lun-ts'ung*, p. 68. Ch'en cites the following from "Han-chien-chung-ti Ho-hsi ching-chi sheng-huo" by Lao Kan:

Item	Total price	Price per roll
Silk, 6 rolls *(p'i)*	2,862 cash	447
Silk, 2 *chang* and 5 *ch'ih*	500	1,008
Silk, 1 roll	1,200	1,200
Silk, 1090 rolls and 35 *ts'un*	354,200	325
Silk, 2 rolls	900	450
Plain silk from Jen-ch'eng, 1 roll	618	618
Plain silk, 1 *chang*, 6 *ch'ih*	268	670
Plain silk, 1 *chang*	250	1,000
Light silk, 1 roll	800	800
Green silk, 1 roll	800	800
White silk, 1 roll	1,400	1,400
Knitted silk, 1 roll	1,000	1,000
Coarse hempen cloth, 3 rolls	1,000	333
Hempen cloth, 1 roll	400	400
Medium hempen cloth, 1 roll	290	290
Medium coarse hempen cloth, 8 rolls [sic]	230 [sic]	?

The Chü-yen slips also occasionally include the prices of garments that were traded between the garrison soldiers. A pair of trousers were sold for 80 cash, a cloth robe was priced at 1,400, and another cloth robe sold for 1,300 cash. In this transaction, the buyer seemed unable to provide the cash as a contract had been drawn and witnessed, agreeing to pay the price later. There were five other contracts indicating either the steps buyers had to follow or agreements to pay the debts later. The prices were: 1,500 for a fur coat, 352 for a black shirt, 1,800 for a heavy coat, 2,000 for a robe, 800 for a pair of black trousers, 2,500 for a heavy black robe, and 1,500 for a robe.[31] The price of clothing at the frontier, where everything had to be shipped from the interior, was undoubtedly higher than comparable items in the producing area. Reduced by half, it still seems expensive if one had to spend more than 1,000 cash for adequate clothing: for example, a set of underwear (176), a pair of trousers (400), and a topcoat (750). For a five-person household, this meant an annual cost of 5,000 or more cash.

Ch'en Chih estimates that the price of one *shih* of unhusked grain at the frontier was a little more than 100 cash.[32] The cost for clothing, therefore, would be the equivalent of some 50 *shih* of grain, or 82.5 *hu* in capacity, more than the amount of grain consumed in one year by two persons who ate three *hu* of grain each month.

The cost for religious activities for an average household is also difficult to estimate. Let us assume that one had to participate at least twice in the community ceremony called *she.* As mentioned in the *Ssu-min-yüeh-ling,* this ceremony was to offer sacrifices to the

---

Ch'en also cites the following miscellaneous items:

Item	Total price	Price per roll
Medium coarse hempen cloth from Kuang-han, 19 rolls, 8-2/3 *ts'un*	4,320 cash	228
Heavy silk from Ho-nei, 8 rolls, 3 *ch'ih*, 4-2/3 *ts'un*	2,978	372
Yellow fine raw silk, 1 *chin*	350	
Fine raw silk, 2 *chin*	434	

A roll *(p'i)* is four *chang* in length; one *chang* is 10 *ch'ih* or 100 *ts'un.* The *chin* is a unit of weight for raw silk.

31. Ch'en Chih, *Liang-Han ching-chi . . .* , p. 61.

32. Ibid., p. 58. The grain price at the ordinary level in the interior provinces should have fallen below the mark of one hundred cash per *shih* in the Western Han period and within the range of 30-80 cash per *shih* in the Eastern Han period (cf. Li Chien-nung, *Hsien-Ch'in Liang-Han ching-chi-shih kao,* p. 196). The price of clothing in the interior provinces also should have been less expensive because the shipping costs were lower. The general ratio of clothing to grain price perhaps would remain within the range of the rate presented by Ch'en Chih according to the market at the frontier.

local guardian deity and was held in the spring and autumn. Both the ceremony of *la* in the winter and the New Year ceremony were also important occasions and even the poorest person had to spend a little for celebration (III, 1: A.1, B.1, H.1, and L.1). Li K'uei allowed 300 cash, the value of 10 *shih* of unhusked grain, for such expenses.[33] The Chü-yen slips included one piece saying, "Contribute to the Autumn *she,* 1,200 cash, on the date of *I-mao,* ninth month, third year of the Yüan-feng reign (78 B.C.)."[34] Unfortunately, this brief statement does not tell us whether one household contributed this money or whether it was the total cost for the ceremony. Using the Eastern Han grain price of 60-70 cash per *shih* and Li K'uei's estimate of the value of 10 *shih* of grain for the religious and social activities, then the total would be 600 to 700 cash per family.[35]

A Han farming household had to pay property, land, and poll taxes, and corvée dues. The land tax was in kind of crops, yet it was levied upon the size of an area instead of on the amount of produce.[36]

The Han land tax presumably was levied according to a ratio of the produce. When the dynasty was founded, the tax was one-fifteenth of the annual produce. Though there was a brief period of rate change, it was restored by 195 B.C. Emperor Wen reduced the rate to one-thirtieth in 168 B.C. and then completely abolished the land tax in the next year. For twelve years, there was practically no land tax until Emperor Ching decreed in 156 B.C. to levy a permanent rate of one-thirtieth on land. At the beginning of the Eastern Han dynasty, the tax rate was once as high as 10 percent; it was soon returned to the old rate of one-thirtieth in A.D. 30. In other words, the official taxation rate was normally one-thirtieth of the yearly produce[37] (I, 17, 18). Nevertheless, the tax was levied according to a fixed amount upon a

33. Swann, *Food and Money*, p. 141.

34. Lao Kan, *Chü-yen Han-chien: k'ao-shih chih-pu*, "K'ao-cheng" (1960), p. 66; "Shih-wen," p. 8 (serial no. 338, tablet no. 280.15). In the "Shih-wen," however, Lao deciphers the character which should be *she* as *fu*. There is clear reference to payment of 900 cash to the *she* in the multistrip document reconstructed by Michael Loewe. See *Records of Han Administration*, pp. 98-99.

35. The estimation of grain prices is based upon Li, *Hsien-Ch'in Liang-Han . . .* , p. 196.

36. Some scholars suspect that the land tax was taxation upon the income drawn from the land; see Liu Tao-yüan, "Shang Yang pien-fa yü Liang-Han t'ien-fu chih-tu," *Shih-huo* 1, no. 3: 1-13. It is also regarded as a taxation on production because of the government control of irrigation; see Kimura Masao, "Shin Kan jidai no denso to sono seikaku," *Rekishigaku kenkyū*, no. 232, pp. 3 ff. For critiques of such an argument, see Hiranaka Reiji, "Kandai no denso to saigai no yoru sono genmen," *Chūgoku kodai no densei to zeihō*, pp. 106 ff.

37. Ma Fei-pai, "Ch'in-Han ching-chi tzu-liao (7): Tsu-shui chih-tu," *Shih-huo* 3, no. 9: 9-33. Yoshida Torao, *Ryōkan sozei no kenkyū*, p. 14.

certain area. Good year or bad, the government demanded the same amount from the farmer, putting a great burden on the farmer during the lean years as well as allowing a good income in good years. At any rate, the Han scholars lamented in the *Discourses on Salt and Iron* that the farmer suffered from the combined blows delivered by famine and taxation (III, 12). Such a deviation from the official taxation rate seems to be the result of the bureaucracy avoiding the strenuous task of evaluating the actual production each year. Thus the record of acreage of each landowner became essential for taxation. As early as 216 B.C., the first emperor of the Ch'in dynasty had decreed that the people should report their own landholdings, and registration of one's own land remained a common practice throughout the Han dynasty.[38] The government had to carry out detailed land surveys repeatedly (see, for instance, I, 19). When the Eastern Han was established, one of its most careful efforts was the resurvey of all the land and in A.D. 39, a decree was issued to make one nation-wide. Numerous grand administrators of commanderies actually were punished or executed because of false reports. For some time there was resentment against the strict measures of the land survey (*HHS* 1B: 12b; 76: 76b). In A.D. 76-81, Ch'in P'eng, a grand administrator, took extreme care to measure the cultivated land and classify the fields into three grades of fertility. Registers were kept in both the local district and prefectural offices. The model case was brought to the court, and it was ordered that all divisions and commanderies adopt the same system of land recording (II, 41).

Had the land tax been exacted according to production there would have been no need to keep an accurate record of each farmer's land. The careful measurement of acreage implies only one thing: taxation at a certain fixed amount on a given unit of area. The problem is that the historical record does not tell what the fixed rates were.[39] It is

38. *SC* 6: 44. This particular incident is missing in the *Shih-chi* text. It is the early commentator Hsü Kuang who preserved in a note this missing section. The term *ming-t'ien,* registration of land, is interpreted in various ways, yet it is generally accepted that the basic function was for registration of ownership. Nishijima Sadao, "Kandai no tochi shoyūsei: tokuni myōden to senden ni tsuite," *Shigaku zasshi,* vol. 58, no. 1.

39. It is difficult to date the beginning of a fixed rate based on area. Equally difficult to decide is the beginning of the gradation system. Yoshida suggests that Ch'in P'eng initiated the gradation. Nakamura feels that gradation had been a standard system of Han taxation. Yoneda cites a Chü-yen tablet of 31 B.C., which includes the term *land record* of three grades and thus reaches the conclusion that the gradation system probably started between 81 and 31 B.C. at the latest, since the *Discourses on Salt and Iron* tells that the single rate system was still in practice in 81 B.C. Cf. Yoshida, *Ryōkan sozei no kenkyū,* p. 18 ff. Hiranaka

also complicated by the change in the size of the mu in about 87 B.C. during the last year of the reign of Emperor Wu from 100 paces to 240 paces[40] (III, 12). Since the changes favored the landowner, it would be logical to assume that on the newly expanded area one still needed to pay only the rate estimated according to the area of the old measurement.[41] It is also possible, however, that the change was made because of plowing land by oxen, a farming technique suitable for working a larger field as well as recognition of the newly reclaimed land. Both were beneficial from the point of view of the farmer.[42]

Fixed rates were probably the effect of bureaucratization, which had a tendency to formalize and standardize specific action. Thus the Han system of taxation on production possibly evolved to one on acreage, and was eventually recognized as a fixed rate. The whole process might have taken more than one generation and there might have been regional differences in both the substance and speed of change, a phenomenon not unusual under the Han local administrative system, as administrators had extremely great discretionary authority over their own regions. Ch'in P'eng's case illustrates this particular local initiation of a new system, which can be regarded as a landmark in the evolution of taxation practices only during the Eastern Han period. The Western Han probably had its own taxation change. Among the Chü-yen Han tablets, one slip reads, " . . . field records; registering the crops in the upper, middle, and lower grades; estimating the rate of harvest; and the calamities thereof, if any." The tablet was accompanied by another one that bears the date of "the second year of the Chien-shih reign," that is, 31 B.C., thus it is possible that the slip on "land record" is of the same date.[43] This is the earliest available record of the three-grades system. The record, however, is too brief and did not say whether the grades judged fertility of the fields or

---

Reiji, "Kandai no denso to saigai no yoru sono genmen," p. 143 ff. Yoneda Kenjirō, "Kandai denso sateihō kanken," *Shiga daigaku kyōiku gakubu kiyō*, no. 17 (1967), p. 120 ff.

40. Hamaguchi Shigekuni, "Chūgoku shijō no kodai shakai mondai ni kansuru oboegaki," *Yamanashi daigaku gakugeibu kenkyū hōkoku*, no. 4.

41. Yoshida, *Ryōkan sozei no kenkyū*, pp. 17-18. Please notice, however, that in the period of 11 years prior to 155 B.C., due to government grace there was no land tax. The seeming reduction of tax from one-fifteenth to one-thirtieth can be interpreted as a compromise (*SC* 10:30).

42. Amano Motonosuke, "Chūgoku bosei kō," *Tōa keizai kenkyū hō*, n.s., no. 3.

43. Lao Kan, *Chü-yen Han-chien: k'ao-shih chih-pu*, "Shih-wen" (1960), p. 85 (serial no. 4171, tablet no. 113.6) and p. 192 (serial no. 9586, tablet no. 113.3).

size of the harvest. This can be cited only as an indication that taxation was differentiated into various grades.[44]

The productivity of Han agriculture is not discussed here; however, the average yield per mu concerns us. Chung-ch'ang T'ung gave such a rate in his critique on Han taxation. He suggested that even at a tax rate of one-tenth of the produce, the farmer who produced an average of 3 *hu* of grain per mu would not be too heavily burdened. An exceedingly light tax, such as the current one-thirtieth of the produce, would not provide the state with sufficient surplus against famines or wars.[45] One-thirtieth of 3 *hu* is one-tenth of a *hu*, or 1 *tou*, as tax on each mu, whereas at the rate suggested by Chung-ch'ang T'ung, the tax would be 3 *tou* per mu (III, 13). The tax paid to a marquis in 32 B.C. from 40,000 mu in his fief amounted to over 1,000 *shih* of grain, or more than 1,650 *hu* of unhusked millet.[46] Thus the tax on each mu would have been four-tenths of a *tou* (or 4 *sheng*) of grain.

Another case in A.D. 87 showed an additional revenue to the government of over 100,000 *hu* of grain from tax on 20,000 *ch'ing* of new arable land.[47] The tax on this, then, would have been about one-half *tou* (or 5 *sheng*).

If a hypothetical family paid tax at a rate of 5 *sheng* per mu, a figure falling within a reasonable range of variation, the total land tax in kind would have been 3.5 *hu* of grain for each 70 mu of land that they tilled.[48]

Besides the grain tax, there was the so-called hay tax, which also provided regular revenue to the government, although probably a relatively minor one[49] (I, 8). Since there is no information other than its name, it is impossible to estimate its actual figure among the expenditures of a farming household.

The property tax, called *tzu-suan,* first appeared in 142 B.C.,

44. Yoneda, "Kandai denso sateihō kanken," pp. 71-72.

45. Wolfram Eberhard estimates the average yield as being 1.0 to 1.5 *shih* per mu; see "Bemerkungen zu statistischen Angaben der Han-Zeit," *T'oung Pao* 36 (1940): 4-5. Very high yields could reach as much as 6.4 *hu* per mu, however (IV, 18, n. 66).

46. *HS* 81: 10b; Ch'ü, *Han Social Structure*, p. 91. A different opinion is held by Ch'ien Ta-hsien, who took this amount as an income of three years. Cf. Chü, *Han Social Structure*, pp. 91-92, n. 95.

47. *HHS* 24: 24a. The acreage is given in the biography of Ma Leng; the amount of increased revenue is given in *Tung-kuan Han-chi* as quoted by the commentator. It is not clear, however, whether the new revenue resulted from newly reclaimed land or from a general improvement in the fertility of fields in the whole region.

48. This rate of five *sheng* is also suggested in Hiranaka, "Kandai no denso to saigai no yoru sono genmen," pp. 160-62. Cf. Yoneda, "Kandai denso sateihō kanken," pp. 66 ff.

49. Ma Fei-pai, "Ch'in-Han ching-chi tzu-liao (7)," pp. 9-33.

when the emperor decreed that the minimum property needed to enter the government service should be four *suan* units instead of ten. The commentator Fu Ch'ien explained that on each 10,000 cash worth of property, which was one *suan* unit, the tax was 127 cash.[50] In 119 B.C., Emperor Wu issued the famous decree in which the property taxes were expanded to cover all the property and income from commercial transactions. The decree clearly listed the items to be taxed, the rate of taxation (which was 127 cash per *suan* unit of property), and the preferential rate granted to officials, local elders, and garrison soldiers. For transacted commodities (or more likely, the profit earned thereon), one *suan* unit would be 2,000 cash; for manufactured goods, one *suan* unit would be 4,000 cash; and for vehicles and boats, each would be one unit[51] (II, 8).

Applying these figures to some of the items mentioned in inventories of Li Chung and Hsü Tsung (III, 5, 6), we can estimate the taxes due on that property (table 5).

TABLE 5

PROPERTY VALUE OF A HYPOTHETICAL FARMING HOUSEHOLD

1 residence	3,000 cash
70 mu of land	7,000
2 oxen	6,000
2 ox carts	4,000
Total	20,000

The annual *tzu-suan* tax, at a rate of 120 cash per 10,000 cash value, should be 240 cash, which is equivalent to 4 *shih* of grain in kind at a hypothetical price of 60 cash per *shih*.[52]

The poll tax was divided into two parts: the *k'ou-ch'ien* for minors and the *suan-fu* for adults. Although not called *suan-fu,* there had been a poll tax in the Ch'in period, which was paid in grain.[53] It obviously was one of the regular revenues of the government; therefore, Tung Chung-shu listed the poll tax along with the land tax.[54] In 203 B.C., Kao-tsu, the founding emperor of the Han dynasty, decreed

50. *HS* 5: 9b-10a. Dubs, *HFHD* 1: 329-30.

51. Swann, *Food and Money*, pp. 280 ff.

52. It should be noted that the price of land in the frontier was less than the fertile land in the well-irrigated fields in the interior, for instance, the land in the capital region. For the rate of property tax, see ibid., no. 531.

53. *SC* 89:6, especially the commentator Fu Ch'ien's note; also *Huai-nan-tzu*, SPPY ed., 13:8b (hereafter cited as *HNT*).

54. *HNT* mentions the poll tax as being transmitted to the Privy Treasury, 13:8b; I, 4.

the collection of *suan-fu* at the rate of 120 cash per head. Theoretically, every subject had to pay it[55] (III, 14). Occasionally the rate fluctuated. In the reign of Emperor Wen (179-157 B.C.), the rate was three times reduced to 40 cash. How long each of these reductions lasted, however, is not clear. By 52 B.C., once again the rate was reduced to 90 cash. And in 41 B.C., the rate was 80 cash. Frequently, some specific groups, such as the aged and the vagabonds, whom the government tried to induce to resettle, were exempted.[56] In summary, an adult male or female during the Han dynasty normally paid a poll tax of 120 cash, in spite of brief reductions and special exemptions.

The *k'ou-ch'ien* was a poll tax levied on minors between seven and fourteen years old. The origin of this tax is not clear; it is suspected that the enormous expenditures during the reign of Emperor Wu caused its creation.[57] It is more likely, however, that Emperor Wu only increased the rate of this tax from 20 to 23 cash per head, while the lower taxable age limit was three years old, until a revision raised it to seven years old (III, 15). The rate of 23 cash seems to have remained constant throughout the Han, beginning in the reign of Emperor Wu.[58]

The poll tax was paid in cash; it was only when the grain price was low that it could be paid in grain[59] (III, 16). In the hypothetical household, the five adult persons had to pay a total of 600 cash annually, the equivalent of ten *shih* of grain if we assume one *shih* of grain to be worth 60 cash.

The last category of expenditures concerned payment to a substitute for military and corvée services. Tung Chung-shu mentioned both in his memorial: one month of corvée in a man's local area and two years of military service (I, 4). In addition, there was the three days' garrison duty each year. The local corvée and three days' garrison

55. *HS* 1A: 39b. Dubs, *HFHD* 1: 93. The Han administration took great pains to see that the census was carefully taken. In the eighth month of each year, everyone, including the handicapped and the sick, had to be counted in person. See Nagata Hidemasa, "Kandai nintōzei no hōkai katei," *Tōyōshi kenkyū* 18, no. 4 (1960): 50 ff.

56. Yoshida, *Ryōkan sozei no kenkyū*, pp. 90-93.

57. Ibid., p. 92, and n. 10, p. 97.

58. *Han-i*, as quoted by the annotator's note to a passage referring to events in A.D. 46, *HHS* 1B: 16b-17a. Also, Wang Ch'ung mentioned it as part of a common saying, "Poll tax of 23 cash at seven years old," *Lun-heng*, SPPY ed., 12: 13a.

59. Yoshida takes the decree of 79 B.C. as an order effective thereafter. Yoshida, *Ryōkan sozei no kenkyū*, pp. 25-26. The decree clearly states, however, that it was to be effective only in that particular year.

duty could be eliminated by payment of fees, which became another tax regarded as one of the regular government revenues (III, 17).

The important issues are: the total number of days one was in service, the age limits of the service, and the amount of the fees paid for a substitute. The first is complicated by the reference to "thirtyfold those of ancient times" in Tung Chung-shu's memorial (I, 4; III, 17). There are scholars who have tried to make an exact sixty days' service of this.[60] It should be noted that however one punctuates the text, Ju Shun's comment gives only two years of military service (III, 17, especially n. 119). The two years of service was confirmed by the *Han-i* (also cited by Ju Shun) whose author said that a youth reaching twenty-three years of age should serve as a regular conscript and spend one year as a guard and one year in the infantry or cavalry (III, 18).

The thirty-three years between ages twenty-three and fifty-six was commonly accepted as the period for military service obligations. Nevertheless, in the Han Chü-yen tablets, some slips say that the draft could begin before twenty-three, probably as early as twenty.[61] The retirement age of fifty-six might also be optional because the verb in the *Han-kuan-i* implies that one would be granted retirement if he was entitled to it and so chose. There is no clear indication of compulsory retirement age in Han sources (III, 18).

Ju Shun mentioned two kinds of fees paid for the substitute: 2,000 cash monthly for the local corvée and 300 cash for the frontier garrison duty. The rate indeed seems high, especially if compared with normal wages.[62] Since wages would have fluctuated according to the money value and the market situation, any fixed rate, whether 300 or 2,000 cash, would not be ideal.[63] Among the labor and military obligations, the regular conscript obligation could not be deferred.[64]

60. For instance, Yoneda Kenjirō, "Kandai yōeki nissū ni kansuru ichi shiron," *Tōhōgakuhō* 27 (1967): 189-212. Liu Tao-yüan, "Shang Yang pien-fa yü Liang-Han t'ien-fu chih-tu," *Shih-huo* 1, no. 3: 1-13.

61. Ch'un-shu Chang, "The Han Colonists and Their Settlements on the Chü-yen Frontier," *Tsing-hua Journal of Chinese Studies*, n.s., 5, no. 2 (1966):194.

62. *HS* 35: 5b, annotator Fu Ch'ien's note, in which the monthly wage rate is given as 300 cash.

63. For instance, the wage of a servant was 1,000 cash monthly by the mid-second century (III, 11).

64. Ho Ch'ang-ch'ün suggests that the garrison duty was translated into another form of tax. Yet one still might go to the front as a garrison soldier if one's turn came up. If he did not want to go, he had to pay a fee of 2,000 cash a month. Ho, "Ch'in-Han-chien ke-t'i-hsiao-nung-ti hsing-ch'eng ho fa-chan," in *Han-T'ang-chien feng-chien t'u-ti-so-yu-chih hsing-shih yen-chiu*, pp. 23-24. Ho's suggestion, however, is not verifiable.

If a man paid the smaller figure of 300 cash to avoid garrison duty and another 300 cash to avoid corvée labor, then the hypothetical household, which had two males, would pay 1,200 cash annually to have substitutes for these duties.

Table 6 shows the expenditures of a farming household, both in cash and in kind.

TABLE 6

EXPENDITURES OF A HYPOTHETICAL FARMING HOUSEHOLD

In kind and self-sufficient	
Food	140 *hu* of grain (unhusked rice)
Land tax	3.5 *hu*
Total	143.5 *hu* of grain
Clothing	5,000 (self-sufficient)
In cash payments	
Religious and social activities	1,600 cash
Property tax	240
Poll tax	600
Labor exemption	1,200
Total	3,640 cash

Converting the grain to cash at an arbitrary rate of 60 cash per *shih* of grain, the grain should be worth (143.5 *hu* is 87 *shih*) 5,220 cash. The clothing, worth 5,000 cash, is provided by women of the family. The self-sufficient items have a cash value of 10,220. The other four expenditures are met with payments totaling 3,640 cash. In other words, a farming household had to find cash to pay 26.3 percent of its basic expenses.

Fluctuations in the price of grain, however, would have changed the ratio: a high market meant a good income; a low market, a poor one. The median range of the price of grain was between thirty and eighty cash per *shih,* although it could be as high as several thousand cash per *shih* in times of calamity and as low as four or five cash in highly productive years.[65] This range perhaps represents both the longer trend of price changes and the seasonal fluctuations of the grain market. The ever-normal granary, designed to control the price of grain, was established in 54 B.C. in the frontier commanderies and

65. Li, *Hsien-Ch'in Liang-Han . . .* , pp. 194-95.

later developed as a regular institution. The idea was for the government to buy grain when the market was low and to sell when it was high, keeping the price stable as well as maintaining a steady supply of grain.[66] The results, however, seem to have been rather ineffective. By A.D. 68, when the court discussed the restoration of the ever-normal granary, the objection was raised that the common people were not benefited by it (III, 19).

The numerous references to price fluctuations in the *Han-shu* and the *Hou-Han-shu* also testify to the ineffectiveness of the ever-normal granary. In the years of good harvests, the farmer found that the unit price of his product dropped and in the years of crop failures he faced starvation. Thus the farmer was often the loser, and constantly subject to pressure. Natural disasters were also a part of his life. During the 462 years of the Han period (including the reign of Wang Mang), there were forty-three droughts, sixty-eight floods, and thirty-seven famines caused by locusts and other insects. Typically, there would be one natural disaster every three years. And man-made disasters such as foreign wars and local disturbances were an added burden.[67] To survive, every possible means had to be pursued. Perhaps it is in this aspect that the Han farmer continuously improved farming techniques, making them more intensive, raising many varieties of crops, and utilizing the land to its maximum efficiency. The advancement in farming techniques became a necessity for the farmer who realized that the selling of his principal crops would not provide an adequate income. What appears in the *Ordinances* and the mock slave contract indeed reflected not just life in a well-to-do farming household, but, more significantly, the style of living of the general farming population that relied heavily upon income from sericulture, vegetable gardening, cottage industry, animal husbandry, horticulture, and marketing activities. These activities will be discussed in chapter 6 (III, 1, 7).

In summary, as discussed in the preceding chapter, Han China had a market economy, in spite of the government's protection of agriculture, and this created the need for cash, which put the farmer in a disadvantageous position if he failed to increase his cash-making production. Meanwhile, a well-developed national economy made regional interdependence necessary. As a result, intensive and diversified agriculture emerged during the centuries of the Han dynasty.

66. Swann, *Food and Money,* pp. 195-96, and n. 292.

67. Ma Fei-pai, "Ch'in-Han ching-chi-shih tzu-liao (3)," *Shih-huo,* vol. 3, no. 1. Li Chien-nung reaches somewhat different conclusions. By Li's count, during Western Han there were thirty-two years of natural disasters within a period of 214 years, while during the Eastern Han, there were 119 years of disasters out of 195 years. Li, *Hsien-Ch'in Liang-Han . . . ,* pp. 162-64. It must be noted, however, that these natural disasters were not always nation-wide.

IV

FARMING RESOURCES

Cultivated Plants

Cultivated plants of people in the Han dynasty included cereals, vegetables, and other cash crops. Five principal grains were recognized, although there is disagreement about which five grains were the most important. In one theory, the top five were common paniculate millet *(Panicum miliaceum),* spiked or Italian millet *(Setaria italica),* beans, wheat, and rice. In another theory, rice was replaced by hempseed as one of the five leading grains.[1] The order of significance of these grains, of course, varied. In ancient times, the millets, being indigenously cultivated, were the most important cereals in the diet of the northern Chinese. By the Chan-kuo period, beans and spiked millet (a choice variety of the foxtail millet) were the basic subsistence. In Han times, the most common grains were wheat and spiked millet, although rice became more important as people migrated to the south, where both the soil and weather were ideal for its cultivation.[2]

In 1956, a group of Western Han tombs were excavated and archaeological findings confirm the literary record. The terra cotta objects discovered in one of the tombs included twenty earthen granary models, consisting of four groups of five models, on which are marked

1. Cf. Cheng Hsüan's comments on the "Chih-fang-shih" in the *Chou-li* and the *Lü-shih ch'un-ch'iu.* In fact, in the "Shen-shih" chapter in the *Lü-shih ch'un-ch'iu,* SPPY ed. (hereafter cited as *LSCC*), six kinds of grain are discussed. Cf. Wan Kuoting, *Wu-ku shih-hua,* pp. 3-4.

2. Ch'ien Mu, "Chung-kuo ku-tai pei-fang nung-tso-wu k'ao," *Hsin-ya-hsüeh-pao,* vol. 1, no. 2 (1956).

the five leading grains: common millet, spiked millet, beans, hemp, and wheat. On each of the granaries there is the sign "ten-thousand *shih* of such-and-such grain," no doubt a symbolic claim.[3] In another site at Lo-yang, a large number of terra cotta granary models are labeled with the names of grains. Among the sixty-seven models, there were eighteen labeled with wheat-barley and their derivatives; seventeen, spiked millet; sixteen, beans, hemp; four, common millet; four, rice; and one, a certain kind of leaven.[4] As life in the other world is a reflection of life in this world, the frequent mention of these grains tells us that wheat-barley, spiked millet, and beans probably were the three most commonly consumed cereals and millet, hemp, and rice were secondary ones. Rice would not appear as a grain consumed daily in Lo-yang in the northern plains.

Investigation of the names of grains and their varieties in the lexicons and agriculture books also confirms the observation on the significant shift from certain cereal crops to others. In the lexicon *Erh-ya,* which was composed to provide annotations to the pre-Ch'in classics, spiked millet and its variations in color and other characteristics were given six names. There were four variations of hemp, three kinds of common millet, two names for rice, two for varieties of oats, and one for soybeans.[5] In the *Chi-chiu-pien,* a primer text for the teaching of reading and writing, composed in the second half of the first century B.C., two kinds of rice, two kinds of millet, two kinds of spiked millet, and one kind of hemp were all given in one sentence, followed by another sentence in which foods of wheat flour and cooked beans are mentioned.[6] The book of Fan Sheng-chih, contemporaneous with the *Chi-chiu-pien,* includes among the cultivated plants: spiked millet, millet, winter and spring barley and wheat, soya and lesser beans, hemp, water darnels, and paddy rice, which was found only in places with adequate water supply[7] (see document V, 4).

In the first century B.C. the Chinese farmer included in his repertoire of cereal grains spring and winter barley-wheat, soybean, and

3. Kuo Pao-chün et al., "Lo-yang Chien-pin ku-wen-hua i-chih chi Han-mu," *K'ao-ku hsüeh-pao,* 1956, no. 1, p. 23.

4. Lo-yang fa-chüeh-tui, "Lo-yang hsi-chiao Han-mu fa-chüeh pao-kao," *K'ao-ku hsüeh-pao,* 1963, no. 2, pp. 48-49, table 2.

5. *Erh-ya,* SPTK ed. (hereafter cited as *EY*), chapter on nomenclature of grasses, passim.

6. *Chi-chiu-pien,* SPTK ed., pp. 30-31; hereafter cited as *CCP.*

7. Shih Sheng-han, *On Fan Sheng-chih-shu, An Agricultural Book of China Written by Fan Sheng-chih in the First Century B.C.,* p. 48. The translation of the original text was rendered by Shih Sheng-han in the same volume. Our translation, though largely based upon the work done by Shih, is a much revised one.

lesser beans; hemp was regarded more as an oleiferous plant, on a par with the newly introduced sesame and Perilla (V, 4: G.8).

In the Eastern Han period, the first dictionary *(Shuo-wen chieh-tzu,* compiled by Hsü Shen) gives eight variations for barley-wheat; seven for spiked millet; six for rice; four for beans; four for hemp; three for common millet; and two for taro.[8] The list itself shows that many different grains were then available to the North China farmer, as demonstrated in the *Ssu-min-yüeh-ling, Ordinances for the Four Peoples.* The farmer could choose among barley, wheat and oats, beans, lesser beans and peas, and early or late spiked millet. There was only one kind of rice or common millet mentioned, however (III, 1, passim).

It seems clear that the most important grains in Han times were spiked millet and wheat. It is for this reason that Tung Chung-shu said that crop failures of either spiked millet or wheat had entered the historical record (IV, 1). The two most commonly cultivated variations of spiked millet were *liang (Setaria italica* Beauv. var. *maxima* Al.) and *su (Setaria italica* Beauv. var. *germanica* Trin.).[9] The grains of the *liang* and the *su,* being somewhat larger than that of the common millet, could be a little more easily husked and they presumably tasted much better. Although spiked millet required a longer growing period, it was a tough plant, tolerating wind and not needing much water—an ideal crop for arid and windy North China. Also, spiked millet could be planted virtually anytime after the first spring rain and before the first frost; hence, Fan Sheng-chih's book stated that no rigid time was required for the sowing of spiked millet (V, 4: D.1). The cultivation of spiked millet nevertheless suffered from a few disadvantages. First, the lightweight grain would easily shatter from the spikelet and thus grow by itself the next year. Without adequate care the plant would produce only a low yield of poor quality. What bothered farmers was the difficulty that self-regeneration created for field management, such as ground dressing, weeding, and so on. This particular problem was explicitly pointed out in the *Ch'i-min-yao-shu,* which, therefore, suggested rotating spiked millet with some other crop.[10] Whether the Han farmer knew the reason for rotation as discussed in the *Ch'i-min-yao-shu* is difficult to determine. Nevertheless, some rotation system for spiked millet and wheat did appear in practice. In the third century B.C. work, *Lü-shih ch'un-ch'iu,* there is the line, "For this year, a good

8. *Shuo-wen chieh-tzu,* SPTK ed., passim; hereafter cited as *SWCT.*

9. The identification of these two grains is suggested by Yu Ching Jang in "Shu-chi-su-liang yu kao-liang," *Ta-lu tsa-chih* 13, no. 4:20.

10. Shih Sheng-Han, annotator, *Ch'i-min-yao-shu chin-shih,* p. 31, 3.3.1 and 3.3.3.

crop of spiked millet; for the next year, a good crop of wheat (*LSCC* 26: 6). The second-century exegetic scholar Cheng Chung told of the practice of planting spiked millet after the wheat was harvested (V, 5).

Spiked millet does not demand good soil, but does demand thorough preparation of the field. Because the seed is relatively small, it needs land that has been carefully cultivated, completely weeded, and on which the moisture is well conserved.[11] This meant constant work in ground dressing and weeding before as well as during the growing season. Many of the general principles of intensive farming developed in China were related to the demand for spiked millet cultivation.

Both wheat and barley were first domesticated in West Asia. Their Chinese names, minor *mai* and major *mai,* respectively, suggest that the former was introduced into China later than the latter. A Japanese scholar suggested that wheat was not in China until the second century B.C. when Chang Ch'ien brought some back from the Western Regions together with a number of foreign plants, among them grape and alfalfa.[12] On the other hand, Chinese scholars note that barley and wheat had their own names, *mou* and *mai,* in ancient times, that the term *mai* in ancient works normally referred to wheat and that wheat actually has been discovered in neolithic sites. Therefore, they said, wheat has long been available in China.[13] In fact, since both barley and wheat were discovered in neolithic sites, perhaps both had been available in China very early.[14] They were both cultivated during the the Han period. Nevertheless, by the second century B.C. wheat had not been popularly accepted by the people, at least not by the people in the metropolitan areas of the capital, the present province of Shensi,

11. Ts'ao Lung-kung, "Chung-kuo nung-shih wen-hsien-shang su-ti tsai-p'ei," *Nung-shih yen-chiu chi-k'an* 2: 95-96.

12. Japanese scholars assert that the references to *mai* in pre-Ch'in works actually refer to barley. Shinoda Sumeru, "Gokoku no kigen," *Shizen to bunka,* no. 2, and Amano Motonosuke, *Chūgoku nōgyōshi kenkyū*, pp. 62, 85-87. Yü Ching Jang suggests that wheat was introduced into China via Mongolia in the third century B.C. Yü, *Tsai-pei chih-wu-k'ao* 2:61-63.

13. Hsia Wei-ying, *Lü-shih ch'un-ch'iu Shang-nung teng-ssu p'ien chiao-shih,* p. 118. Hu Hsi-wen, "Chung-kuo hsiao-mai tsai-p'ei chi-shu chien-shih," *Nung-yeh i-ch'an yen-chiu chi-k'an* 1 (1958): 56, and Wan Kuo-ting's letter dated September, 23, 1960, to Amano Motonosuke, in Amano, *Chūgoku nōgyōshi kenkyū,* p. 86. A recent book touching on this problem is P.T. Ho's *The Cradle of the East;* see pp. 73-76, 352-53.

14. For barley, Tokunaga Shigeyasu and Naora Nobuo, *Manshu teikoku Kichirin-shō Kōgōton dai-ikkai hakkutsu hōkoku*, pp. 36-37, and for wheat, Anhwei-sheng po-wu-kuan, "An-hui hsin-shih-ch'i shih-tai i-chih-ti tiao-ch'a," *K'ao-ku hsüeh-pao,* 1957, no. 1, p. 23. Note that in the latter case the archaeological report mentions only *mai,* without specification of being *hsiao mai* (wheat). Its identifica-

which is now a wheat-producing area. Not only has Tung Chung-shu suggested that the people near the capital should be urged to plant wheat, but also a specialist was dispatched there in the first century B.C. to teach people the technique of wheat and barley cultivation[15] (IV, 1). This phenomenon, however, probably was a local one. At any rate, the book of Fan Sheng-chih recorded the farming of wheat in great detail. And the Han tablets at Chü-yen, dated about mid-first century B.C., disclosed the provisions of *mai* in the western frontier military settlements.[16] Wheat and barley became so common in the countryside by the second century A.D. that both are mentioned in children's rhymes (IV, 2). A tablet dated just after the end of the Han period at a western frontier military settlement listed wheat, barley, and millet as three crops that were planted there (IV, 3).

Since wheat and barley can grow either in winter or in spring, these plants made ideal partners for millet, which also has a flexible growing season. Millet has a shallow root, but the long roots of wheat and barley can help to loosen the soil underneath, especially desirable in the compact loess soil in North China. This makes wheat-barley and millet compatible rotation crops.

The legumes were of considerable significance in Chinese agriculture. Soybean was first portrayed in the ancient bronze inscriptions as a plant with a few dots around the roots. The dots are interpreted by modern scholars as indicating that the ancient people were aware of the root nodules, which bear the function of nitrogen fixation.[17] Although people of that time could not have known of the existence of nitrogen, it is possible that the farmers of ancient days had empirical knowledge of the morphological characteristics of the leguminous crops and their value as good green manure. One of the beans, probably soybean, was named *jung-shu*; *jung* is generally acknowledged as a term the Chinese used for foreigners (or barbarians) from the Western Regions.[18] At least one of the beans, the broad bean, was introduced

---

tion as wheat is provided by Wan Kuo-ting (see n. 13).

15. *Chin-shu*, SPPY ed., 26: 8 (hereafter cited as *CS*), and Lien-sheng Yang, "Notes on the Economic History of the Chin Dynasty" in his *Studies in Chinese Institutional History*, p. 182.

16. Lao Kan, *Chü-yen Han-chien: K'ao-shih chih-pu*, "K'ao-cheng" (1960), pp. 59-61. The dates of such tablets ranged from the earliest one of 102 B.C. to the latest of A.D. 33. The majority are of 81-25 B.C.; see Ch'en Kung-jou and Hsü P'ing-fang, "Kuan-yü Chü-yen Han-chien-ti fa-hsien ho yen-chiu," *K'ao-ku*, 1960, no. 1, p. 47. Lao, however, suggests that the grain *mai* is mainly barley.

17. Hu Tao-ching, "Shih-shu-p'ien—Shih-lun wo-kuo ku-tai nung-min tui ta-tou ken-liu-ti jen-shih," *Chung-hua wen-shih lun-ts'ung* 3: 111-12.

18. For the definition of *jung* in the ancient days, see Ying-shih Yü, *Trade and Expansion in Han China*, p. 5. An annotator, Sun Yen, however, suggested

into China during the Han period by Chang Ch'ien, an envoy, who in 139-126 B.C. and 115 B.C. went to the Western Regions, today the province of Sinkiang, to establish ties with the states there.[19]

In Fan Sheng-chih's book, soybean was presented as a safe crop even in adverse years, and it was suggested that five mu of soybeans be planted for each person in the household (V, 4: D. 6). The lesser beans, however, were considered extremely difficult to grow (V, 4: D. 7).

In the pre-Ch'in period, beans were one of the basic foods, on a par with the better variety of spiked millet *(su),* which replaced common millet and foxtail.[20] Therefore, Ts'ui Shih in his *Ordinances for the Four Peoples* listed beans and lesser beans among crops to grow and for marketing (III-1: B. 8, E. 6, for instance).

Beans were ideal for rotating with millet, since soybean could be sown any time between early spring and early summer and lesser beans could be sown in late summer.[21]

Wheat took precedence over beans in the Han period, possibly because it was more flavorsome and also because it preceded millet as a rotating crop, thus taking over part of the otherwise exclusive role of leguminous crops. Although beans remained a food, their value was more as a green cover to ameliorate the soil. That perhaps was why the lesser bean, considered a less safe crop, was still cultivated. In the *Ch'i-min-yao-shu,* green lentils were plowed under after blossoming to prepare the field for melons, onions, and so forth (IV, 7).

Considering just the three principal crops, millet, wheat-barley, and legumes, the Han farmer of North China had a fairly good choice for developing a multicrop system.

The farmer in the South had rice culture to develop. Although paddy rice by nature is a southern staple, it did appear in North China. At Ho-nan, in neolithic sites of the Yang-shao culture, imprints of rice on earthen vessels were discovered. That rice has been identified as a

---

that *jung* here simply meant "large," hence not necessarily related to the Western barbarians. *EY,* "Shih-ts'ao."

19. Shih, *Ch'i-min-yao-shu chin-shih,* p. 78. For Chang Ch'ien's mission, see Yü, *Trade and Expansion,* pp. 135-37.

20. Ch'ien Mu, "Chung-kuo ku-tai pei-fang nung-tso-wu k'ao." Tsou Shu-wen, "Shih-ching shih-tai su-chi-pien," p. 41; P'an Hung-sheng and Yang Ch'ao-po, "Chan-kuo shih-tai-ti liu-kuo nung-yeh sheng-ch'an," p. 52, both in *Nung-shih yen-chiu chi-k'an,* vol. 2. For the translations of passages regarding beans and millet in ancient works, see IV, 4, 5, 6.

21. Shih, *Ch'i-min-yao-shu,* 6.2.1 and 7.1.2.

a kind of short grain variety *(Oryza sativa* L. var. *japonica).*[22] Specimens of the same grain have been discovered in neolithic sites in the Huai River valley, the Yangtse River valley, and the Yangtse delta.[23] Since there is evidence of the existence of wild rice in South China, it is suggested that paddy rice in China is perhaps the domesticated descendant of the indigenous wild species.[24] It then spread to southeast China as well as to the Huai River valley. Rice grain discovered in a Han tomb at Lo-yang was also identified as the same variety as that found in the neolithic sites.[25]

In the Han, other than regular rice, there was also dry rice and the glutinous variety. The latter, used to brew wine (technically, rice wine, Japanese *sake,* is a beer, but we follow tradition in calling it wine), perhaps had been in China quite early; possibly the rice mentioned in the *Shih-ching* was the glutinous type.[26] The time to sow glutinous rice, as given in the book by Fan Sheng-chih, was one month later than regular rice (V, 4: D. 5a). Dry rice was rare in China. It is mentioned only twice in the ancient works, both dated not earlier than the Han period. Etymologically, the character denoting dry rice indicates that it was originally a water plant and the anatomy of the

22. Amano, *Chūgoku nōgyōshi kenkyū,* pp. 95-96. Ch'en Tsu-kuei, "Chung-kuo wen-hsien-shang-ti shui-tao tsai-p'ei," *Nung-shih yen-chiu chi-k'an* 2:65.

23. Ting Ying, "Chiang-Han p'ing-yüan hsin-shih-ch'i shih-tai hung-shao-t'u-chung-ti tao-ku-k'o k'ao-ch'a," *K'ao-ku hsüeh-pao,* 1959, no. 4, pp. 31-33. Anhwei-sheng po-wu-kuan, "An-hui hsin-shih-ch'i shih-tai i-chih-ti tiao-ch'a," *K'ao-ku hsüeh-pao,* 1957, no. 1, pp. 23, 27, and Kiangsu-sheng wen-kuan-hui, "Chiang-su Wu-hsi Hsi-shan kung-yüan ku-i-chih ch'ing-li chien-pao," *Wen-wu ts'an-k'ao tzu-liao,* 1956, no. 1. Kiangsu-sheng wen-wu kuan-li wei-yüan-hui, "Chiang-su Wu-hsi Hsien-li-tun hsin-shih-ch'i shih-tai i-chih ch'ing-li chien-pao," *Wen-wu ts'an-k'ao tzu-liao,* 1955, no. 8. A new discovery of great significance is in Yu Hsiu-ling, "Tui Ho-mu-tu i-chih ti-ssu-wen-hua ch'eng ch'u-t'u tao-ku ho ku-ssu ti chi-tien K'ang-fo," *Wen-wu,* 1976, no. 8, pp. 20-23.

24. Ting Ying, "Chung-kuo tsai-p'ei-tao-chung-ti ch'i-yüan chi ch'i yen-pien," in *Tao-tso k'e-hsüeh lun-wen hsüan-chi,* ed. Chung-kuo nung-yeh k'e-hsüeh-yüan, p. 8.

25. Ting Ying, "Chiang-Han p'ing-yüan," pp. 31-33. A Japanese scholar, however, identified the grain as *Oryza sativa* var. *Indica,* as he noted the length of the grain. See summary by Amano, *Chūgoku nōgyōshi kenkyū,* p. 96. Moreover, in one neolithic site at Sung-tse, near the vicinity of Shanghai, rice grain was discovered that had been identified as *hsien* rice *(Indica)*; see Wu Shan-ch'ing, "Lüeh-lun Ch'ing-lien-kang wen-hua," *Wen-wu,* 1973, no. 6, p. 57. Yet most cultivated rice in China before the Sung dynasty was still that of the *keng* variety *(japonica)*; see Ho Ping-ti, *Huang-t'u yü Chung-kuo nung-yeh-ti ch'i-yüan,* p. 180. The *keng,* of course, is a mutative variation of *hsien* rice. Yu Hsiu-ling, "Tui Ho-mu-tu i-chih ti-ssu-wen-hua ch'eng ch'u-t'u tao-ku ho ku-ssu ti chi-tien K'ang-fo," pp. 20-23.

26. Ch'en Tsu-kuei, "Chung-kuo wen-hsien-shang-ti," p. 71.

plant also demonstrates the survival of the structure of a water plant.[27]

Even if dry rice is included as an alternate crop, the Han farmer in the areas of easy access to water had a good choice of crops. Since millet and leguminous crops are adaptable to most climates, the rice cultivator could arrange crop rotation. In certain areas, such as the Huai River valley, even wheat and rice can be organized into rotation systems. In fact, rice cultivation in the Han period could be found as far north as the present provinces of Shansi and Hopei[28] (II, 40, for instance).

Rice cultivation in Han times had already acquired many of the basic farming techniques that were to become typical in China. Most significant was the preparation of a rice paddy with a flat and level bed, which limited the size of the individual field. The painstaking job of weeding in water and the transplanting of young sprouts from the nursery to the field are both extremely labor consuming (V, 4: D. 5 and III, 1: E. 5). The heavy demand for water probably also helps to explain the irrigation systems in Nan-yang commandery and other areas along the transitional zone between the north and the south.[29]

As well as the cultivation of cereals, the Han farmer invested a good proportion of his labor and land on vegetable cultivation. A second century B.C. work[30] gives the ratio of 20 percent of total yield for the growth of vegetables, as shown in table 7.

TABLE 7
DISTRIBUTION OF PRODUCTION

Kinds	Amount *(shih)*	Ratio
Total	50	100
Cereals	30	60
Vegetables	10	20
Fodders and "six animals"	10	20

SOURCE: Yu Yü, "Yu Hsi-Chou tao Ch'ien-Han," *Nung-shih yen-chiu chi-k'an* 2: 14, and IV, 8.

27. Ibid., p. 72.
28. Ibid., p. 66.
29. See discussion on irrigation, pp. 99-107 in this chapter.
30. The book, attributed to Commandant Yin, allegedly contained chapters on cultivation of garlic, onion, mallow, melons, and mustard. See Shen Ch'in-han's comment quoted by Wang Hsien-ch'ien, *HS*, 30: 49a.

From the newly opened interflow of materials with the states of Central Asia, mainly the Western Regions, the Han farmer acquired many new types of plants, among which there were watermelons, cucumber, broad bean, shallot, garlic, pepper, sesame, grape, and alfalfa.[31] Judging from the vegetables mentioned by Ts'ui Shih in the *Ordinances for the Four Peoples,* nevertheless, the list of vegetables was fairly short: melon, gourd, mallow, hibiscus, alfalfa, mustard greens, turnip, taro, ginger, wild ginger, onions, shallot, garlic, leeks, water pepper, and thyme (III, 1, passim). Note the large number of seasoning spices and the garlic-onion group, which were mainly for culinary use. The remaining are far from a rich variety; even if elm pods and soybean leaves are added to the list, the Han diet did not have a fraction of the vegetables that Chinese diets do today. To grow these relatively few vegetables still kept the Han farmer busy virtually eight months a year, according to the *Ordinances* and as illustrated in table 8.

The principal cash crops were mulberry, hemp, sesame, indigo, and gourd; mulberry leaves were used only as food for silkworms. Male hemp was split and the tissue used for making linen and other fabrics. Female hemp, sesame, and Perilla were oleiferous crops, although hempseed was also considered a cereal food.[32] By the time of Ts'ui Shih, hempseed was obviously no longer a cereal since it is not included in his list at all; wheat was substituted for hempseed in the autumn (III, 1: H. 5). Indigo was the principal dye for blue color and in Ch'en-liu indigo replaced millet as the main crop during the late second century A.D. (IV, 9).

In summary, cereal crops had been developed to include millet, wheat, barley, oats, rice, and beans; each of these crops could be planted at various times. In addition, there were a good number of vegetables and cash crops so that the Han farmer had a broad range from which to plan a profitable farming system. Nevertheless, the better the quality of his crop, the more labor it demanded. For instance, spiked millet presumably tasted better than millet, yet it also required more care and more water. Wheat was even more appreciated than spiked millet; its demand for water and for better conditions in the field is also greater. Rice eventually became the principal crop of China. In the Han period one can trace the change from a crop pattern based upon rather primitive to highly intensive

31. Wan Kuo-ting, et al., *Chung-kuo nung-hsüeh-shih,* p. 230; Burton Watson, trans., *Records of the Grand Historian* 2: 279-80.

32. Shih Sheng-han, *On Fan Sheng-chih-shu,* p. 48. Hempseed was offered to ancestors as a newly harvested crop in fall; see *LSCC* 7: 1, for instance.

TABLE 8

DISTRIBUTION OF CROPS IN THE YEAR

Crop	Month							
	1	3	4	6	7	8	9	10
Melon	P	P		store				S
Gourd	P			fruit		H		
Mallow	P			P		dry	preserve	
Hibiscus		SS		P				
Alfalfa	P				P	P		
Taro	P							
Ginger			P				S	
Wild ginger		nursery					S	
Turnip			SS	P	P			H
Mustard	P/T		SS	SS	SS	P		
Onion								
Large	P			T	P			T
Small		T	T		P			
Shallot					P			
Garlic								
Large						P		
Small			H	P	P	P		
Variant	P							
Leek								
1	P				S	H		
2	P/T				T			
Water pepper	P							
Thyme	P							

SOURCE: *Chung-kuo nung-hsüeh-shih*, p. 230, table 30.

NOTE: P = plant; T = transplant; S = separate; H = harvest; SS = select seed.

farming. The development of farming techniques, however, will be discussed in the next chapter.

Soil and Its Modification

Soil cultivation appeared in at least two ancient Chinese sources, the "Yü-kung" chapter in the *Shu-ching* and the "Ti-yüan" chapter in the *Kuan-tzu*. The dates of these sources are far from their alleged ones. The "Yü-kung" allegedly reflects tributes received from the nine regions by the great culture hero, Yü, who saved China from the deluge in the twenty-fourth century B.C. This work, however, is generally regarded as having been written between the sixth and third centuries B.C.[33] In the "Yü-kung," color, texture, and the fertility of the soil of nine regions are described and graded (see table 9).

TABLE 9

Soils of Nine Regions

Region	Location	Soil	Tax (Grades)	Fertility (Grades)
Chi	Shansi, northern Honan, and northern and western Hopei	white loam	1-2	5
Yen	Western Shantung, part of Honan, and Hopei. Lower reach of Yellow River	black, rich	9	6
Ch'ing	Shantung Peninsula and part of Liaotung Peninsula	white, rich	4	3
Hsü	Huai River valley and part of Shantung	red, clayey, rich	5	2
Yang	Lower Yangtse River and the coastal area	muddy, miry	3	8
Ching	Central reach of Yangtse River and South China provinces	muddy, miry	3	8
Yü	Honan and southern Hopei	loam; lower parts: rich dark, with lumps	1-2	4
Liang	South Shensi, Szechwan, Yunnan	blue black	7-8-9	7
Yung	Northwest provinces	yellow loam	6	1

Source: IV, 10.

33. The earlier date, twenty-fourth century B.C., was indiscriminately accepted

Note that the regions with the best soil were in Yung, Hsü, Ch'ing, and Yü, the area extending along the whole Yellow River from the loess plateau to the lower plain. The soil of the lowest grades was distributed in Liang, Ching, and Yang (the Yangtse River valley), and the mountainous provinces in the southwest. In other words, the soil was graded according to the criteria for dry-land farming, possibly for cultivation of such crops as millet, which needs very little water. Strangely, tax grades did not coincide with fertility of the soil. Chi, Yü, Ching, and Ch'ing were most heavily taxed. These regions were occupied by the most populous states, Han, Chao, Wei, Ch'i, and Ch'u, during the Chan-kuo period, and remained areas of high population in the Han, although the crowded metropolitan area of the Ch'in-Han capital, Yung in the "Yü-kung," did not appear as one of the most highly taxed regions. The correlation between grades of taxes and population density points to two probable interpretations. First, the poll tax constituted a significant proportion of the pre-Ch'in tax reservoir. Second, the actual production rate was determined by the labor put into farming rather than the quality of the soil. The availability of labor, in turn, was correlated with the size of the population. Since the original text gives the criteria of taxation in terms of products from the soil in each region and since no mention is made of the poll tax, the second interpretation, therefore, seems more likely. If that were the case, then the "Yü-kung" author perhaps only listed the soil grades and the tax rates as a matter of record rather than stressing the quality of the soil.

The *Kuan-tzu* also includes some interesting information on soils. In the "Ti-yüan" chapter the level of underground water, the quality of soil, and the native flora are discussed. Hsia Wei-ying asserted that this chapter dated from the Chan-kuo period and that it covered the eastern part of China, excluding the state of Ch'in. Yu Yü, on the other hand, argued that the work was written during the reign of Emperor Wu and that the area covered was mainly the present province

by some agricultural historians; see Milton Whitney, *Soil and Civilization.* On the other hand, the iconoclast Ku Chieh-kang pulls the date down to the early third century B.C.; see his annotation on "Yü-kung" in *Chung-kuo ku-tai ti-li ming-chu hsüan-tu,* ed. Hou Jen-chih et al., pp. 1-6. Ch'ü Wan-li, however, puts the "Yü-kung" date in the mid to late Ch'un-ch'iu period; see his "Lun Yü-kung chu-ch'eng-ti shih-tai," *Bulletin of the Institute of History and Philology* 35 (1964): 53 ff. Professor Lao Kan disagrees with both Ku and Ch'ü by dating the "Yü-kung" to the mid or late Chan-kuo periods. See his "Huang-t'u yü Chung-kuo nung-yeh-ti ch'i-yüan pa" postface in *Huang-t'u yü Chung-kuo nung-yeh-ti ch'i-yüan,* by Ho Ping-ti, p. 193.

of Shensi, the Kuan-chung region.[34] Yu Yü, by drawing upon modern soil records, presents the more convincing case.

According to the "Ti-yüan," the soil can be classified first by terrain: plains, hills, and mountains. The chapter begins with a description of the five kinds of soil, probably in the capital region, and how far down the underground water was. First, the irrigated alluvial soil, with the water table 35 Han *ch'ih* below the surface, was the most fertile of all soils, and any of the five grains could grow on it. The second was a kind of red loam, which is described as "loose as well as strong, a very fertile soil, possibly of a balanced portion of clay and sand." The underground water was 28 Han *ch'ih* below the surface; this was good soil for growing any of the five grains. Third, a kind of yellow sand, was regarded as poor soil, too loose to maintain adequate moisture, with the underground water 27 *ch'ih* below the surface. It was even hard to erect walls and buildings on this soil. Only millets could grow here. The fourth was also a red soil, perhaps brown clay, on which beans and wheat grew. The underground water was salty and was 14 *ch'ih* below the surface. Finally, there was a dark soil, whose color probably indicated heavy deposits of decomposed organic materials, that could grow wheat. The underground water, 7 Han *ch'ih* below the surface, tasted bitter. This soil seemed to be in the lowland of the river valleys or adjacent to riverbanks.

The "Ti-yüan" chapter gives fifteen kinds of soil on hills; in such areas the water level ranged from 42 to 126 Han *ch'ih* below the surface. Four of the last five were found at higher elevations where underground water was virtually unreachable. The gradual elevation in height of land, where the underground water lay deeper and deeper, indicates a pattern of terraces between the high mountains and the river valley. No flora data is given in the text for hilly areas.

Strangely, the areas in the mountains are described with their local flora and the availability of water, which seems much easier to reach than the water in the hilly terraced land. For neither the hills nor mountains is soil condition described, nor fertility rates given.

The latter half of the "Ti-yüan" appears to be written by another person. It starts with a review of the different heights to which plants grow and how that height affects the distribution of each of the twelve kinds of plants. The first category of soils consists of three subdivisions, as shown in table 10. These classifications reflect the notion that fertility of soil is determined by a combination of texture and water supply.

34. Cf. Hsia Wei-ying, *Kuan-tzu Ti-yüan-p'ien chiao-shih,* and Yu Yü, "Kuan-tzu Ti-yüan-p'ien yen-chiu" in *Nung-shih yen-shiu chi-k'an* 1 (1959):17ff.

TABLE 10

CLASSIFICATION OF SOILS

Soil	Grade	Colors	Characteristics	Crops Grown
Wu-shu	A	5	Moist, yet not sticky; hard, yet not crisp; no mud to catch vehicle wheels; no dirt to soil limbs	All grains
Wu-wu	B	5	Tight and compact, yet with small holes that allow insects and worms; moist and mellow	Grains of large and small sprouts; all varieties of hemp
Wu-wei	C	5	Neither solid nor crunchy; fine grain; good moisture in both high and lowlands	*Ta-wei-wu, Hsi-wei-wu,* 2 kinds of grain, possibly millet
	Fertility (in percentage of top grade)			
Wu-yin	80		Black-green, fertile, powderlike	2 varieties of rice
Wu-jang	80		Powder-like, moist, heaped-up	2 varieties of paddy rice
Wu-fou	80		Like broken grain, preserving moisture	2 varieties of millet
Wu-chü	70		Hard, yet moist	2 varieties of S.I.*
Wu-lu	70		Hard, yet strong	2 varieties of millet
Wu-chien	70		Yellow, fluffy, like husks	2 varieties of millet
Wu-p'iao	60		Purple-blue, fluffy, like ground seashells	2 varieties of black P.M.**
Wu-sha	60		Sand, dusty	2 varieties of P.M.
Wu-ke	60		Sticks together; clusters of large particles	2 kinds of P.M.
Wu-yu	50		Like excrement	2 kinds of P.M.
Wu-hung	50		Like "mouse liver"	2 variations of S.I.
Wu-chih	40		Very wet, yet sandy	2 kinds of rice
Wu-hu	40		Loose, tolerates neither flood nor drought	2 kinds of beans
Wu-fu	30		Hard, yet not strong	Dry-land rice
Wu-chieh	30		Salty and bitter	White rice

SOURCE: *Kuan-tzu,* SPPY ed., 19: 1-7. Cf. Miao, *Chung-kuo nung-hsüeh-shih,* pp. 144-50; Yu Yü, "Yu Hsi-Chou tao Ch'ien-Han," pp. 35-36.

* S.I. = *Setaria italica*; ** P.M. = *Panicum miliaceum*

Although the classifications of soil in the *Kuan-tzu* are based mainly upon texture and availability of water, it is worth mentioning that the authors of the "Ti-yüan" chapter noted the impoverishment of the loose soils, which were described as "fluffy," "loose," "cluster of particles," and so forth. Such kinds of soil very likely were the result of leaching, the downward percolation of plant nutrients. In the valleys of the Yellow and Huai rivers, the soils are largely brown, gray-brown podsolic, chestnut, and light chestnut, which is calcareous throughout because of recent deposits of loess. Such soils are ordinarily prone to leach.[35]

Lying behind the discussion of soil in the "Ti-yüan" chapter, of course, is the recognition that the crop had to be selected that was most suitable to the soil. This is perhaps a very common notion; nevertheless, the discussion presented in the *Lü-shih ch'un-ch'iu* also challenged this basic assumption by suggesting various soil modifications.[36] A remnant of the *Fan Sheng-chih-shu* declared that land productivity need not be determined entirely by formation of the soil, rather, that fertility could be created by fertilizers (V, 4: G.1a).

Agrology recognizes that soil changes in time.[37] This includes not only the natural process of chemical weathering and mechanical translocation, but also the control over soil exercised by human efforts, such as cultivation and fertilization. Man can virtually transform soil so that it differs completely from its original texture.[38]

The *Fan Sheng-chih-shu* contains specific discussions of soil modification. Hard soil was to be softened by early pulverization and light soil was to be hardened by treading grasses and leaves into it or even by firm treading without the grasses and leaves (V, 4: G.1 and G.4a). The treatment of seeds with animal or silkworm excrement or bonemeal was a way of enriching the soil in the immediate vicinity of germination by providing the seed with optimal nutritional supplies *hoc tempore hoc loco*[39] (V, 4: C.1 through D.1b). Han scholars had

35. For the distribution of the soil, see T. H. Shen, *Agricultural Resources of China*, pp. 20-21. For the process of leaching, see Brian T. Bunting, *The Geography of Soil*, pp. 90-97, 163-78.

36. *LSCC* 26: 6, and our introductory chapter.

37. Bunting quotes V. V. Dokuchayev, "Soils are not only very variable in space but also comparatively unstable in time. Actually we do not know of any soil which has retained its character forever." Bunting, *Geography of Soil*, p. 78.

38. Ibid., pp. 19, 130-31.

39. See the comments by Shih Sheng-han in *Fan Sheng-chih-shu chin-shih*, pp. 58-60. Shih points out that silkworm excrement contains a significant amount

identified the treatment of seed as one method of ameliorating the soil. Commenting on one paragraph in the *Chou-li,* the Han annotator Cheng Hsüan stated that the ways to modify the soil consisted of matching the crop and the soil and treating the seed with animal bone decoction. And Cheng Hsüan explicitly declared that such a soil amelioration process was comparable to the method of Fan Sheng-chih[40] (IV, 11).

In fact, even new soil was modified by human effort, beginning with the opening of virgin land for cultivation. In both woody areas and grasslands, the original vegetation was burned and the residue allowed to decompose before the soil was cultivated. The decomposed ash, of course, would provide the soil with valuable nutrient salts[41] (IV, 12). This land reclamation process, although similar to slash-and-burn farming, was aimed at limited cultivation in the year after the felling of trees and the creation of a good field in the third year and thereafter. Such a practice had its origin as early as in the Shang period and is vividly depicted in Western Chou literature.[42] By using fertilizers and irrigation, of course, man manages constantly to modify and change the character and quality of the soil. To some extent soil is under the control of human manipulation, instead of determining the crop man must plant.

Fertilizers in Han China were in the main the two manures, green and animal. The conscious use of green manure most likely appeared in China as early as the eleventh century B.C. and not later than the fifth century B.C.[43] A passage on green manure, which perhaps is the first recorded, appears in the *Li-chi.* One section of the "Yüeh-ling" is a discussion of how weeds should be cut, burned, and then submerged in water during the late summer when excessive heat could

---

of potassium, nitrogen, phosphates, auxins, and vitamins, together with complicated groups of microbes, both active and resting, and that bone decoctions add to the soil soluble nutrient salts, organic matter, and collagen that support microbic activities.

40. Cf. Shih, *Fan Sheng-chih-shu,* pp. 56-57.

41. A problem closely related to the one discussed here is the "fire cultivation and water hoeing" method, which is to be discussed in chap. 5.

42. Chang Cheng-lang, "Pu-tz'u p'ou-t'ien chi ch'i hsiang-kuan chu-wen-t'i," *K'ao-ku hsüeh-pao,* 1973, no. 1, pp. 93-105. Cf. *MSCI,* 16-4:3 (James Legge's translation, *The Chinese Classics* 4: 448). Such reclamation process is not to be confused with the fallow system; see Yu Yü, "Yu Hsi-Chou tao Ch'ien-Han-ti keng-tso chih-tu yen-ke," *Nung-shih yen-chiu chi-k'an* 2:3-4.

43. Ch'en Liang-tso, "Wo-kuo li-tai nung-t'ien shih-yung-chih lü-fei," *Ta-lu tsa-chih* 46, no. 5:1-2, and "Chung-kuo ku-tai nung-yeh shih-fei chih shang-chüeh" (as appendix to Hsü Cho-yun, "Liang-Chou nung-tso yü chi-shu," *Bulletin of the Institute of History and Philology* 42, pt 4 [1971] :829-42.

accelerate their decomposition to "fertilize the field and ameliorate the soil" (IV, 13). The Han agricultural treatise, the *Fan Sheng-chih-shu,* also contains a passage on the use of green manure. The author warned that the field should not be pulverized until the weeds began to grow so that the field would be ameliorated after the weeds were turned under and had decayed in the soil (V, 4: 5a).

"Animal manure" includes both human and animal excrement. An agricultural historian recently argued that waste from human settlements seems to have been used as fertilizer in horticulture in the vicinity of such settlements before fertilizers were consciously applied to agriculture on an extensive scale,[44] and in present times night soil from the cities of China is still shipped to vegetable gardens and rice paddies. The toilets included in the terra cotta models of houses, farms, and so forth, found in the Han tombs speak for their significance as being more than merely one part of the residence.[45] Apart from night soil, pigs and poultry provided the major source of animal manure (see plate 1). This is in contrast to the European sources of fertilizer, which drew largely on cattle and horses. European farmers had a large amount of fallowed land to provide their animals with extensive grazing space, while the Han Chinese farmer simply could not afford to fallow, for the land had to be cultivated continuously. An early Han estimation of the income of a farm household asserted that 20 percent of the total production was related to the "six animals" (chickens, dogs, pigs, horses, oxen, and sheep). This particular category included "husks and chaff" (IV, 8), indicating that it embraced both the value of feed and the value of domestic animals. The proportion of either one, therefore, is even smaller than 20 percent (IV, 8).[46] This figure certainly is far below the proportion that animal husbandry occupied in farm household incomes in Europe.

The strenuous effort by the Han farmer to improve and modify the soil shows that land was precious and that the farmer could not afford large acreages. The minimal animal husbandry also emphasizes the necessity of maximum utilization of land for production of food and cash crops. Furthermore, smaller farms and the two problems stated above constituted a vicious circle. The more effort one put

44. Ch'en, "Chung-kuo ku-tai nung-yeh shih-fei chih shang-chüeh."

45. In the Han tombs discovered at Shao-kou, Lo-yang, for instance, twenty-six clay models of pigpens are found in twenty-four graves. The pen, often with models of pigs in it, is built adjacent to the toilet chamber. Under both the toilet and the pen there are pipe openings leading toward a lower clay plate, which obviously is a depository for night soils and pig feces. Lo-yang-ch'ü k'ao-ku fa-chüeh-tui, *Lo-yang Shao-kou Han-mu,* p. 141.

46. Yu, "Yu Hsi-Chou tao Ch'ien-Han," pp. 13-14.

PLATE 1. MODELS OF ADJOINING PIGPENS AND TOILETS FROM HAN TOMBS AT LOYANG.

These terra cotta models have pipe openings leading toward a lower clay plate representing facilities for the easy collection of manure (Lo-yang-chü k'ao-ku fa-chüeh-tui, *Lo-yang Shao-kou Han-mu*, p. 141).

into intensive farming, the smaller the possibility one could work an extensive area. The smaller the source of animal manure, the less likelihood there was of satisfactorily improving the soil quality of large areas. It is apparent that modification of the soil with manure, treatment of seed, and restricted animal husbandry all indicate that the Han farmer had to adapt to a small farm operation.

IRRIGATION

Irrigation, one of the major utilizations of natural resources, can be implemented by controlling water through at least three means: the reservoir, the canal ditch, and the well. The purposes are also threefold: to water crops, to improve the quality of saline soil, and to control flooding. In Han China these utilizations of water each met the needs of different regions.

In pre-Han times, extensive irrigation systems were developed in the present provinces of Shensi, the middle and lower reaches of the Yellow River, the Huai and Han valleys, and the Szechwan Basin. It was in these areas also that most of the Han irrigation projects were constructed.[47]

Each of three early projects met the demands of the local soils and crops. The oldest project is the Shao Pei, or the Shao Reservoir, in northern Anhwei Province, along the tributaries of the Huai River. The construction of the Shao Reservoir was attributed to a chancellor of Ch'u during the reign of King Chuang (ca. 613-591 B.C.). In the first century after its construction, the whole system was repaired after a period of deterioration. The historical records available to us, especially the *Shui-ching-chu,* an ancient historical geography of the fifth century, appear to be an account of the system after extensive repair. Nevertheless, the basic features of the Shao Reservoir probably remained very much the same—a means to store water from two rivers to irrigate 1,000,000 mu of rice fields. The reservoir had five water gates to receive and release water and its circumference was more

47. The Yellow River loess region and the Huai and Han valleys are regarded by Ch'ao-ting Chi as the first key economic areas in Chinese history. The Szechwan Basin, though somewhat peripheral, also had strategic significance after the irrigation network was constructed there in the third century B.C.; see Ch'ao-ting Chi, *Key Economic Areas in Chinese History*, pp. 66-70, 75-80, 96-97. A recent statistic on the irrigation works recorded in the *Shui-ching-chu* reveals that 48 percent of the sixty-one pre-Han reservoirs were in the Yellow River valley while all of the fourteen canals but one were in the three regions mentioned above. Huang Yaoneng, "Shui-ching-chu shih-tai so-ch'u-hsien-ti Chung-kuo ku-tai ch'ü-pei fen-pu chi ch'i so-tai-piao i-i," *Yu-shih yüeh-k'an* 43, no. 5:56 ff.

than 120 *li* (IV, 14, 15). The Huai River valley is within the rice-growing region, so the reservoir insured the supply of water to the rice paddies. The primary function of this irrigation, therefore, was stabilization of the water supply.

The Chang River valley irrigation system was for a different purpose. Hsi-men Pao in the reign of Marquis Wen of Wei (424-387 B.C.), or Shih Ch'i in the reign of King Hsiang of Wei (334-319 B.C.), or both of them, are credited with tapping water from the Chang River to irrigate fields so that "the soil, saline ever since antiquity," was made fertile enough to grow rice and the *liang* variety of millet *(Setaria italica* Beauv. var. *maxima* Al.) (IV, 16, 17). Similarly, the Cheng Kuo Canal, which was named after the engineer who masterminded the project in the present province of Shensi, was also designed to change saline soil to good fertile land. An extensive irrigation system rechanneled water from the muddy Ching River to carry silt to make a layer of fertile topsoil on the poor quality fields (IV, 18). Both Yeh and Kuan-chung, where the Cheng Kuo Canal was constructed, are in the loess region where a good supply of water would both serve as a solvent of excess salt and would provide a vehicle for capillary action, bringing up the minerals that lay in the deep formation of loess. This same texture also makes the soil subject to leaching. The annual deposit of river silt, which is suspended in water during the inundation, is the best available remedy for such soil.

The third irrigation system was the Tu-chiang Dam built by Li Ping in the Szechwan Basin during the fourth century B.C. (306-251 B.C.). The earlier sources state that the primary motivation for this water-control project was to improve water transportation by dividing into two channels the course of a river that was blocked by rocks and cascades. Tapping water to the fields was only a by-function. Today, however, irrigation has become the principal function of Tu-chiang Dam, which is well maintained and still in full use, surviving some twenty-three centuries (IV, 19). The climate and soil in the Szechwan Basin are suitable for growing paddy rice and the dam facilitated supplying water, emerging from stony mountains and carrying no mud or silt, to the paddy fields in the Ch'eng-tu plain. This project, therefore, is more like the Shao Reservoir than the network in Yeh.

In the Han period, water-control projects served the same purposes as they had earlier, although, strangely, field irrigation does not seem to have been the initial or primary concern. The first major enterprise in the Han period was undertaken in 130 B.C. when a canal was dug to shorten the distance grain was transported from the eastern part of the empire to the densely populated capital region. Several tens of thousands of laborers worked for three years to open up the three-hundred *li* canal. Meanwhile, more than ten thousand *ch'ing* of fields near the canal benefited from easy access to water (IV, 20). A few years later, the difficulty of shipping massive amounts of grain from the east to the capital again demanded some alternative. The governor of Ho-tung, a commandery adjoining the capital district across the Yellow River, proposed the irrigation of five thousand *ch'ing* (or half a million mu) of marginal land so that grain could be produced within easy reach of the capital district, eliminating or reducing grain transportation from the east. In his proposal, canals were to be dug to bring water from both the Yellow River and its tributary, the River Fen, into the deserted land along the riverbank, which was being used for grazing and growing fodder. However, the river shifted its course and the project was a failure (IV, 21). It is strange, in this case, that the land on the riverbanks should have needed special irrigation. One possibility is that the land was actually on a former riverbed, resulting from an earlier change of the river's course. Another change then caused the canals to be useless. Another possibility is that the leach-prone soil had made it necessary to supply silt for a layer of topsoil. Five years later, another section of salinated land was to be irrigated. The water reached the fields through a series of underground canals (or wells) (IV, 22), which would not provide silt to make topsoil. In this case, the high yield expected from the project would have been met not only by washing away alkalinity but also by bringing water to some sort of water-loving high-yield crop, such as rice. Rice did grow here during the Han period, but today Shensi is no longer a significant rice-producing area (IV, 23 and see IV, 30).

In 111 B.C. six auxiliary canals were added to the Cheng Kuo Canal to irrigate fields on the high lands (IV, 24). And in 95 B.C. one more major canal was dug to add 4,500 *ch'ing* of irrigated fields to the arable acreage in the capital district. The crops were specified as two variations of millet. The advantage was attributed to the heavy silt carried by the Ching River (IV, 25).

The emphasis on construction of irrigation canals in the capital district before and during the first century B.C. (IV, 26) changed from the need for transportation to the need to reclaim land, and eventually to the irrigation of crops. By the time the six auxiliary canals were dug, a decree was issued to emphasize the benefit of water on grain crops, and rice was especially identified as a crop in the capital district (IV, 24). To assure an abundant food supply for the capital area population, the original approach was first to import grain from the east, then to increase local production by expanding the productive area, and finally, to improve the yield per unit in a given area through irrigation.

The popularization of some water-loving crops, such as rice and wheat, ought to bear a significant correlation to the rising demand for irrigation. Ho Ping-ti cited the difference in the efficiency of transpiration among various plants and his findings are given in table 11.

TABLE 11

WATER DEMAND OF CROPS

	First test		Second test	
	Index	Efficiency	Index	Efficiency
Setaria italica	274	3.65	224	4.46
Sorghum vulgare	285	3.51	268	3.73
Triticum vulgare	550	1.82	403	2.48
Zea Mays L.	361	2.77	—	—
Glycine soja	646	1.55	—	—

SOURCE: L. J. King, *Weeds of the World,* table 6, p. 180, quoted by Ho Ping-ti, *Huang-t'u yü Chung-kuo nung-yeh-ti ch'i-yüan*, p. 131, table 6.

In this table, dry-land crops, such as wheat *(Triticum vulgare)* and soybean *(Glycine soja),* show more than double the capacity for water than millet and sorghum.[48] Rice, a waterweed, needs even more water. Earlier in this chapter, we noted that wheat and soybean, with the addition of one kind of millet *(Panicum miliaceum)* made up the important cereal grains of the Han period, and *Setaria italica* occupied a less significant position than it had in previous times. Increasing demand for water perhaps would help to explain the ostensible increase of droughts in the records of natural disasters. The frequency of major droughts in the pre-Ch'in and Ch'in-Han periods, up to the early part of the first century B.C., is shown in table 12.

The pattern of increasing floods, and especially the sharp break between periods 3 and 4, however, differs from that of the drought

48. The uneven annual distribution of precipitation in North China makes the situation worse. During April and May the wheat needs water badly, yet the rainfall in these two months usually is insufficient. Chu K'o-chen, "Lun wo-kuo ch'i-hou-ti chi-ke t'e-tien chi ch'i yü liang-shih tso-wu sheng-ch'an-ti kuan-hsi," *Ti-li hsüeh-pao* 30, no. 1 (1964): 4-5. Today in Sian, the annual rainfall is 22.6 inches, but in April and May, when crops need the most water, the average is less than 4 inches. *Rand McNally Illustrated Atlas of China* (1972), pp. 54-55.

record.[49] One of the plausible explanations of the rising frequency of floods is that the marginal land near the flooding zone had been inhabited as a result of population growth, or the overexploitation of water resources caused poor conservation (IV, 27).

TABLE 12

FREQUENCY OF DROUGHTS

Periods	Length of period (in years)	Number of Occurrences	Interval between (in years)
1. 722-480 B.C.	242	17	14.23
2. 479-247 B.C.		no record	
3. 246-180 B.C.	66	7	9.4
4. 179-141 B.C.	39	5	7.8
5. 140-87 B.C.	53	13	4.07
Average interval during 3 through 5			6.32

SOURCE: The data are derived from Yu Yü, "Kuan-tzu Tu-ti-p'ien t'an-wei," pp. 8-9. There is no way to test the reliability of these figures.

Spectacular population growth in the capital district in the Western Han period put tremendous pressure upon agricultural production. The capital district, which consisted of the capital and two neighboring commanderies in the Western Han dynasty, had a population of several million. People of outstanding prestige, wealth, influence, and rank in the government hierarchy were occasionally ordered to move to the capital region. Ordinarily when an emperor began to build his mausoleum in preparation for his death, a new county would be formed to house the elite households of that particular generation.[50] Four to six million *shih* of grain had to be imported to feed the regular population, in addition to the rotating garrison troops stationed there.

49. The frequency of floods also increased as follows:

Periods	A Duration (in years)	B Occurrences	C Interval years (A/B)
1.	242	11	22
2.		no record	
3.	66	3	22
4.	39	4	9.75
5.	53	6	9.38

SOURCE: Yu Yü, "Kuan-tzu Tu-ti-p'ien t'an-wei," pp. 8-9.

50. Lao Kan, "Liang-Han hu-chi yü ti-li-chih kuan-hsi," *Bulletin of the Institute of History and Philology* 5 (1935): 198-201.

It seems that the local population pressure alone would and could spur the improvement of agriculture, including the building of expensive and extensive water-control projects and the adoption of new crops.[51] As internal disturbances ended the Western Han dynasty and the old capital district near Ch'ang-an ceased to be the hub of activities, the population decreased, and the old irrigation systems deteriorated from neglect. A stele inscribed in A.D. 181 tells the story of the malfunction of one water reservoir and of the resulting flood suffered by a town (IV, 28).

New key economic areas gradually developed. The middle and lower reaches of the Yellow River continued to be plagued by frequent inundations of this muddy river, but people still exploited the fertility of its heavy silt. The floods of the Yellow River became more acute as inhabitants tended to farm too close to the riverside, because irrigation caused the flow to slow down (IV, 27, 29, 30). The most effective irrigation systems were developed along the tributaries of the Yellow and the Huai rivers where, as early as the second century B.C., canals had been dug to bring water into the fields[52] (IV, 31). According to Ch'ao-ting Chi's statistics, among the fifty-six water-control projects during both Han periods (206 B.C.-A.D. 220), eighteen were in Shensi, nineteen in Honan, four in Shansi, five in Hopei, and one in Anhwei.[53] Most of the works in the Yellow and Huai river valleys were done in the Eastern Han period.[54] One of the most ambitious efforts for flood control was in A.D. 69 when several hundred thousand laborers were mobilized to build dikes and dig canals along a stretch of one thousand *li* (IV, 32).

The area of key significance is in the Nan-yang-Ju-nan region, the valleys of the Han and Huai rivers. Here the precipitation is more abundant than in Kuan-chung, the old capital district, and the general climate is on the edge of the rice-wheat transition line. Irrigation was important in determining the selection of crops,[55] and a number of reservoirs of various sizes were in existence in this region in the Han

51. Kimura suggested that the creation of new arable land was correlated with the intention of the imperial government to grasp direct control over the farmer. Kimura Masao, *Chūgoku kodai teikoku no keisei*, p. 199.

52. Tsou Pao-chün pointed out that the foundations of ancient *Chinese* civilization were found along such tributaries. "Chung-kuo wen-hua ch'i-yüan ti," *Tsing Hua Journal*, n.s., vol. 6, nos. 1-2 (1967).

53. Chi, *Key Economic Areas in Chinese History*, p. 36.

54. Kimura noted that a few of them actually had been built in the pre-Ch'in period. Kimura, *Chūgoku kodai teikoku no keisei*, pp. 159 ff. Huang Yao-neng, "Shui-ching-chu shih-tai," pp. 56 ff. The dates, however, appear to be uncertain.

55. For weather information, see *Rand McNally Illustrated Atlas of China* (1972), pp. 54-59.

period. The first extensive operation was the system built up by Shao Hsin-ch'en in the second half of the first century B.C. In the Nan-yang commandery several tens of places benefited by canals. Although the total irrigated area is an impressive one—thirty thousand *ch'ing*—each of the scattered individual projects seems to have been modest. The first historical data we have on this is a stele upon which the regulations for water distribution were inscribed in order to prevent disputes. It seems that the commandery authority served as arbiter rather than as a directly involved manager (IV, 33).

Another important water reservoir is the Hung-hsi in Ju-nan. It is not known when the reservoir was first built, but around 15 B.C. it was demolished because of the inundation it caused (*HS* 84: 20a-b). People complained that they had only a diet of taro and beans because of the destruction of the pond, indicating that the pond served to supply water to a crop that demanded more water than beans and taro, possibly rice. Some fifty years later, the reservoir was restored. It is said that Ju-nan remained an area of extensive irrigation systems throughout much of the Eastern Han period. Canals, reservoirs, and water gates constituted the main features of the network, which was probably for irrigation of rice fields (IV, 35, 36).

The construction of irrigation projects spread in all directions. In the Huai River valley, the Yangtse valley, and further to the south, as far as today's Vietnam, where rice was presumably the main staple, canals were dug and reservoirs were built for irrigation (IV, 37, 38, 39, 40, 41). The millet-and-wheat producing areas, nevertheless, were not ignored as old irrigation systems were repaired from time to time. For example, those in today's Shansi were repaired by the central government during the second century (IV, 42). The need for irrigation projects in the north could be for enriching the alkaline soil as well as for compensating for the shortage of precipitation (IV, 43, 44).

Water-control techniques of the people in the Han period were quite sophisticated. For example, the Po Canal included a conduit at least fifty *li* in length before the water reached the dam at which the water head was built so that it caused the water to jet a considerable distance while the heavy silt stayed in the canal.[56] The sophistication of Han hydraulic engineering is shown in the designs of the Dam of the Heavenly Well, which consisted of a series of cascades by which the water could return to the original stream (IV, 45). In another system, during heavy rains the flow from a reservoir was stopped and

56. Yu, "Kuan-tzu Tu-ti-p'ien t'an-wei," *Nung-shih yen-chiu chi-k'an* 1:10-11. We accept Yu's position that the "Tu-ti" chapter of *Kuan-tzu* was compiled in the reign of Emperor Wu and that it refers to the Po Canal system.

PLATE 2

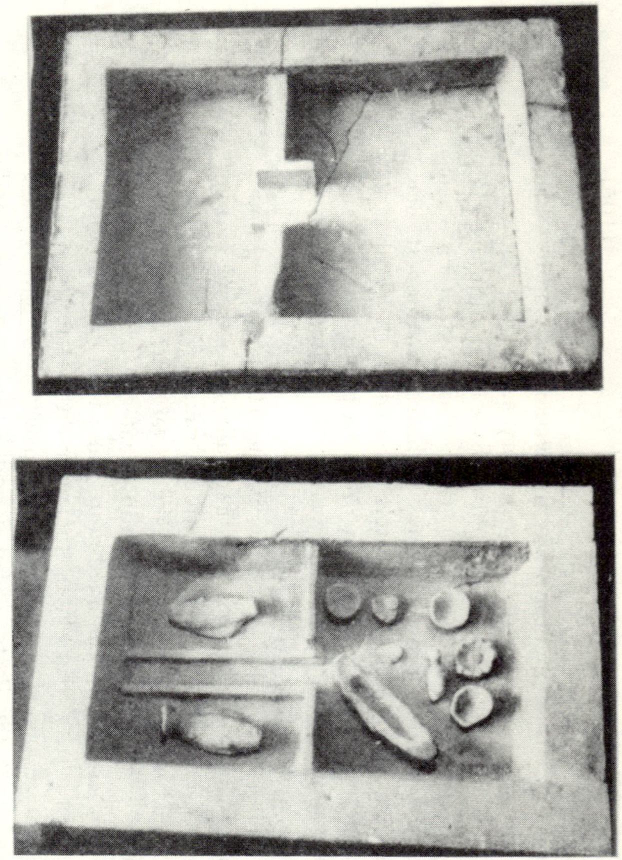

PLATE 3

PLATE 2. MODEL OF WELL-OPENING WITH A PAIR OF BUCKETS FROM HAN TOMBS AT SHAO-KOU, LOYANG, Photograph of original model and line drawing that includes models of irrigation troughs (Lo-yang-chü k'ao-ku fa-chüeh-tui, *Lo-yang, Shao-kou Han-mu,* p. 141).

PLATE 3. MODELS OF RICE PADDIES AND PONDS FROM TOMBS AT CHENGTU. These models include fish and boats, even a frog (Liu Chih-yüan, "Ch'eng-tu T'ien-hui-shan yai-mu ch'ing-li-chi," *K'ao-ku hsüeh-pao,* 1958, no. 1, p. 97).

and the irrigation channels were used to drain excess water from the fields (IV, 46). Even the water distribution in the field was accomplished by a fairly complicated network of trenches and troughs (IV, 47). Mechanical devices were also developed, as when water wheels and siphons were used where water had to be raised from low to high levels. Both kinds of devices are still visible in the countryside of China today (IV, 48).

The size of the irrigation works, of course, varied greatly. The large ones were constructed by imperial decrees, with several tens of thousands of people working for several years, and lesser ones built by local administrations. One construction project required 40,000 man-days of labor (IV, 49). It is not improbable that some local projects were sponsored by the administration but financed locally (IV, 50). In the Western Han period the offices of water control were organs of the central government, from which agents were assigned to local levels. The founder of the Eastern Han, however, decreed that the Office of Water Control should be under the commandery administration.[57] This change of jurisdiction perhaps indicates that rather than a bureaucratic reshuffle, water controls had become so numerous and commonplace that initiative for their construction had to lie with local officials.

Actually, it was not unusual for private sectors to undertake construction projects. Wealthy individuals built reservoirs to irrigate their own land as did the Fan clan of Nan-yang. Although Fan Chung was the father of the empress, the reservoir was built before the establishment of the Eastern Han; thus, it was an individual enterprise.[58] A stele inscription records another interesting case in which a small community collectively dug a public well to reach underground water. Ninety-one persons, each making contributions, jointly raised several hundred thousand cash. None of the contributors listed was identified as an officeholder. Thus utilization of water reservoirs was not just the concern of the government, but also that of the ordinary people, and they jointly took the initiative, shared the cost and the burden, and enjoyed the benefits (IV, 51).

Less than two decades ago, archaeologists excavated ninety-seven clay models of wells at Loyang. Most of them were one-of-a-kind items from eighty-six excavated Han tombs. Such phenomena show that in Han times each farming household could have had a well at its disposal. The terra cotta wells were usually elongated in shape

57. *Hou Han-chih* 26: 11b and 28: 9b. Cf. Kimura, *Chūgoku kodai teikoku no keisei*, pp. 212-13.

58. T'ung-tsu Ch'ü, *Han Social Structure*, pp. 286-87 and n. 166.

and equipped with pulleys, rollers, and buckets. The water tank was built right under the well frame and at the other end of the tank there was an opening. The whole system, therefore, seems more than adequate for drawing drinking water. The pair of buckets, which were side by side in the elongated well opening, could be rolled up and down in rotation, and the tank with the opening was most likely the beginning of an irrigation trench[59] (see plate 2).

Small ponds also served irrigation purposes. In the Han tombs terra cotta models of rice paddies in Szechwan are often immediately connected to a little pond in which there are fish, boats, and even frogs[60] (see plate 3).

In summary, the Han farmer counted on several ways to utilize water. The tendency, however, gradually shifted from government-sponsored efforts to use water to increase arable acreage by modifying the soil toward numerous small-scale private enterprises to increase the productivity of fields by supplying water where precipitation was uneven, or by supplying water abundantly to grow high-yielding and/or high-valued crops. The former practice, indeed, is related to the creation of secondary arable land, as Kimura suggests. The latter practice, however, ought to be looked upon as an advancement of farming techniques, perhaps related to the farmer's incentive to reach optimum production on his own farm.

My efforts in this chapter have been to investigate probable conditions under which the Han farmer developed his farming techniques. He had a fairly rich assortment of crops available for planting, either singularly or in rotation as a group. He had soil of a broad range of texture and conditions in different climates. Nevertheless, he took the attitude that man's efforts could modify the soil and make it more adaptable for cultivation. The experiments in harnessing water resources also made the farmer less vulnerable to the happenstance of the availability and distribution of this most precious factor in agriculture. Without the confidence that man could control at least some natural conditions, the Han farmer would not have asserted himself so persistently to find ways of enhancing his productivity.

59. Lo-yang-chü k'ao-ku fa-chüeh-tui, *Lo-yang Shao-kou Han-mu*, pp. 125 ff., 241.

60. Okazaki Satoshi, "Kandai meiki deishō ni arawareta suiden suichi ni tsuite," *Kōkogaku zasshi* 44, no. 2 (1958): 65 ff. Szechwan-sheng po-wu-kuan, "Szu-ch'uan sheng Mu-ma-shan Kuan-kai-ch'ü ku-mu ch'ing-li chien-pao," *K'ao-ku*, 1959, no. 8, p. 423; Szechwan-sheng po-wu-kuan wen-wu kung-tso-tui, "Szu-ch'uan Hsin-chin-hsien Pao-tzu-shan yai-mu ch'ing-li chien-pao," *K'ao-ku t'ung-hsün*, 1958, no. 8 Liu Chih-yüan, "Ch'eng-tu T'ien-hui-shan yai-mu ch'ing-li-chi," *K'ao-ku hsüeh-pao*, 1958, no. 1, p. 97.

V

FARMING METHODS AND TECHNIQUES

FARM SIZE

AT no time during the Han period was the average farm large. Agricultural historians estimate that by the eve of Ch'in unification, an average farmer's holding in the state of Ch'in was much smaller than the ideal of 108 small mu[1] (see document V, 1). In the eastern states the farmers in the more crowded areas had already opened up marginal land such as swamps and fields on the northern slopes of the hills, but even then their average acreage of arable land, because of the pressure of population, must have made their average holdings smaller than those in the state of Ch'in.[2] The Han work attributed to Kuan-tzu noted the disparity of land-use conditions between the crowded areas and the less crowded ones (V,2). Kuo Wen-t'ao estimated that, from the *Kuan-tzu* date, at best a farming household could have an average of 150 small mu.[3] The arable land and population ratios in A.D. 2

1. Here the mu is measured 100 paces in length, smaller than the 240-pace mu in the reign of Emperor Wu; cf. III, 12.

2. Yu Yü, "Yu Hsi-Chou tao Ch'ien-Han-ti keng-tso chih-tu yen-ke," *Nung-shih yen-chiu chi-k'an* 2 (1960): 6.

3. Kuo Wen-t'ao, "Chung-kuo ku-tai nung-tso-chih chih-shih-ti k'ao-ch'a," *Chung-kuo nung-pao,* 1963, no. 9, pp. 43-44. Much larger figures are derived by Yu Yü from another chapter of the same work. His interpretation, however, seems somewhat indirect and ambiguous. Cf. Yu, "Yu Hsi-Chou tao Ch'ien-Han," p. 12.

and the second century give us gross figures of about 70 mu per household, or the equivalent of about 168 small mu.[4]

In 1973, a group of Han tombs were excavated at Feng-huang-shan of Chiang-ling, Hupei, from which some tablets regarding the sizes of tenant farms were unearthed:

Designates of household heads	A	B	C	D	E	F	G	H	I	J	K	L	M	N	O	P	Q	R	S	T	U	V	W
Working hands	1	1	2	4	2	2	2	2	3	2	3	2	4	3	2	3	3	4	4	4	4	4	4
Population	1	3	4	8	2	3	3	4	4	6	6	6	7	6	3	6	4	4	5	6	6	7	5
Farm size (in mu)	8	10	12	15	18	20	20	20	20	20	21	23	23	27	30	30	30	30	30	32	33	37	54

The median farm size was 20 to 30 mu per household, 5 to 7.5 mu per person, or 6.6 to 10 mu per working hand.[5] These figures are considerably smaller than the *Kuan-tzu* figures, probably because the tenant farm was smaller than the independent holdings. It is also possible that the southern paddy farm, as is the case even today, was smaller than its northern counterpart. According to Ch'ao Ts'o's statement, in general one household tilled 100 mu only.

The demand for cash for the inescapable expenditures, such as clothing and taxes, probably pushed the farmer to use the land to its fullest. Since the land tax, although supposedly levied according to production, was exacted according to the size of the plot, the farmer would not have been inhibited from producing as much as he could. The result was a development toward repeated use of the land, either for the same crop or for different crops in rotation. Farming also evolved from a single crop annually to multicrops annually.

4. Cf. table 3 on arable land and population ratio, p. 21. The conversion to the small mu by multiplying 2.4 should be as follows: A.D. 2—163.2; A.D. 105—189.6; A.D. 122—172.8; A.D. 144—165.6; A.D. 145—168.0; A.D. 146—177.6. Cf. Ma Ch'eng-feng, *Chung-kuo ching-chi-shih* 2:352. Yu, "Yu Hsi-Chou tao Ch'ien-Han," pp. 12-13; Kuo, "Chung-kuo ku-tai nung-tso-chih," pp. 43-44.

5. Huang Sheng-chang, "Chiang-ling Feng-huang-shan Han-mu chien-tu chi ch'i tsai li-shih ti-li yen-chiu-shang-ti chia-chih," *Wen-wu*, 1974, no. 6, pp. 71-72, 76; and Hung I, "Chiang-ling Feng-huang-shan shih-hao Han-mu chien-tu ch'u-t'an," ibid., pp. 81-82. The date is estimated as possibly in 153 B.C. Two households are excluded because of incomplete information.

Multicrop System

The first mention of crop rotation is in the *Lü-shih ch'un-ch'iu*, a work completed before the unification of China. The line reads, "This year it is the good millet; in the coming year it is the good wheat."[6]

Double crops were first discussed with "Now, as far as the growth of the five grains on this land is concerned, if one works well, one can have several *p'en* per mu and in a single year one can harvest twice" (*HT* 6:5). In the early Han period, the lower Yellow River valley used a system of repeated cultivation, which yielded five crops in four years[7] (V, 3).

The remnants of the earliest available agricultural work, the *Fan Sheng-chih-shu,* mention the planting of wheat right after spiked millet was reaped. The work thus refers to the practice of crop rotation in the late first century B.C.[8] (V, 4: G. 3a). In the Eastern Han period, wheat was planted immediately after spiked millet, and soybean or millet was planted immediately after the wheat was harvested. This constituted three crops, one after the other, perhaps within two years[9] (V, 5, 6). Some interesting observations can be derived from the work schedule given in the *Ssu-min-yüeh-ling, Ordinances for the Four Peoples.* The farmer depicted in the *Ordinances* sold as well as purchased grains at different times of the year. At least on one occasion, the eighth month, wheat was sold and millet was purchased. If the farmer bought millet at a low price during harvest, the sale of wheat can be explained as advantageous to meet the demand for seed grain[10] (III, 1: H.7).

Besides rotation of crops, land utilization was so intensified that several vegetables were interplanted. The *Fan Sheng-chih-shu* registered the tripartite interpolation of melon, scallions, and beans (V, 4: G.5). Millet was sown with mulberry seed to provide the young mulberry trees with fertilizer, which was the ash of the burned millet stalk, as well as to utilize the space between the young trees when they needed little room (V, 4: D.13a-D.13c).

6. See the discussion of the agricultural level of the pre-Ch'in empire in the introduction of this work. Cf. Yu, "Yu Hsi-Chou tao Ch'ien-Han," p. 10; Kuo, "Chung-kuo ku-tai nung-tso-chih," p. 43.

7. Kuo, "Chung-kuo ku-tai nung-tso-chih," p. 44.

8. Ibid.

9. Ibid. By the time of *Ch'i-min-yao-shu,* the sixth-century agricultural work, the system of having three crops in two years had been well established. For a conclusive work on this, see Yoneda Kenjirō, "Seimin yōjutsu to ninen sanmō-saku," *Tōyōshi kenkyū* 17, no. 4 (1959): 407 and passim.

10. Yoneda, "Seimin yōjutsu to ninen sanmō-saku," p. 423.

Such intensive use of land, of course, would be relatively common in the agriculturally well developed Kuan-chung region and the eastern plain along the Yellow River during the Western Han, or the region between the Yellow and the upper Han rivers. In the less developed areas there could still be extensive farming. For instance, the fertile land in the south along the Yangtse River had not changed during the Western Han period from simple slash-and-burn farming. Its tremendous productivity would not be exploited until more people migrated there.[11]

New Farming Method—Tai-t'ien

Repeated cultivation and rotation of various crops was aimed at the most economic utilization of land by arranging the plantation in sequence. Parallel to the maximum temporal arrangement was the optimum use of space so that plants were close enough to permit the best growth within a given unit, yet not so crowded as to deprive them of proper nutrition and space to mature. Early in the first century B.C., a farming method called *tai-t'ien* was brought to the attention of the Han court by Chao Kuo. This method was summarized by Pan Ku in the *Han-shu*; however, the text is so ambiguous and brief that different interpretations are presented by historians.

With the *tai-t'ien* method, a mu of land was divided into furrows and the soil plowed up from the trench was piled beside the furrow to form a ridge one *ch'ih* high. The seed was planted in the furrow and soil from the ridge was constantly pushed down to surround the root of the plant as it was growing. Eventually the soil on the ridge was all returned to the trench. The next year, new furrows were made between the old ones (see fig. 1). Chao Kuo also improved the implements to meet the needs of this new method, which was put into practice on government land as an experiment. The result was a much higher yield than the old method of broadcasting the seed in fields without trenches (V, 7).

The *tai-t'ien* method was misinterpreted as one variation of the "three-field" system (or, more properly, a "two-field" system), which allowed one part of the land to lie fallow. Since the soil on the ridge had to be shoveled back into the trench gradually to provide support as the plant grew, it is inaccurate to say that the space between the furrows or trenches lay fallow.[12]

11. A more detailed discussion of the slash-and-burn method and its relationship with rice cultivation is given on pp. 119-20.

12. The main advocates of the hypothesis of laying half of the space fallow are T'ao Hsi-sheng and Ch'en Hsiao-chiang; see T'ao, *Hsi-Han ching-chi-shih*, p. 67, and Ch'en, "Kuan-yü i-mou-san-ch'üan wen-t'-ti shang-chüeh p'ing-ta Yang-chün,"

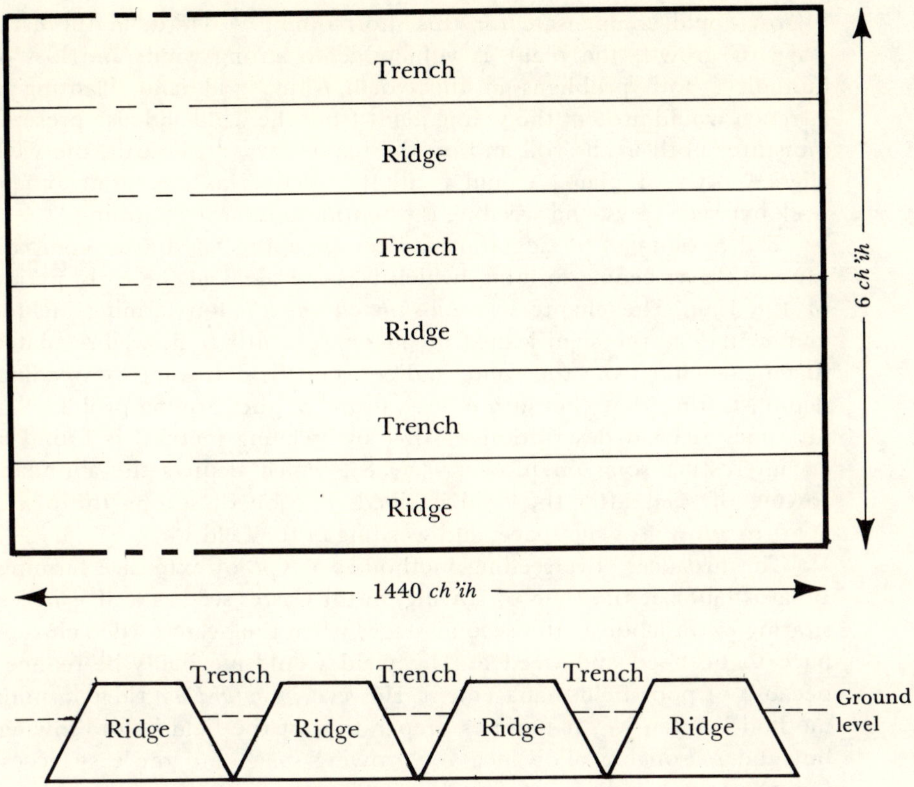

Fig. 1. *Tai-t'ien* (ridge method) farming. *Above,* top view; *below,* cross view.

The *tai-t'ien* method was a type of ridge farming. The advantages of regulating the furrows are fourfold. First, the well-regulated alignment would permit better ventilation and more sunlight than where plants grew in clusters, the concomitant of overseeding. The importance of good ventilation had been noted even in pre-Han farming, at least in principle.[13] Second, the Han plow permitted only a limited depth in plowing. The constant and gradual building up of soil around the root of the growing plant by pushing soil from the ridge into the

Shih-huo, vol. 2, no. 1; cf. Yu, "Yu Hsi-Chou tao Ch'ien-Han," pp. 11 ff., and Yang Chung-li, "Tsai-lun i-mou-san-ch'üan sui-tai-ch'u," *Shih-huo*, vol. 2, no. 4.

13. The relevant passage in the *LSCC* is discussed in the introduction of this book, pp. 7-9. Also compare my article, "Liang-Chou nung-tso yü chi-shu," *Bulletin of the Institute of History and Philology*, vol. 42, pt. 4.

furrow could compensate for this shortcoming.[14] Third, in the early stage of growth the plant is vulnerable to strong winds and loss of humidity, both problems in the North China arid land. Planting in a trench would protect the young plant from the wind and also preserve moisture, both in the soil and on the plant leaves.[15] Fourth, the well-aligned rows of plants would facilitate weeding as the farmer could walk between rows, and weeding is essential to intensive farming.[16]

The advantages of ridge farming can be appreciated after comparison with overseeding farming, which was prevalent in the early period of the Han. The chapters in *Lü-shih ch'un-ch'iu* on farming include comments on the significance of adequate ventilation, well-regulated spacing, shelter from the wind, and constant field dressing by weeding, demonstrating that the author was aware of the farming problems in his time. A vivid description of that overseeding method is found in an interesting song on plowing (V, 8), which depicts the abundant sowing of seed after the field is tilled, the job of setting the plants apart to allow growing space, and weeding in the field.[17]

The broadcast overseeding method is a type of extensive farming. It saves labor at the time of sowing, but it wastes seed as well as necessitating extra labor at the second stage, when the overcrowded clusters have to be hoed and weeded. The yield would inevitably be reduced because of poor field management. However, the *tai-t'ien* ridge farming method is intensive. It requires preparation of the field before sowing, but the reasonable allowance for growing space and the easy access for constant weeding requires less labor and increases the yield per unit.

Another contribution of the *tai-t'ien* method was the improvement of implements, especially the plow, used with either oxen or men (see plates 4, 5). Plowing was the only means of cutting deep and straight trenches and of turning the soil to form ridges. The *Han-shu* includes a comment that plowing and sowing implements were improved. Integrated into the *tai-t'ien* ridge farming method was the alignment of plants, which required a regulated spacing of seed. Ts'ui Shih, about two centuries later, pointed out the advantages of Chao Kuo's plow, where three plows were pulled by one ox and all the work of tilling and sowing was carried out by one person (V, 9). This plow probably did not have three plowshares: more likely it was a seeder with one seeding chute, which released seed at measured intervals, and two light

14. The characteristics of the plow are discussed later in this chapter.

15. Chung-kuo nung-yeh-i-ch'an yen-chiu-shih, *Chung-kuo nung-hsüeh-shih,* p. 151.

16. Nishijima Sadao, *Chūgoku keizaishi kenkyū,* p. 98.

17. Ibid., pp. 95-98.

PLATE 4. DETAIL OF RUBBING FROM LATER HAN TOMB IN SZECHWAN. This stone carving shows what may have been a new implement, the long-handled sickle (Ch'ung-ch'ing-shih po-wu-kuan, ed., *Ch'ung-ch'ing-shih po-wu-kuan-ts'ang Ssu-ch'uan Han-hua-hsiang-chuan hsüan-chi* [Peking: Wen-wu, 1957], p. 8).

PLATE 5. RUBBING FROM A TOMB IN SHENSI PROVINCE. This stone carving illustrates an ox-drawn plow; note the decorative grain pattern at the top (Shensi-sheng po-wu-kuan et al., "Mi-chih Tung-Han hua-hsiang-shih-mu fa-chüeh chien-pao," *Wen-wu*, 1972, no. 3).

plows, which turned the soil over and covered the seed. One Han tomb mural depicts a tripod seeder drawn by an ox. The excavator of this first-century tomb reports that the three "legs," the plowshares, were unmistakably visible.[18] Hence the plow was significant in the alignment during both tilling of the field into trenches and seeding along the furrow.

There is a disagreement among scholars concerning the government's motivation in promoting the *tai-t'ien* method. One school held that the purpose of government promotion was to benefit the large landholders, especially the politically privileged group who had access to government land. Much of this argument is built upon the need for using draft animals, which were quite expensive. Part of the argument is based upon the interpretation of the phrase in the *Han-shu*, which gives the possible reading that the privileged landholders were urged to adopt this new improvement in farming.[19]

Opponents of this view point out that Han policy during the reign of Emperor Wu was one of encouraging the small farming household to increase production, while large landowners of that period were generally suppressed. They interpret the phrase in another possible reading to establish an hypothesis that the ordinary farmers were taught to adopt the new farming method[20] (V, 7, n. 55).

Draft animals were not required in the *tai-t'ien* method since the text makes it explicit that manpower could be substituted if the farmer did not possess an ox. The farm did not have to be large; the experimental field, which was a side lot inside the palace compound and under the walls, was small. The farms that relied upon the exchange of labor among farmers to pull the plow were small; otherwise, the men would never have been able to finish the job on time.

Larger farms could also benefit from *tai-t'ien*. The suggested standard farm of 500 mu was fairly large, and the application of *tai-t'ien* on government land is now verified in the Chü-yen Han wooden strips, on which the terms the *tai-t'ien granary* and *tai-t'ien* frequently appear as one part of the frontier garrison organization. It took only two years from the date of the proposal being submitted to the court to

18. Shansi-sheng wen-wu kuan-li wei-yüan-hui, "Shan-hsi P'ing-lu Tsao-yüan-ts'un pi-hua Han-mu," *K'ao-ku,* 1959, no. 9, p. 463, and plate I, 4. Cf. Chung-kuo nung-yeh-i-ch'an yen-chiu-shih, *Chung-kuo nung-hsüeh-shih,* p. 154. A note to the *Ch'i-min-yao-shu,* where this information was quoted, relates this kind of plow with the seeder of his own time; see Shih Sheng-han, *Ch'i-min-yao-shu chin-shih,* p. 13.

19. Ito Norio, "Daidenhō no ichi kōsatsu," *Shigaku zasshi,* vol. 69, no. 11 (1960). Also *Chung-kuo nung-hsüeh-shih,* pp. 155-56. Cf. V, 7, n. 55.

20. Nishijima, *Chūgoku keizaishi kenkyū,* pp. 116 ff.

establish the *tai-t'ien* method at the frontier.[21] Since the evidence relates to both small, private farms owned by individual households and large government farms operated by military units in the frontier colonies, it can be assumed that the *tai-t'ien* method could be adopted on farms of all sizes.

I believe the Han court promoted the *tai-t'ien* method to increase agricultural output around the capital and in the nearby northwest frontier. At the time of Emperor Wu, the Kuan-chung area was heavily populated as a result of voluntary and compulsory migrations. Providing an adequate supply of grain for the populace already constituted a serious problem.[22] Emperor Wu himself was tired of the costly foreign wars, which occurred mostly northwest and north of the capital and had drained much of the resources of Kuan-chung. Hence, the *tai-t'ien* method perhaps was adopted to spur agricultural production rather than to benefit any particular social group. Although labor demand for *tai-t'ien* was very high, it was used in such a well-planned pattern that efficiency was greater than labor used in overseeding farming and resulted in a higher crop yield. The *tai-t'ien* method may have been practiced as early as the third century B.C., considering the similarity between some of the basic principles of the *tai-t'ien* method and those discussed in the *Lü-shih ch'un-ch'iu,* but its spread through government promotion was important in agrotechnological advancement.

NEW FARMING METHOD—OU-CHUNG

Further elaboration on the foundation of the *tai-t'ien* method is in the *ou-chung,* or pit-farming method, found in the *Fan Sheng-chih-shu.* Pit farming consisted of an intensive concentration of labor and fertilizer within small squares, or pits, which were evenly spaced either beside a long strip or in an area arranged as a matrix. The size and depth of the pit varied according to the needs of different crops. For example, wheat or millet could grow in pits of six *ts'un* across, six *ts'un* deep, and nine *ts'un* apart on first-class soil, while melons needed pits three *ch'ih* across, five *ts'un* deep, and so far apart that there could be only twenty-four pits in a mu (see fig. 2). Whatever the arrangement of the pits or whatever the crop, the farmer had to keep the pits well irrigated and fertilized. The purpose, as given in the text, was to counter-

21. Ch'un-shu Chang, "The Han Colonists and Their Settlements on the Chü-yen Frontier," *Tsing Hua Journal,* n.s., 5, no. 2 (1966): 161, 251. Ch'en Chih, *Liang-Han ching-chi shih-liao lun-ts'ung,* pp. 52-53.

22. Nancy Lee Swann, trans., *Food and Money in Ancient China,* pp. 260 ff., 300.

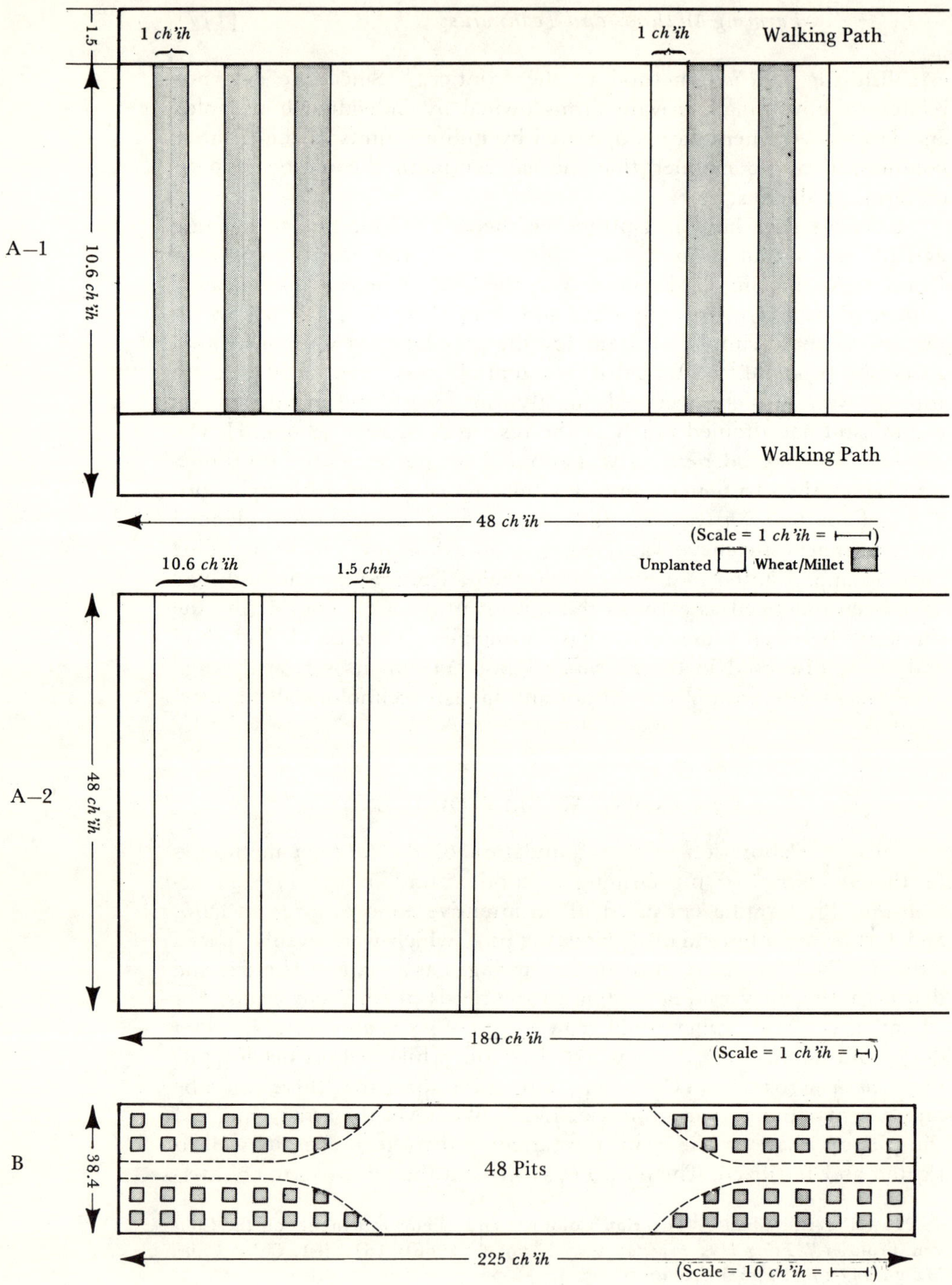

Fig. 2. Pit farming. A–1 and A–2, arrangement in strips; B, arrangement in matrix.

act drought as well as to utilize marginal land too small or inconvenient for regular plowing (V, 4:G.1).

The alleged per unit yield as given in the text was incredibly high. It is stated that two persons working on ten mu of land using pit farming would be able to feed themselves for twenty-six years! However, figures in different parts of the text show discrepancies (V, 4: G.1h, G.1i). Nevertheless, experiments in pit farming carried out in recent centuries, and even in recent decades, produced exceedingly high yields. The disadvantage of pit farming is its high labor consumption: the division of land into small plots precludes the use of animals or machines.[23]

Fan Sheng-chih, the acknowledged author of this first century B.C. book on agriculture, probably did design a way of farming that required little land and no animal power, but a great deal of labor and care, to meet the needs of peasants farming very small patches of land. Pit farming includes ground dressing, weeding, irrigation, and so on—much work on a small area. However, a good part of the work can be shared by women and minors, and Fan's method allowed the entire potential labor force in a household to be used, thus transforming the liability of the poor peasant to an asset. The brief reference to Fan Sheng-chih in the *Chin-shu* states he was once assigned to teach people in metropolitan districts the cultivation of wheat.[24] Actual field experience perhaps led Fan to devise a farming technique that would help the land-short farmer overcome the disadvantages deriving from social injustice.

Wet-Land Farming

Both *tai-t'ien* ridge farming and *ou-chung* pit-farming methods are for dry-land crops. Of the fourteen crops mentioned in Fan's work, only rice is a wet-land product; he discusses the cultivation of paddy rice. The significant feature is water in the paddy, which was kept at an even depth and temperature by controlling its flow into the fields. (Small fields are ideal in that they permit a fairly even depth of water.) Opening gaps on the mud banks directly facing each other allows the water to flow only a short course, with the least

23. There have been numerous attempts to apply pit-farming techniques with certain modifications. Ten records, mostly by "gentleman farmers" of the seventeenth and nineteenth centuries, were collected by Wang Yü-hu in a single volume entitled *Ou-chung shih-chung.* For experiments in 1958 in Hupei and in Honan, see *Chung-kuo nung-hsüeh-shih*, p. 178. The disadvantages given here are pointed out by Shih Sheng-han, *On Fan Sheng-chih-shu*, p. 64.

24. During the reign of Emperor Ch'eng (32-1 B.C.), Fan received such an assignment; see *CS* 26:8 and *HS* 30:49a, note by the annotator Yen Shih-ku.

effect on water temperature. Thus the water can be kept relatively warm. On the other hand, opening gaps in the banks diagonally to each other creates a longer watercourse, allowing more movement and hence cooling. The lowered water temperature is necessary for the rice paddy during warm weather or excessive sun. By using this simple yet effective device, the farmer created the most favorable conditions for the growth of rice (V, 4: D.5, D.5c).

What seemed a common practice in rice cultivation, especially in southern China and in the Yangtse River valley, was water weeding, which probably had its origin in very primitive slash-and-burn farming.[25] The phrase "to till with fire and to hoe with water" appears regularly in Han literature describing rice agriculture in the Yangtse River valley.[26] Although there are different interpretations, the Han annotator Ying Shao (fl. late second century) probably provided the best. According to Ying Shao, the grass in the field was first burned to the ground, then water was tapped into the paddy where the rice seed was sown. Weeds and rice grew side by side until the rice seedlings reached seven or eight *ts'un* high, when the grass was cut down and more water was tapped into the paddy to cover the grass roots, which soon suffocated. Both the cut grass and the roots became fertilizers and nourished the rice (V, 10 and n. 63).

Cheng Hsüan (A.D. 127-200), a contemporary of Ying Shao, also mentioned the use of water to get rid of weeds. In his annotation to the *Chou-li,* he pointed out that the grass should be cut in midsummer when summer showers would flood the field and prevent its regrowth. In the fall, when the field was dry, the grass should once again be cut down to prepare the field for rice the next year[27] (V, 11, nn. 65, 66). What is pertinent is that the rice field prepared this way lay fallow one year, or at least from midsummer to spring.[28] Unlike

25. An excellent discussion of water-weeding farming is in Nishijima, *Chūgoku keizaishi kenkyū,* pp. 185-234. Also cf. Amano Motonosuke, " 'Kakō-suidō' no ben: Chūgoku kodai Kōnan suitō gijutsu kō," *Shigaku zasshi,* vol. 61, no. 4 (1952). Yoneda Kenjirō, "O Shō 'Kakō-suidō' chū yori mitaru Gokan Kō-Wai no suitō-saku gijutsu ni tsuite," *Shirin,* vol. 38, no. 5 (1955), and Shih, *Ch'i-min-yao-shu chin-shih* 1:110-18.

26. For instance, *SC,* 30:36, 69:27; *HS* 6:18a, 28B(2): 65a, and E. M. Gale, trans., *Discourses on Salt and Iron,* pp. 18-19. The translation of this term given in our text is the one adopted from Lien-sheng Yang, "Notes on the Economic History of the Chin Dynasty," *Studies in Chinese Institutional History,* p. 175; cf. V, 13.

27. Cheng Hsüan was not only a scholar but also a tenant farmer; see *HHS* 35: 11b-12a.

28. Lien-sheng Yang implies that this method was mainly for reclaiming "virgin soil" (Yang, "Notes on Economic History," p. 175, n. 91). I believe that this

the repeated—even interlacing—cultivation discussed in Fan's book, the rice farming discussed by Ying and Cheng was land-extensive rather than labor intensive. One explanation for the different approach is that even as late as the second century, the sparsely populated regions south of the Yangtse River provided relatively more arable land. Such a hypothesis can be supported by comparing the south three quarters of a century later. In A.D. 278, Tu Yü, a general who conquered the south for the Tsin dynasty, noted in his memorial that the "fire tilling and water hoeing" method of cultivation was no longer applicable in the south, as there were no more large areas of unused land and the population was no longer small (V, 12).

This rather primitive method of rice farming probably was in practice only in the south. Rice farming in the Yellow River valley and the Han metropolitan districts followed another path. Rice had long been cultivated in North China: rice kernel impressions and even rice kernels have been discovered in Neolithic sites. In the Classics, the areas of the present provinces of Shantung, Shansi, Shensi, and Honan all had rice. In the Han period, water reservoirs were constructed in the northern commanderies as far as present-day Hopei, Shensi, Honan, and Shansi, making rice cultivation possible.[29] The repeated use of arable land and the shorter growing period in the north probably forced the farmer to use the land more economically and to develop better farming techniques. One device was the transplanting of seedlings from the nursery to the field, a method still used in China. The advantage is obvious: while other crops are still in the field, seeds are already beginning to germinate in a nursery. In the *tai-t'ien* and pit farming methods, the crops were to be spaced in a well-planned pattern. In rice cultivation it is difficult to make furrows or pits and only the planting of seedlings of a certain height can make an orderly arrangement possible. Perhaps that was why transplanting of rice is mentioned as a chore in the fifth month in the *Ssu-min-yüeh-ling*[30] (III, 1:E, 5). The sixth century *Ch'i-min-yao-shu* includes an interesting comparison of rice cultivation in the north and in the south. In the south, the seedlings were not transplanted, while in the north the seedlings were transplanted at the height of seven or eight *ts'un.* However, by then, much of the water weeding technique had been incorporated into the rice cultivation process in North China (V, 13).

---

method was not merely for reclaiming virgin land, but was a regular annual practice on fields to be turned into rice paddies.

29. Ch'en Tsu-kuei, "Chung-kuo wen-hsien-shang-ti shui-tao tsai-p'ei," *Nung-shih yen-chiu chi-k'an* 2: 64-66, and Okazaki Fumio, "Shina kodai no tōbei inasaku kō" in his *Nanbokuchō ni okeru shakai keizai seido.*

30. Cf. Nishijima, *Chūgoku keizaishi kenkyū,* pp. 208-9.

These accounts reveal that farmers who settled in the south, as long as they had extensive land, practiced less intensive farming than they had in the crowded north. The southern way of ground burning, however, did have the advantage of providing fertilizers in the water-bed of the rice paddy, which dry-land farming hardly could match. Fire-tilling and water-hoeing could therefore be easily appreciated for rice cultivation in the north. There is, however, the possibility that the practice of slash-and-burn farming from the primitive stage of agriculture in the north did survive in this particular aspect, but it is dubious that after several centuries of an elaborated system of repeated cultivation such an ancient practice would come to the mind of the farmer unless there was some new stimulus. Very likely such a new stimulus emerged as knowledge of the southern methods became known in the north. There is at least one case in which farmers of the southern ethnic minority Yüeh were brought to the north to cultivate the lowland on the riverside after the northerners failed to produce anything on land newly reclaimed from the river (V, 14).

There is a brief discussion on the selection and preservation of seed grain in Fan's book. The strong, large, and high-yielding single ears were selected for next year's seeds. Chinese farmers have continued this practice to this day, with the consequence that numerous varieties of specimens exist side by side. Breeding had not been adopted to develop ideal specimens; selections right from the field seem to have been the only method[31] (V, 4: F.1). Further improvement of selected seeds by isolated cultivation in well-cared-for fields, from which the best were again selected, had yet to be developed, as registered in the *Ch'i-min-yao-shu.*[32]

To protect the seed grain from excessive heat and humidity, the seed was stored with great care. It was first thoroughly dried and then put in either bamboo or pottery vessels to which herbs that were good for pest control were added. Interestingly, Fan Sheng-chih rather superstitiously suggested that insects could be prevented if a horse munched a few mouthfuls of the seeds (V, 4: F.2, F.3).

IMPLEMENTS

The implements used by Han farmers were made mainly of iron and wood, as they are today. Some of the wooden implements, such as the digging stick, were actually survivals of the most primitive ones

31. Ch'en Tsu-kuei, "Chung-kuo wen-hsien," pp. 89-90, and Ts'ao Lung-kung, "Chung-kuo nung-shih wen-hsien-shang su-ti tsai-p'ei," *Nung-shih yen-chiu chi-k'an* 2: 105-6.

32. Shih, *Ch'i-min-yao-shu chin-shih,* p. 21 (2.3.1-2.3.3).

that had been in use possibly as early as the Neolithic period. The digging stick, *lei,* appears at the head of a category of wooden implements (*SWCT* 14a:2), each of which had a very specific function (V, 15). The same lexicon also has a list of wooden implements, including various kinds of hoes, plows, smoothers, harvesters, threshers, and so on (V, 16). The iron implements included were the spade, mattock, plow, double-pronged hoe, garden hoe, sickle, and scythe (V, 17). Spades had no less than four variations, each bearing a distinctive name.

In another Han glossary, the *Shih-ming,* the etymologist provides the readers with descriptions of various implements; surprisingly, many of them resemble even in detail their descendants used in China today. The etymologist also states that the farming method determined the morphology of the tool, so that a narrow hoe was specifically for weeding along a trench, which probably was related to ridge farming (V, 18).

The evolution of the plow demonstrates the diversification determined by its specific functions. The prototype of the plow was simply a larger digging stick. Ancient texts frequently mention "paired tilling" *(ou-keng).* The most sensible interpretation of paired tilling is that one person pulled the improved digging stick while another person stepped the stick into the soil. As the stick acquired cutting blades to penetrate the soil it had become virtually a plow. The blades gradually would develop into more efficient plowshares. The back ridge that eventually evolved into the moldboard would also help to turn over the soil. By then the plow would have been too large to be pulled by a person and would have been used with oxen or horses. Plows made entirely of wood could not match the pulling force of draft animals, and an iron edge added to the blade was a significant improvement. The Chan-kuo plowshare excavated at Hui Hsien is precisely that, wood with an iron blade attached to it.[33]

The wide V-shape that exceeded 120 degrees of the Chan-kuo plow could hardly cut deeply into the soil, nor was the cast-iron blade of such a wide angle strong enough to sustain the pressure needed for deeper tilling.[34] The plow of the Han period appears to have been an improvement. In Han archaeological sites a few very large plows with sharp triangular shares and heavy ridges have been excavated.[35] Bring-

33. For a more detailed discussion of this, see Hsü, "Liang-Chou nung-tso yü chi-shu," pp. 801-10.

34. Hsia Nai, *Hui-hsien fa-chüeh pao-kao,* p. 82.

35. Tung-pei po-wu-kuan, "Liao-yang San-tao-Hao Hsi-Han ts'un-lo i-chih," *K'ao-ku hsüeh-pao,* 1957, no. 1. One of the pieces discovered at Shih-chia-chuang, however, is of rather dubious date. See Huang Chan-yüeh, "Chin-nien ch'u-t'u-ti Chan-kuo Liang-Han t'ieh-ch'i," *K'ao-ku hsüeh-pao,* 1957, no. 3, p. 100, n. 1.

ing the plowshare to a more acute angle increased its penetration and made deep plowing possible. The plows of Eastern Han show remarkable improvement over those used in the Western Han.[36]

The Han plows, nevertheless, were not uniform—there were some so small that it seems unlikely that a draft animal was used. At least one small piece, excavated at Ch'in-ho in Hopei, is clearly a leg of a seeder rather than a plowshare.[37] The different plows point out the inevitable local variations within the vast territory of China as well as indicating that the purpose of the work dictated the shape and size of the tool. The big plow could be very useful in the breaking of new ground, and perhaps it is not coincidence that the large ones were found in Liao-yang, Ku-lang, and Chen-te, all of which were outlying frontier lands in the Han period. The middle-sized ones, light enough to be pulled by two persons, fit the description of the small implements that Chao Kuo advocated for ridge farming. And the very small ones, if they were seeder's legs, certainly served to regulate the spaces between plants, essential if the farmer were to reach the optimum combination of full utility of the land, good ventilation, and a reasonable allowance of working space for good management (V, 7 and 9).

The use of the plow was promoted by the Han government, particularly in the peripheral regions, where officials encouraged people to adopt ox-drawn plowing.[38] The wide distribution of Han archaeological sites where plows were found and the many representations of plowing in Han murals and brick carvings confirm the information in the literary records.[39] The iron foundries under the government monopoly agency were said to have had a notorious inclination towards casting "large implements," and in comparison with most other tools, even a small plow was large. The credit-conscious agency of the iron monopoly probably did regard the manufacturing of plows an easy task in fulfilling its assignments and the government manufacture and sale of the plow probably helped promote its use (V, 19).

The farm implements of daily use that made up the greater part of the Han remains consisted of spades, hoes, and the sickle-sythe group. Spades had the longest tradition, for they were a direct descen-

36. Tseng Jung, "Han-tai-ti t'ieh-chih kung-chü," *Wen-wu,* 1959, no. 1, p. 16.

37. Huang, "Chin-nien ch'u-t'u-ti," and Amano Motonosuke, *Chūgoku nōgyōshi kenkyū,* pp. 746-50. For the Ch'in-ho plow, see *Ch'üan-kuo chi-pen chien-she kung-ch'eng-chung ch'u-t'u wen-wu chan-lan t'u-lu* (1954) plate 2; T'ien Ch'ang-wu, "Han-Wu-ti," *Jen-min Chung-kuo,* 1960, no. 8, p. 56.

38. Amano, *Chūgoku nōgyōshi kenkyū,* pp. 751-72.

39. Huang, "Chin-nien ch'u-t'u-ti," and Amano, *Chūgoku nōgyōshi kenkyū,* pp. 753-56. A recently published piece is the stone relief from a Han tomb, *Wen-wu,* 1972, no. 3, plate 11.

dant of the ancient digging stick.[40] The Han period remains included spades with a straight edge and those with a curved edge, those with a shovel shape and those that were flat, some that were double pronged and some that were single and narrow, some straight-handled and some angle-handled, and so on. The heavy *chü* was designed to break ground, the narrow *ch'ien* was designed to cut weeds between furrows. Spades made of wrought iron became fairly common in the Han period, whereas the Chan-kuo iron specimens are all cast.[41]

The hoes had similar diversification: heavy ones to break ground, light to pulverize the earth clods, and narrow to weed. The handles were fixed to the iron head in various ways to get the best results. Wrought-iron pieces were also common among the Han samples.[42]

The sickle-scythe group was comprised of scythes with long handles, sickles with short handles, and crescent-shaped hand sickles. The long-handled scythe appears clearly in a harvest scene in a Han tomb brick carving, where two persons are swinging scythes in a swampy field while three other persons are using hand sickles to cut the head stalks.[43] The scythe was regarded as suitable for overseeding farming as the grain was not planted in rows; the crescent-shaped hand sickle, which is possibly a survival of a cutting knife developed as early as the Neolithic period, was for cutting off only the heads, leaving tall stalks in the field.[44] The short-handled sickle was an improved implement. It cut the stalk a few inches above the ground so that the harvest could be collected much faster. There would be no necessity to clean the field again, because the short stem left in the field could easily be turned under as fertilizer. On the other hand, the most efficient use of short-handled sickles demanded neat and straight rows. Ridge farming, therefore, is correlated to the best use of sickles.[45]

The site of a Han foundry of the late Western Han period was discovered in 1958 at Kung-hsien, Honan. There were facilities to produce a fairly high grade of wrought iron, with a low content of carbon, by means of coal-burning furnaces of high temperature. The implements unearthed in the foundry were mostly of cast iron; most

40. Hsü, "Liang-Chou nung-tso yü chi-shu," p. 812. Sun Ch'ang-hsü, *Lei-ssu-ti ch'i-yüan chi ch'i fa-chan*, pp. 13 ff.

41. Tseng, "Han-tai-ti t'ieh-chih kung-chü," pp. 16-18. Li Wen-hsin, "Ku-tai t'ieh-nung-chü," *Wen-wu ts'an-k'ao tzu-liao*, 1954, no. 9, p. 81.

42. Tseng, "Han-tai-ti," p. 16; Li, "Ku-tai t'ieh-nung-chü," p. 32-35.

43. K'ao-ku-yen-chiu-so, *Hsin Chung-kuo-ti k'ao-ku shuo-huo*, plate 218.

44. Li, "Ku-tai t'ieh-nung-chü," pp. 85-86; Liu Hsien-chou, *Chung-kuo ku-tai nung-yeh chi-chieh fa-ming shih*, pp. 58-62.

45. Liu, *Chung-kuo ku-tai nung-yeh*, pp. 59-62; Li, "Ku-tai t'ieh-nung-chü," p. 86.

of the carpenters' tools, however, were made of good wrought iron.[46] The invention of the water-driven bellows probably had considerable effect on the production of iron in the Eastern Han period (V, 20). These discoveries should confirm the possibility that there was a turning point in iron technology at the end of Western Han and that the Eastern Han people were able to produce wrought iron to be used for all tools and implements.[47] It is important to note, moreover, that the iron implements in the Western Han period, as represented by the Liao-yang San-tao-hao remains, were of cast iron and small and fragile, on much the same technological level as the Chan-kuo pieces.[48] The Eastern Han implements were of wrought iron, larger, and of good quality.[49]

With all this discussion of Han implements, of course, there still must have been some regional variations as well as individual differences (V, 21). Furthermore, the poorer farmer perhaps could afford to use only inferior implements (V, 22).

Summary

Agrarian technology made these advancements in the Han period: well-planned arrangement of plantation areas in the fields replaced the broadcasting, overseeding method; multicrop systems were gradually developed; rotation of crops and fertilization made repeated cultivation a normal pattern, unless land lay fallow, or for instance, where there

46. Honan-sheng wen-hua-chü wen-wu kung-tso-tui, *Kung-hsien t'ieh-sheng-kou Han-tai yeh-t'ieh i-chih.* Also, cf. Nan-ching po-wu-yüan, "Li-kuo-i ku-tai lien-t'ieh-lu-ti tiao-ch'a chi ch'ing-li," *Wen-wu,* 1960, no. 4.

47. Huang, "Chin-nien ch'u-t'u-ti," pp. 105-8. Some scholars have suggested that the use of iron had started as early as in the Western Chou or even in the Shang dynasty. Yang, *Chung-kuo ku-tai yeh-t'ieh,* pp. 12-23. T'ung Shu-yeh, "Ts'ung Chung-kuo k'ai-shih yung-t'ieh-ti shih-tai wen-t'i p'ing Hu Shih-p'ai-ti chih-hsüeh fang-fa," *Wen-shih-che,* 1955, no. 2. A few iron pieces, dated back to very early times, seem to be of meteorite iron rather than of processed iron ore. A compromise is suggested by Amano Motonosuke that iron began to be used for implements in 1,000-500 B.C., "Chūgoku ni okeru suki no hattatsu," *Tōhōgakuhō* 26 (March, 1956): 105 ff. I feel that whatever the beginning date was for the use of iron, its significance would not have been apparent until the late Ch'un-ch'iu period.

48. Huang, "Chin-nien ch'u-t'u-ti," p. 98. Tung-pei po-wu-kuan, "Liao-yang San-tao-hao Hsi-Han ts'un-lo i-chih," *K'ao-ku hsüeh-pao,* 1957, no. 1.

49. Li, "Ku-tai t'ieh-nung-chü." Precise statistics on the comparison are very difficult because of the ambiguity and inconsistency of the terminology applied by individual field workers in their original field reports. Cf. Huang, "Chin-nien ch'u-t'u-ti," pp. 104, 107-8; Yang K'uan, *Chung-kuo ku-tai yeh-t'ieh chi-shu-ti fa-ming ho fa-chan,* p. 43.

had been failure of a harvest for two consecutive years or in the south where arable land was plentiful before it was highly populated. Intensive farming made ground dressing, weeding, and constant pulverizing standard features of Chinese farming, which meant that intensive care of each individual plant had been brought almost to the level of gardening. The variety of implements also indicates the significance of such activities. Such farming demanded a concentration of labor and was associated with smaller farms. Also the farmer's incentive was crucial to guarantee the thoroughness of the field work, which made extensive use of slaves or contract labor unreliable. All these factors help to explain the maintenance of independent farmers in the Han period and the preference of large landholders for tenancy rather than slavery.

There is always the temptation to estimate the rate of productivity per unit or per capita. For instance, Wan Kuo-ting tried to compare the labor output per capita in terms of area (mu) between ordinary digging-stick tilling and oxen-plow tilling. Wan's comparison is given in table 13.

TABLE 13

OUTPUT BY FARMING METHOD

Period	Method	Tilled/laborer (in mu)
Reign of Emperor Wen (179-157 B.C.)	Unspecified	20.5
Reign of Emperor Wu (140-87 B.C.)	Digging stick	10
(Same)	*Tai-t'ien*	166.7

SOURCE: *Chung-kuo nung-hsüeh-shih*, p. 154, table 21.

Care should be taken, however, in estimating and comparing the growth of productivity of different periods, since the sparse data on Han agricultural productivity are subject to many variables: time, kinds of grain, location,[50] and so on. A general advancing trend of agricultural output, nevertheless, still can be visualized by drawing upon a number of clues. Amano, in spite of his own precautions, took a few items found in archaeological sites to imply that there were indeed changes in several aspects: The size of residences and wells

50. Amano Motonosuke, "Chūgoku kōdaishika no shosetsu o hyōsu," *Rekishigaku kenkyū*, 180:33.

in the Western Han period were larger than those of Eastern Han; the smaller residences and wells in the later period indicate smaller households being sustained on smaller, intensively operated farms; bricks were more extensively used in the wells of the Eastern Han period, indicating a higher standard of living; granaries in Western Han sites were smaller than those of Eastern Han; Eastern Han pottery was normally larger; iron implements and tools, even wrought-iron ones, were widely used in Eastern Han, but excavated sites of the Western Han period revealed little more than one-half of the iron found in Eastern Han sites.[51] The appearance of an animal-powered or water-powered pounding device for husking grain, an improvement over the simple hand pounding mortar, is probably related to the increase of productivity (V, 23).

Thus we can speculate that the Eastern Han farmers who were members of smaller households could produce more grain and lived a relatively better life with more food in jars and bowls than their predecessors in the Western Han. Advancement in agriculture was owing not only to the better implements made of iron, but also to the development of intensive farming, where a significant portion of the work, including weeding, irrigation, and fertilizing, as well as work carried out in the house, such as treating the seed, was done by women and minors. Thus the small farming household was an integrated production unit.

51. Amano Motonosuke, "Kandai gōzoku no daitochi keiei shiron," *Takikawa hakushi kenreki kinen ronbun shū* 1:16-17; Kuo Pao-chün, "Lo-yang hsi-chiao Han-tai chü-chu i-chih," *K'ao-ku t'ung-hsün,* 1956 no. 1; Huang, "Chin-nien ch'u-t'u-ti."

VI

THE NONAGRICULTURAL ALTERNATIVES

THE Z ACTIVITIES

IN the preceding chapters, production methods in Han farming were discussed. Work in the field, as noted in the *Ssu-min-yüeh-ling,* varied with the season. During spring, summer, and fall, the farmer was busy with cultivation and harvest. After the ninth month, field work was put aside and the winter was given over to repair jobs (see document III, 1: I.1, J.1, L.3). Intensive farming, which included light yet time-consuming labors such as fertilizer spreading, weeding, and insect control, involved women and minor members of the farming household during the busiest season. At the less busy times, however, those who were released from field activities became idle, and to keep them still productive the farming household had to reallocate its energy to activities other than direct farming. Denoted here as the "Z activities," they span a broad range: processing of food and fuels, spinning and weaving of textiles, metalworking, dressing and tanning leather, manufacture and repair of tools, implements, pottery, and ceremonial objects, as well as housebuilding, fence repairing, and activities such as recreation, protection, transportation, and distribution.[1]

1. The term *Z activities* is adopted from Stephen Hymer and Stephen Resinick, "A Model of an Agrarian Economy with Nonagricultural Activities," *American Economic Review* 59, no. 4, pt. 1 (Sept., 1969): 403, n. 1.

The Han farmer, as demonstrated in chapter 3, was market-oriented because many of his expenditures had to be in cash. There were ways to make up the difference between his expenditures and his cash income: produce goods that he could sell, or produce goods that he had to otherwise buy. The Z activities embraced both categories, but some of them served the dual purpose of producing goods for both home consumption and for marketing.

Textile production, including the various steps from spinning and weaving to fabrication and tailoring, had been women's jobs in rural households since antiquity. "Man as tiller, woman as weaver" remained a standard description of labor division in China until modern times. One of the earliest associations of woman and weaving perhaps is the star name of Weaving-girl (Vega, the brightest star of Lyra) (*MSCI* 13:1-7; 18: 5-6.). Chia I, by citing the ancient saying, showed that he regarded weaving by women as significant as farming by men. Han local officials often made it their primary responsibility to see not only that the farmers were doing well in the fields to produce food but also that mulberry trees and hemp were planted to keep raw materials for textiles in good supply (VI, 1, 2, for instance). In fact, the Han people regarded the combination of farming and textile production as a distinctive characteristic of Chinese agrarian culture, in contrast to the way of living on the frontiers in the north and northwest (VI, 3). Pan Ku, depicting the ideal life of the farmer, also regarded weaving and spinning as part of production. During the winter months women wove and spun in the evening as well as during the day, working the equivalent of one and one-half days, by modern standards (VI, 4). A woman could produce an average of 5 *ch'ih* of ordinary fabric each working day in the Western Han period and 13 1/3 *ch'ih* daily in the Eastern Han period (VI, 5, 6).

Although the winters could be as long as four months in North China, spinning and other preparations might require two months, the same as the time spent weaving[2] (III, 1: A.5, F.2, J.5). The hypothetical household (as given in chapter 3) provided two female laborers who together worked 120 days a year.[3] The annual output from weaving then should be fifteen bolts (*p'i*) (for 5 *ch'ih* output as the daily average) or forty bolts (for 13 1/3 *ch'ih* output as the daily average). The annual clothing needs for the household members themselves has been estimated (chapter 3) as one piece per head. Let us

2. Nancy Lee Swann suggests a four-month winter working schedule, *Food and Money in Ancient China*, p. 129, n. 72. *The Ordinances for the Four Peoples* gives two winter months and one summer month for weaving.

3. In order to make a conservative estimate, we use thirty days for each month instead of the forty-five days suggested in VI, 4.

make a generous assumption that each person had two pieces of clothing (one piece for spare), then the marketable amount of fabric produced should be ten to thirty pieces. Using the same estimated price for pieces of fabric as suggested in chapter 3, the household's revenue from weaving could be between 5,000 to 8,000 cash for low production, or 15,000 to 24,000 for a high production rate, an income almost equal to that from food crops, or even more.

Clothing prices, of course, varied with the quality of the material. A coarse cloth, called seven *tsung*, was the poorest quality and probably worth the least amount of money. For economy, convict laborers had to wear clothing of this grade of cloth.[4] The price of eight-*tsung* fabric was recorded in a Han tablet at Chü-yen as 230 cash per piece (*p'i*). The price of seven-*tsung* fabric, then, ought to have been lower. The most expensive fabric could be worth as much as 3,900 cash per piece. Between these extremes, other prices listed range from 300 to 1400 cash, with ordinary fabrics falling between 500 and 800 cash.[5] The range, indeed, was very broad and not only represented differences in quality, but also fluctuations in price. The income from fabric production was subject to enormous market changes.

Raising domestic animals could also provide a substantial income for the Han farming household. Meat, although not always in the daily diet, was not uncommon at festivals or other occasions worthy of celebration (VI, 8). On the birth of the founding emperor of Han, Kao-tsu, and of the boy born to his neighbor, other neighbors congratulated the two households with presents of sheep and wine. As the two boys grew up and developed a close friendship, neighbors did the same once again (VI, 7).

By the first century B.C., meat served at meals was popular in the countryside. The price of a hog is not clear in the text; nevertheless, it probably was relatively high. It indeed was expensive from the viewpoint of the consumer, but would also provide a good income for those who raised such animals for market. Terra-cotta pigpens, with pigs and piglets in them, are common in Han tombs. The San-tao-hao Han village site consisted of six complexes of farming houses, to each of which was attached a pigpen. The outlet of the pen was shared with that of the toilet, indicating a design to facilitate the

4. *SC* 11:14. *Tsung* is the unit of measure of the density of the woof or filling thread in cloth. Each *tsung* consisted of eighty threads, according to the annotator Chang Shou-chieh. The loom width of Han cloth was 22 Han inches (*ts'un*). The seven-*tsung* density, therefore, should be 560 threads within the width of 22 inches, just about as coarse as fine medical gauze.

5. Ch'en Chih, *Liang-Han ching-chi shih-liao lun-ts'ung*, pp. 67-69. Also cf. our chap. 3.

collection of manure[6] (see plate 1). Hog raising could be done at a specialized animal farm, as in the case of Liang Hung, who raised pigs for a living after he finished studying at the Imperial Academy[7] (VI, 9).

The special pig farm could be a large enterprise—worth as much as a million cash (*SC* 129: 31). It could also provide jobs for the poor, who herded the animals in swamps, on seashores, or on other available unarable land (VI, 10, 11). Nevertheless, the pigpen and the house toilet being connected to the same manure storage illustrates that the farmer kept pigs for selling (and consuming) and for securing manure. The terra-cotta pigpen ordinarily was large enough to keep four or five pigs, which probably was about the number a farming family was able to afford.

As with pig raising, sheep breeding could be profitable. Pu Shih multiplied his property simply by breeding sheep on grazing land (VI, 12). Since sheep are mentioned with cattle as an index of the general well-being of the people while China was largely an agrarian country, it seems that sheep were also often kept in ordinary farming households (VI, 13).

The price of pigs was probably about the same as that of sheep, and a head of sheep could be sold for 900 to 1000 cash, the monthly salary of a laborer at the official rate[8] (VI, 14).

Most of the domestic animals could be taken care of by minors, who also could be hired by several households or an entire community to look after the animals in herds. The pay of a shepherd or swineherd, who were (just as in China today) most likely boys of the community, hardly could have been high (VI, 16, 17).

Winemaking was another profitable auxiliary activity. The liquor was worth twice the price of the grain that was used.[9] Han wine was not distilled; rather, it was served with the lees settled to the bottom. The volume of original grain would be at least doubled with the addition of water and the profit would be great (VI, 18, 19, 20).

Manufacturing tools, utensils, and even construction materials

6. Wang Chung-shu, "Han-tai wu-chih wen-hua lüeh-shuo," *K'ao-ku t'ung-hsün,* 1956, no. 1, p. 58, and Tung-pei po-wu-kuan, "Liao-yang San-tao-hao Hsi-Han ts'un-lo i-chih," *K'ao-ku hsüeh-pao,* 1957, no. 1, pp. 121 ff.

7. In this case Liang Hung owned neither land nor other property; he obviously was not a farmer then. The government park, where he kept the pigs, was at that time opened to the people for gathering fuel and for fishing, as well as for use as pasture.

8. Cf. Lao Kan, "Han-chien-chung-ti Ho-hsi ching-chi sheng-huo," *Bulletin of the Institute of History and Philology* 11: 70-71. Poultry, especially chickens, were also raised. The price, however is not known (VI, 15).

9. Lao Kan, *Chü-yen Han-chien: K'ao-shih chih-pu,* "K'ao-cheng," pp. 62-63.

added to the farmer's income. In the Han village site at San-tao-hao, a number of kilns for firing tiles were found, but their rudimentary structure and limited capacity suggests that such production was not a full-time professional job. The tiles unearthed at the residential site were the same as the ones produced so it is apparent that local production and local consumption were united. There was, however, at least one complex of two identical kilns, demonstrating the possible development toward full-time production from two kilns in rotation. It may be deduced that their products were also meant to be marketed outside the local area.[10]

Production and Marketing

Han government factories produced a limited amount of consumer goods. First, there were not many government factories in the whole empire. The "Geographical Treatise" ("Ti-li-chih") of the *Han-shu* gives the sites of the offices of manufacturing *(kung-kuan)* under eight commanderies: Ho-nei, Ho-nan, Ying-ch'uan, Nan-yang, T'ai-shan, Chi-nan, Kuang-han, and Shu. There were also three offices for textile manufacturing *(fu-kuan)* in Ch'i, as well as others in Ch'en-liu and Hsiang-i. In contrast, the distribution of offices for salt production *(yen-kuan)* and offices for iron production *(t'ieh-kuan)* spread to virtually all quarters of China,[11] and the iron and salt agencies had to be widespread also to provide people with their needs. The other factories, which existed only in limited areas, probably produced very little for ordinary consumption. Secondly, even though the government factories were of considerable size, their products probably were costly luxury goods, such as works of lacquer, gold and silver, fine embroidery, and so on, for use by the imperial house and closely related people (VI, 21, 22).

In the Eastern Han period, the function of the Office for Manufacturing has been described as the collection of manufacturing taxes in the commanderies where there were numerous artisans.[12] This change from the supervision of production to the collection of taxes is perhaps an indication that the government agencies in the Eastern Han period did not operate their own workships. Associated with such a change are two possibilities. One is that there could have been more large-scale private enterprises taking the place of government operations. However, business suffered a heavy blow during the reign of Emperor

10. Tung-pei po-wu-kuan, "Liao-yang San-tao-hao," p. 125.
11. Li Chien-nung, *Hsien-Ch'in Liang-Han ching-chi-shih kao,* pp. 169-72.
12. *Hou Han Chih Chi Chieh* ed., 28:9b.

Wu, who intervened to curtail the growth of private capital, which seemingly challenged the monolithic power of the state.[13]

State jealousy of mercantile practices and the light taxes on agriculture jointly pushed capital into safe and profitable investment in land,[14] a trend that would not be favorable for the emergence of large-scale private producers.[15] Hence, another possible reason for the establishment of the office of tax collection was the flourishing of small-scale manufacturing. Such manufactured products were sold within a relatively limited area for local consumption, as, for example, were the tiles produced at San-tao-hao. The tax on these could be levied only locally.

In the *Ordinances for the Four Peoples,* numerous activities to produce materials both for household consumption and for marketing are listed. Not only does the extant text include accounts of the selling and buying of items such as grain, clothes, cloth, and so on, the *Hou Han-shu* biography of the author, Ts'ui Shih, also explicitly mentions that the Ts'ui family supplemented their income by producing and selling wines and such preserved food as pickles and soybean paste. These products are frequently noted in the text of the *Ordinances.* The Ts'ui household, indeed, did not produce the miscellaneous items simply to supply their own needs in the manner of a manorial operation. In fact, the title of the *Ordinances for the Four Peoples,* in which the "Four Peoples" refer to the Scholar, Farmer, Artisan, and Merchant, implies an orientation toward a marketing economy, and away from the self-sufficiency assumed by a number of scholars.[16]

Marketing activities are well reflected in the mock contract composed by Wang Pao. Both selling and buying at the markets are mentioned as routine, and markets existed in large and small cities, towns, and in fairs along both the major and the side roads. The trade routes even reached areas of non-Chinese tribes[17] (III, 7). Utsunomiya points

13. Cho-yun Hsu, "The Changing Relationship between Local Society and the Central Political Power in Former Han," *Comparative Studies in Society and History* 7, no. 4 (1965):358 ff.

14. John A. Harrison, *The Chinese Empire,* pp. 143-49.

15. Comparing the tone in the *Shih-chi* and the *Han-shu,* one can appreciate the difference between author's attitudes toward mercantile enterprises held in the time of Emperor Wu and those in the Eastern Han *SC* 129 and *HS* 91.

16. For the activities in the *Ordinances for the Four Peoples,* see III, 1. For the Ts'ui family's brewing and preserved food trade, see *HSS* 52: 18a. John Harrison, drawing upon the study by Martin Wilbur, suggested that Han agriculture was basically that of a natural economy; see Harrison, *Chinese Empire,* p. 148, and cf. C. Martin Wilbur, *Slavery in China during the Former Han Dynasty,* p. 246.

17. Utsunomiya Kiyoyoshi, *Kandai shakai keizaishi kenkyu,* pp. 349-51, especially the map on p. 350.

out that the outer ring for trade stretched as far as two hundred kilometers from Wang's home, while his inner circle, the one frequently visited, covered a fifty-kilometer radius.

The hierarchy of the marketing places and its differentiated relation to the farming population bears great resemblance to the marketing system that William Skinner noted was characteristic of preindustrial nineteenth-century China.[18] As Skinner suggests, Chinese peasantry followed a cyclical trend. While external conditions were favorable, the peasants actively participated in a marketing system that at its zenith could be a nation-wide network of very complicated interdependency between localities and regions. In a period of chaos, the peasants would gradually withdraw and finally reduce their contact with the external world to such a degree that the village virtually was shut in.[19] The concept of Z activities is helpful to explain the characteristic of an open-and-shut rhythm. It is possible to readjust the economic pattern only if the farmer is able to switch back and forth between the production of consumable and marketable goods. The Z-activity products are the ones interchangeable as either F, which are consumed goods, or Z, some of which are marketable.[20]

The situation in which the farmer made his decisions is not hard to discern. The products of his cottage industry were likely to be local staples. As transportation posed a problem or road conditions became hazardous, the staple would not be shipped to its normal market. Local surplus of these items would cause the price to drop and profit would be trimmed or even disappear. The farmer also noticed that certain goods that were produced in other places were no longer available in the nearby marketplace because of the same conditions that cost him his distant market. He had to find means to produce substitutes for the unattainable goods with the resources he had and so time, energy, labor, and capital were all switched from production of the goods he could market to that of goods his household needed.

This seems to have been the situation that from time to time, during dynastic decline and nation-wide disturbances, turned an active marketing economy into a self-sufficient manorial economy. Fan Chung's farm, most frequently cited by historians, demonstrates a closed manorial livelihood pattern (III, 2). Fan Chung lived in the last quarter of the Western Han period, during the chaotic years of Wang Mang. For half of his life there were national disturbances and

18. G. William Skinner, "Marketing and Social Structure in Rural China," Part I, *Journal of Asian Studies* 24, no. 1 (1964): 3-43.

19. G. William Skinner, "Chinese Peasants and the Closed Community," *Comparative Studies in Society and History* 12, no. 3 (1971): pp. 270 ff.

20. Hymer and Resinick, "A Model of an Agrarian Economy," pp. 403 ff

it is likely that Fan took advantage of peaceful years to trade in manufactured products and hence became wealthy. As the empire was torn by war and inappropriate governance, the Fan household turned inward, even manufacturing its own furniture with lacquer and timber produced on its own land. The final phase of the development toward a closed community is the fortification of the village, which Fan Chung's son, Hung, and his relatives guarded with arms. Such self-defense organizations were often formed around kinship groups that were already in existence. During the disturbances, at the time of the downfall of the Wang Mang regime, such organizations consisted of both relatives and nonrelatives (VI, 23, 24). Similar self-defense organizations appeared once again in the third century as China fell into disorder at the end of the Han period (VI, 25, for instance). Such communal self-defense efforts were often led by the most prominent families of the areas. It is therefore understandable to read in the *Ordinances for the Four Peoples* that the potential leader should pay much attention to nourishing the goodwill of his kinsmen (III, 1: A.1 and I.1).

Migration

During famines or other calamities, even cash income from nonagricultural activities could not forestall starvation. Some of the population then had to move to other areas of the empire. The frontiers of the northwest and northeast and the vast open land in the south below the Yangtse and Huai rivers provided a haven for the migrants from the crowded key economic areas. Sometimes the government ordered the organization of large-scale migrations; at times provisions were supplied by the government, such as in 119 B.C., when some 725,000 poor persons from the eastern provinces were resettled in the northwest and in the north (I, 31). Nevertheless, other than such government-ordered massive migrations, every famine would have driven people from their homes to seek temporary relief.[21] Many of those who wandered perhaps went back to their homes, but many settled in new places where land was open and the soil was fertile, and where the arm of the government had not extended.

The northwestern region, of the three new areas, was largely managed as a number of military colonies. The northeast, roughly the

21. Wang Chung-lo listed fifty major population movements between 119 B.C. and A.D. 188 as starving people wandered around the empire; see Wang Chung-lo, "Kuan-yü Chung-kuo nu-li she-hui-ti wa-chieh chi feng-chien kuan-hsi-ti hsing-ch'eng wen-t'i," *Chung-kuo ku-tai-shih fen-ch'i wen-t'i t'ao-lun-chi*, pp. 450-52.

present-day provinces of Liaoning and Jehol and part of North Korea, was an integrated part of China through most of the four centuries of the Han. Lao Kan points out that the region beyond Chihli Bay had been an asylum for Chinese refugees since the Ch'in period.

Investigating the demographic data in the *Han-shu* and *Hou-Han-shu,* Lao Kan also noted the increase in population through the Western and Eastern Han periods in individual commanderies. Although the total figure for the whole region was somewhat less in the Eastern Han than in the Western Han period, Lo-lang, for instance, had an influx of over one hundred thousand persons during the Eastern Han period.[22] The most significant changes in population distribution were in the commanderies in the Yangtse valley and the coastal areas, with their fertile soils and mild climates.[23] The population of the thirteen commanderies thereof made up 5 percent of the national total in A.D. 2 and leaped to 15 percent in A.D. 140.[24] Lao Kan's population tabulation is given in table 14.

In a previous study we noted that a good number of the migrants to the south were not included in the demographic record as they had settled deep in the mountain areas where they were free from heavy taxes and corvée burdens; at least the immigrants could enjoy a short period of respite until the law reached their new homes.[25] To colonize new land, however, often required capital, which an ordinary farming household did not have. Group operations therefore were necessary. In the *San-kuo-chih,* there are frequent references to the "*tsung* people" and "leaders of the *tsung.*" T'ang Ch'ang-ju is correct in that the term *tsung* refers to organizations on a kinship base. The larger ones could consist of several ten thousand households dwelling inside fortified camps that paid only those taxes specified by themselves and that refused to accept agents dispatched by the local government[26] (VI, 26).

22. Lao Kan, "Liang-Han hu-chi yü ti-li-chih kuan-hsi," *Bulletin of the Institute of History and Philology* 5 (1935): 202-8.

23. A recent study by a geographer notes that the weather in China during the reign of Emperor Wu (140-87 B.C.) was warmer than today. During the Eastern Han period the weather began to turn colder and by the period of the Three Kingdoms and thereafter, the weather was one to two degrees colder than it is now. Chu K'o-chen, "Chung-kuo chin-wu-ch'ien-nien-lai ch'i-hou pien-hua-ti ch'u-pu yen-chiu," *K'ao-ku hsüeh-pao,* 1972, no. 1, p. 17. It is possible that the cold weather in the north pushed some vegetation as well as people out of the north.

24. Liu Ping-jen, "Liang-Han-ti jen-k'ou yü shih-liang cheng-ts'e," *Yü-kung* 7, nos. 1-3: 177-78.

25. Hsü Cho-yun, "San-kuo Wu-ti-ti ti-fang shih-li," *Bulletin of the Institute of History and Philology,* vol. 37, pt. 1.

26. T'ang Ch'ang-ju, *Wei-Chin Nan-pei-ch'ao shih lun-ts'ung,* pp. 3-7.

TABLE 14

POPULATION CHANGES IN SOUTHERN COMMANDERIES

Commanderies	Western Han	Eastern Han	Increase or Decrease	Relation of Population Eastern: Western (In percentages)
Nan Chün	718,540	747,604	+29,064	104
Chiang-hsia	219,218	265,464	+46,246	121
Lu-chiang	457,333			
&Liu-an	+178,616	424,683	-211,266	67
Chiu-chiang	780,520	432,426	-348,094	55
Lin-huai &	1,237,764	611,083	-357,213	74
Kuang-ling	+140,722	+410,190		
K'uai-chi	1,032,604	700,782	149,374	114
		+481,196		
Tan-yang	405,170	630,545	+225,375	156
Yü-chang	351,965	1,668,906	+1,316,941	474
Ch'ang-sha	235,825	1,059,372	+823,547	449
Kuei-yang	156,488	501,403	+344,915	320
Wu-ling	185,758	250,913	+65,155	135
Ling-ling	139,378	1,001,578	+862,200	718
Han-chung	300,614	267,402	-33,212	89
Kuang-han	662,249	509,438	+52.841	108
		205,652		
Shu Chün	1,245,929	1,350,476	+580,176	146
		475,629		
Chien-wei	489,486	411,378	-40,921	92
		37,187		
Yueh-sui	408,405	622,418	+214,013	152
I Chou	580,463	110,802	+1,427,683	346
		1,897,344		
Tsang-k'e	153,360	267,253	+113,893	174
Pa Chün	708,148	1,086,049	+377,901	153
Nan-hai	94,253	250,282	+156,029	266
Ts'ang-wu	146,160	466,975	+320,815	319
Chiu-chen	166,013	209,844	+43,831	126
Ho-p'u	78,980	86,617	+7,637	110
Jih-nan	69,485	100,676	+31,191	145

SOURCE: Lao Kan, "Liang-Han hu-chi yü ti-li chih kuan-hsi," pp. 208-9.

NOTE: Liu-an was abolished and its territory subsumed under Lu-chiang; the name of Lin-huai was changed to Hsia-p'ei and prefectures of the eastern part of the commandery were assigned to Kuang-ling; hence, the two commanderies have to be treated as one unit; the northern part of K'uai-chi was set off from it as the commandery of Wu; a part of Kuang-han was made administratively distinct as the "dependent state of Kuang-han" (Kuang-han shu-kuo); a part of Shu Chün was made administratively distinct as the "dependent state of Shu Chün (Shu Chün shuo-kuo); a part of Chien-wei was made administratively distinct as the "dependent state of Chien-wei" (Chien-wei shu-kuo); the western part of I-chou was made into Yung-ch'ang commandery (*HHC* 22:41a; 21:22a, 47b; 23A:24a, 25b, 26a, 23a). Chiao-chih and Yueh-lin have been omitted from Lao's table for want of adequate data.

The best illustration of the organization of a new colony is the case of T'ien Ch'ou. A native of the northern frontier, T'ien Ch'ou led a few hundred of his kinsmen into the mountain areas along the border because his native province had been occupied by a warlord. More people joined him at his new settlement and in a few years there was a population of five thousand families. T'ien Ch'ou was elected the leader, basic laws were agreed upon, schools were established, and a Chinese community was formed (VI, 27).

The impact of such group migrations upon both the farmer and farming was pertinent. An ordinary farmer who joined a group to settle at a new place would likely become a dependent of the leader who was either strong charismatically or powerful because he already commanded a core group, that is, his kinsmen, who thence became retainers.[27]

T'UN-T'IEN

The system of *t'un-t'ien* contributed to the transformation of ordinary independent farmers into private retainers. *T'un-t'ien,* military colonies, were originally developed along the north and northwest frontiers to maintain an ample supply of provisions by producing grain right on the frontiers, by soldiers themselves or by other conscripts, such as government-sponsored migrants[28] (I, 28). The tablets found in Chü-yen give firsthand information on the actual conditions in military colonies operated by the Han garrison units along the northwest defense line.[29] In the frontier colonies, there were independent

27. Chin Fa-ken traces the change of status of *pu-ch'ü* ("personal troops") from subordinates to retainers in the Han and the Three Kingdoms periods; see his *Yung-chia-luan-hou pei-fang-ti hao-tsu,* pp. 16-26. T'ang Ch'ang-ju noted that the strong local leaders of Wu had large numbers of private retainers, many of whom were recruits from the mountain dwellers; *Wei-Chin Nan-pei-ch'ao,* pp. 14-26. It appears that the constituents of the local colonial group had declined to the status of retainers at the time when the Wu generals took them over into their private armies.

28. The first proposal to move some of the population to the frontier was made Ch'ao Ts'o in 165 B.C.

29. The most extensive English treatise on the Chü-yen documents is Michael Loewe, *Records of Han Administration.* Ch'un-shu Chang has a study on the garrison system; see "The Colonization of the Ho-hsi Region—A Study of the Han Frontier System" (Ph.D. diss., Harvard University, 1963). A pertinent part of Chang's work is published as "The Han Colonists and Their Settlements on the Chü-yen Frontier," *Tsing Hua Journal of Chinese Studies,* n.s., 5, no. 2 (1966): 154 ff.

soldier-farmers who owned property and paid taxes, with their own dependents and sometimes even slaves living under one roof (III, 5, 6). There were also the soldier-farmers who tilled government land and paid heavy rent. One record gives the information that a soldier-farmer had to yield some 40 percent of his production to the garrison authority. His status appears to have been at the same level as that of a tenant. The land is recorded as being registered under the title of an officer and it is not unlikely that the conscript referred to was regarded as a private subordinate to that particular officer. If so, the conscript could also have been the officer's private retainer[30] (VI, 28). There are, of course, numerous records among the Chü-yen tablets indicating that the conscripts and their dependents received their daily rations from the garrison granary.[31] Under such circumstances the production of the field probably went completely to the granary.[32] It is very likely, in such cases, that the conscripts or the garrison soldiers actually worked in the fields as gang laborers under the supervision of an officer[33] (VI, 29).

As soldiers farmed under the command of their own officers, the rigid discipline of military organization would create a situation in which the officers would be inclined to transform the soldier-farmers into their private retainers who were then at the total disposal of their leaders. Ma Yüan's case is a good illustration. As he led his private army back to the capital, he requested permission to establish *t'un-t'ien* on the vacant land of the imperial park. The production then was equally divided between Ma Yüan, the general, and the producers, his soldiers.[34]

The large-scale transformation from *t'un-t'ien* to private tenancy, however, occurred in the third century A.D. The collapse of the Han central government resulted in prolonged chaos and disturbances at the end of the second century. The people in the key economic areas, such as southern Shensi and southern Honan, left their land when warlords used it for battlegrounds. Ts'ao Ts'ao, one of the warlords, after winning control of the central plain, put the surrendered peasants of the rebel Yellow Turban forces as *t'un-t'ien* units on the

30. Loewe, *Records of Han Administration* 2: 295-97. Lao, "Han-chien chung-ti," p. 67. Ch'en Chih suggests that the rent exacted from the conscript was on the basis of the number of persons on the farm; see *Liang-Han ching-chi shih-liao lun-ts'ung*, p. 51.

31. Loewe, *Records of Han Administration* 2: 64-93.

32. Lao Kan cites a Tun-huang tablet in which the whole yield of a plot went to the granary; Lao, "Han-chien chung-ti," p. 67.

33. Ch'en, *Liang-Han ching-chi*, p. 51.

34. *HHS* 24: 3b. Cf. Wang, "Kuan-yü Chung-kuo nu-li," p. 481.

land deserted by the refugees. The peasants were then given fields to till and even oxen were "leased" to them by the Ts'ao regime. Instead of paying taxes, they paid rent. Because they were actually government tenants, they were exempt from military service[35] (VI, 30).

More and more Offices of Commandants for Farming were established to put unclaimed land back into production. Nishijima, searching through the *San-kuo-chih* and related documents, noted the distribution of such offices, which virtually covered the Wei River valley and the middle reach of the Yellow River valley—the same key economic areas that were the main basis of support and seat of power during the two Han periods.[36]

As soon as the zenith of the Wei regime was over, however, powerful people in the government gradually controlled the *t'un-t'ien* population, who then were transformed from government "tenants" to private retainers.[37] Once general social order was restored, farming productivity again soared because of intensive farming and because the opportunity to market the surplus facilitated the return of nonagricultural activities. Thus the Office of Commandant for Farming engaged in commerce by having one portion of the *t'un-t'ien* population actually work full time in pursuit of nonagricultural profit[38] (VI, 30). This restoration of business activities shows that Han rural economy was flexible enough to respond to the marketing system as soon as conditions became stable and that the F products, which were produced for immediate consumption, could be transferred to marketable items.

PEASANT UPRISINGS

Violence was the last resort of peasants who could not survive by farming, either as a result of exploitation by others, such as landlords, government officials, and the mercantile elements, or because of adverse conditions such as wars, natural calamities, or poor soil. Hunger would drive peasants out of the fields and turn them into vagrants, dissident elements, and eventually rebels against the establishment. The Han dynasty was founded by a leader of such a peasant uprising. One period of great turmoil, when peasants rebelled here and there under the banners of Red Eyebrow, Bronze Horse, and so on,

35. T'ang, *Wei-Chin Nan-pei-ch'ao*, pp. 39-40.

36. Nishijima Sadao, *Chūgoku keizaishi kenkyū*, pp. 302-4, and map on p. 308. Cf. Ch'ao-ting Chi, *Key Economic Areas in Chinese History*, p. 95.

37. Nishijima, *Chūgoku keizaishi kenkyū*, pp. 328 ff. T'ang, *Wei-Chin Nan-pei-ch'ao*, pp. 39-42.

38. Nishijima, *Chūgoku keizaishi kenkyū*, pp. 326-30.

marked the transition from the Western Han to the Eastern Han. Another peasant rebellion at the end of the second century, known as the uprising of the Yellow Turbans, led to the collapse of the dynasty and to a long period of chaos.[39] The commonly held theory concerning the eruption of large-scale uprisings is that land concentration in the hands of a few deprived people of their means to survive.[40] As discussed previously, land concentration indeed created a serious social crisis, which gravely concerned Han writers and thinkers. However, since the small farmer of the Han period, either independent or a tenant, could derive a considerable portion of his revenue from auxiliary crops and nonagricultural activities—land concentration in the hands of landlords, who still had to parcel the land to tenants, then may not have been a major factor in driving the peasant off his land and into total desperation. An investigation of locations where uprisings originated perhaps can help to clarify the situation.

Let the proposition start with a positive association between peasant uprisings and widespread latifundia. The places where there were the heaviest concentrations of land among the smallest number of landlords ought to be the key economic areas where the most politically powerful persons converged around the court and where land was most valuable because of its proximity to the capital market, irrigation works, and so on. In fact, the famous conversation between Emperor Kuang-wu (A.D. 25-57) and his crown prince gives a good summary of this situation: "In Honan, the Capital, the close courtiers are numerous; in Nan-yang, the home of the Imperial House, the imperial relatives are numerous. [Both groups] possess real estate beyond regulation" (*HHS* 22: 8b). The key economic areas were the commanderies along the Wei River valley, the Yellow River plain, and part of the upper Huai River, namely, the metropolitan district of the capital (the central part of present Shensi), Ho-nei (northern Honan), Ho-tung (southwestern Shansi), Shang-tang (southeastern

39. Good narrations in English of these two periods of peasant uprisings are Hans Bielenstein, "The Restoration of the Han Dynasty," *Bulletin of the Museum of Far Eastern Antiquities,* no. 26 (1954); Howard Levy, "Yellow Turban Religion and Rebellion at the End of Han," *Journal of the American Oriental Society,* vol. 76 (1956); and Rafe de Crespigny, trans., *The Last of the Han.*

40. The most simplistic explanation is found in virtually all of the Marxist-toned historical studies published in China where "exploitation by feudalistic landlords" is an officially endorsed line; for illustration, Ch'i Hsia et al., *Ch'in-Han nung-min chan-cheng-shih,* pp. 62 ff. and 129 ff. A similar suggestion can be found in such celebrated works as Etienne Balazs, *Chinese Civilization and Bureaucracy,* pp. 192 ff.

Shansi), Chao Kuo (western central Shansi and Eastern central Hopei), and T'ai-yüan (central Shansi).[41]

The places of origin of peasant uprisings during the Western Han period are given in table 15; the places where rebellions began in the Eastern Han period are given in table 16.

These cases, with three exceptions (two in the metropolitan district and one in Ch'en-liu), occurred outside of the key economic areas. Most prone to uprisings were the regions of the present provinces of Hopei, Shantung, Anhwei, and Hupei—all were on the periphery of the Shensi-Honan heartland. Such a mutual exclusion of uprisings and extreme land concentrations possibly point to the interpretation that livelihood was supplemented in key economic areas, where the largest urban centers are located and where the largest number of consumers dwelt, because the farmer had access to an excellent marketing system.

The demographic changes in commanderies from the Western Han to the Eastern Han periods, as discussed by Lao Kan, show that the commanderies in the present provinces of Hopei, Shantung, and northern Kiangsu (Lu, Wei Chün, Chü-lu, Ch'ang-shan, Chung-shan, Ch'ing-ho, Chao, Tung Chün, Tung-p'ing, T'ai-shan, Shan-yang, Chi-yin, Tung-hai, Lang-ya, P'eng-ch'eng, Hsia-p'i, Chi-nan, Pei-hai) lost a substantial part of their population.[42] If the outflow of population from these areas was steady throughout the Han period, it indicates a general decline in economic conditions. Therefore, the marketing system in these northern and eastern commanderies probably would not have been active, which would be unfavorable to the production of nonagricultural goods for additional income. Kimura suggests that these peripheral areas, being "secondary arable land" in the mountain regions, could not afford the deterioration of irrigation works from negligence, while, he implies, the "primary arable land" in the valleys and plains would have favorable natural conditions.[43] The decrease in population might be associated with the deterioration of irrigation facilities; poor governance could be the factor starting the deterioration of the general economy.

The Ching-chou area (Hupei and Hunan today), in the south, however, gained in population during the Eastern Han period.[44] Yet

41. Chi, *Key Economic Areas in Chinese History*, p. 95.

42. Lao Kan, "Liang-Han chün-kuo mien-chi-chih ku-chi chi k'ou-shu tseng-chien-chih t'ui-t'se," *Bulletin of the Institute of History and Philology* 5 (1935): 224 ff. passim.

43. Kimura Masao, "Zengo-Kan kōtaiki no nōmin hanran," *Tōkyō kyōiku daigaku bungaku kiyō* 16 (1967): 1-2.

44. Lao, "Liang-Han chün-kuo mien-chi-chih ku-chi chi k'ou-shu tseng-chien-chih t'ui-ts'e," pp. 224 passim.

TABLE 15

PEASANT UPRISINGS OF WESTERN HAN
(Including Wang Mang's reign)

Year	Place	Scale
First century B.C.	T'ai-shan	Unknown
	Nan-yang	"
	Ch'u	"
	Ch'i	"
	Yen and Chao	"
30 B.C.	Metropolitan district	Small
22 B.C.	Ying-ch'uan	"
18 B.C.	Kuang-han	"
A.D. 17	Lang-ya	Medium, then large
	K'uai-chi	Medium
	Chiang-hsia	Medium, then large
	Nan Chün, Chiang-hsia	Large
A.D. 18	Lang-ya	" (Red Eyebrows)
	Tung-hai	
A.D. 21	Metropolitan district	Numerous small ones
	Nan Chün	Large
	P'ing-yüan	Medium
A.D. 23	Tung-p'ing	Large
	Tung Chün	Small
	P'ing-yüan	Large
	Chü-lu	" (Bronze Horse)
	Ch'ing-ho, Wei Chün	Large
	Wei Chün	Medium
	Chen-ting, Chung-shan	Large

SOURCE: Various chapter of the *Han-shu*, collected by Ch'i Hsia et al., *Ch'in-Han nung-min chan-cheng-shih,* pp. 79-94. Scale: hundreds = small; thousands = medium; tens of thousands = large.

the population density in Ching-chou was still below that of the northern and eastern commanderies[45] and was still to develop its active economy. Tada points out that the disturbances in the south occurred much later than in the north and east. In fact, he proposes that the

TABLE 16

PEASANT UPRISINGS OF EASTERN HAN

Year	Place	Scale
A.D. 108	P'ing-yüan	Unknown
109	P'ing-yüan, Po-hai	Medium
110	Chao-kuo	"
111	T'ien-shui	Unknown
132	K'uai-chi	Medium
132	Yang-chou	Unknown
134	I-chou	"
137	Chiang-hsia	"
138	Chiu-chiang	Small
	Wu Chün	"
	Ching-chou	"
139	Yang-chou, Hsü-chou	Medium
141	T'ai-shan	Unknown
142	Kuang-ling	Large
144	Nan Chün, Chiang-hsia	Unknown
147	Ch'en-liu	"
154	T'ai-shan, Lang-ya	Large
162	Ch'ang-sha	Medium
167	Lu-chiang	Unknown
170	Chi-nan	"
172	K'uai-chi	Large
184	Nationwide	" (Yellow Turban)

SOURCE: Ch'i Hsia et al., *Ch'in-Han nung-min chan-cheng-shih*, pp. 149-53.

45. Ibid., pp. 220-21. The Ching-chou commanderies had a density of 3-14 persons a square kilometer; figures for Ch'ing-chou and Chi-chou are approximately 30-130 persons a square kilometer.

migration from the disturbed areas to the relatively underdeveloped south caused the south to break down too.[46]

In either case, with a declining trend or with a developing economy, peasants would have relatively little contact with a marketing economy that would offer them an alternative to subsistence farming. As the margin of subsistence farming was small, peasants, for the sake of survival, had the last resort—violent action against established order.

SUMMARY

The Han farming household, in conjunction with the highly uneven distribution of seasonal labor for intensive farming, developed a heavy dependence upon the production of nonagricultural activities. A self-sufficient economy, which bore a resemblance to a manorial economy, however, emerged when the marketing system was closed to the farming population, which needed only to transfer its energy from producing marketable items back to materials for its own consumption.

The same pattern would be reversed in the case of migrants who had to live in a closed community in a new place until the local marketing network developed enough for them to participate. The dependence upon organization in the new places, however, tended to make individual peasants subservient to strong leadership, which often reduced them to the status of retainers.

The *t'un-t'ien* system was, in actuality, only another form of the migration pattern. The population, instead of being under private leadership, was settled on unclaimed land by the government. Perhaps because some of the *t'un-t'ien* colonies were in the former key economic areas in which conditions and facilities for intensive farming were still available, high productivity was soon restored, leading to a late phase in *t'un-t'ien* development in which commercialization occurred. *T'un-t'ien* were also experienced by the migrants to the south and to frontier colonies, but because conditions were favorable to intensive farming in the key economic areas, the whole process leading to commercialization was more accelerated, and therefore more revealing, than the same process in the development of, for instance, the southern marketing system.

The flexibility of intensive farming is apparent again when peasant uprisings are examined, for they generally occurred outside of the key economic areas. Superior farming facilities and the active marketing economy in the key areas could have compensated for some

46. Tada Kensuke, "Kōkin no ran zenshi" *Tōyōshi kenkyū* 26, no. 4 (1968): 172-74.

of the disadvantages peasants suffered under the latifundia phenomenon, while peasants at the periphery or outside of it would have very little leeway in such calamities as drought, flood, or corrupt governance, which pressed them into rebelling.

EPILOGUE

IN this work, I have tried to establish that agricultural advancement in Han China was related to the development of small-farm operations and that its result, intensive farming, was associated with a market economy that, however, formed in China a particular rural economic pattern. We shall start this summary with a discussion of the impact of policies made by political power upon agricultural development.

The Ch'in government, from its days as one of the seven states contending for dominance, developed steadily toward bureaucratization. The abolition of the feudal system was completed when Ch'in Shih-huang-ti approved the proposal made by Li Ssu that the empire should be governed under the *chün-hsien* system, with no prince enfeoffed as ruler of a vassal state.[1] This measure removed the holders of power that intervened between the emperor and the mass of commoners. It, indeed, was the realization of the thought of Han Fei, the great master of the Fa-chia, who warned that powerful households could provide shelter for ordinary citizens who comprised the pool from which came both producers and soldiers.[2]

The earlier reforms carried out by Shang Yang in the fourth century B.C. encouraged the establishment of independent households as soon as a male had reached adulthood. This reform reduced the potential of forming, particularly within kinship organizations, another possible bloc between imperial authority and the individual subject.

The Ch'in rulers made a successful effort to assert direct state

1. Derk Bodde, *China's First Unifier*, pp. 133-46; also, *SC* 6:25-27.
2. *HFT* 17:13-14, 18:10-11, 20:4.

control over producers. After the unification of China, Ch'in Shih-huang-ti repeatedly stressed in stele inscriptions that he had succeeded in mobilizing the people to engage in agricultural production.[3] It is obvious that Ch'in Shih-huang-ti viewed the farmers not only as providers of the manpower to produce the only real goods in an agrarian economy but also as his stable subjects, simple and earthbound.

The Han Empire, though it came into existence as a result of the overthrow of the Ch'in dynasty, remained as bureaucratic a society as the Ch'in had been. The Han rulers continuously did their best to reduce the influence of various groups in society to minimize their potential menace to imperial power. By the time of Emperor Wu (140-87 B.C.), the Han court had utilized several means—monopoly of crucial commodities, promotion of state enterprises, discriminatory taxation, and other financial and monetary measures—to extend direct imperial control into the economic sphere. These measures coincided with the establishment of monolithic control over the whole society by means of political maneuvering. Most enduring, however, were the constant attempts by the Han government to make a flourishing agriculture the foundation of the bureaucratic empire.

The measures for promotion of agriculture consisted of reduction of the production tax, which was once even completely waived, of repeatedly opening government-controlled land to resettle the farmers who had lost their own lots, and of frequent loans of operating capital to those who were stricken by natural disasters or driven from production by poverty. In addition, magistrates were instructed to take promotion of agricultural production as their assignment of top priority. It should be noted that much of the imperial grace toward the farming population did not occur until at least one full generation after the dynasty was founded. An edict issued by Emperor Wen in 176 B.C. declared that agriculture was the very foundation of the empire, a statement that was echoed in subsequent imperial decrees many times. The treadmill pattern reveals that there was a persistent problem that could not be solved by political manipulation.

3. He visited the east coast area several times between 219 and 212 B.C. Long poems were inscribed at seven locations. The one inscribed in 219 B.C. at Lang-ya, for instance, includes such sentences as: "The Emperor succeeded in making the people work for the basic production" (i.e., agriculture); "Agriculture is upheld, while the unimportant profession [commerce] is discouraged, so that the common people are enriched"; "The small and the big all assert their utmost effort, daring not to be lazy and neglectful"; "The production in all lines is plentiful and multiplied as the works are carried out timely and promptly." Or, the one in 215 B.C. inscribed at Chieh-shih includes "The male works happily in the field; the female fulfills her own share"; "Everything is in order; the benefit is felt in all professions"; "People converge to cultivate the fields as [the subjects of the empire] and all are satisfied with their lives." *SC* 6: 34-46.

The time factor has to be taken into consideration. One generation or two after unification, the Han Empire had a steady population growth possibly at a rate of 1.6 percent. The growth could reflect a surge in the birth rate during those peaceful years or a more effective registration of population under a stable government, or—more likely—a combination of both. At any rate, the population increase must have posed a serious problem—a marked decrease in per capita acreage of arable land. The speed at which land was brought under cultivation hardly seemed to match the demographic growth rate. As E. A. Wrigley points out, the subtle and dynamic interrelationship between demographic characteristics and economic changes may help to explain why in so many preindustrial societies in Europe and elsewhere the agricultural base of the economy slowly and gradually broadened.[4]

Ester Boserup, in her provocative work, *Conditions of Agricultural Growth,* boldly turned Malthusian theory around. She argued that population growth appears to be the motivating force for agrarian change. In typical cases, the cultivator would find it profitable to shift to a more intensive system of land use *only* when a certain density of population was reached. And if such a density were not reached, he may not use intensive farming methods, even if they were known to him, until the critical point of high population pressure was such that he had to adopt labor-intensive methods, that is, to accept a decline of output per man-hour. Obviously, with an abundant population, labor supply would not be a problem. Under such pressure, the farmer strives for an increase of yield per given area of land, namely, an increase in the production rate.[5]

There are always two options regarding farm sizes: either individual farming households working on small plots, or large-scale farming operations under manorial, plantation, village commune, cooperative, or other arrangements. The Han government felt it initially necessary to maintain farmers as the foundation of the empire and to allow no part of the private sector to control human resources. This attitude (which I think had its roots in both Confucian and the Fa-chia ideologies) prevented the appearance of manorial plantations under the control of private landlords. Meanwhile, the motivation for personal gain would make tenancy a better system, even from the viewpoint of the landlord, than the use of gang laborers. Therefore, although there was good opportunity for the privileged social group to seize land from independent owners, their farms were still generally tilled

4. E. A. Wrigley, *Population and History,* pp. 46-50. Cf. T. R. Malthus, *First Essay on Population.*

5. Ester Boserup, *The Conditions of Agricultural Growth,* p. 41.

by tenants on small plots. The large population made labor costs so cheap that slavery was not an institution of production in the Han period. The tenants, nonetheless, still had to carry the burden of paying rent to landlords. The farmer who worked on a small farm simply had to make as much as he could from his limited land to make both ends meet.

What a large population could afford was labor-intensive production. Intensive farming in the Han had these characteristics:

I. *Field preparation*
1. Planting fields in patterns, no broadcasting of seed
2. Cultivating fairly deeply
3. Utilizing the characteristics of the terrain

II. *Seed*
1. Selecting the best specimens for propagation
2. Treating seed before sowing

III. *Cultivation*
1. Timing precisely to take advantage of optimum atmospheric conditions
2. Constant, diligent weeding and insect control
3. Irrigating to provide adequate moisture

IV. *Soil Amelioration*
1. Using manure to fertilize the soil
2. Rotating crops to shorten or to avoid fallow
3. Using leguminous crops to improve soil quality

V. *Land Utilization*
1. Multicropping; on occasion, interpolating crops
2. Growing vegetables on marginal land

VI. *Implements*
1. Using animal power
2. Using specialized tools for various functions

These characteristics show that the Han farmer did command technical knowledge of intensive farming at a level that allowed continuous and rational utilization of land, although some regional variations in actual practice should be allowed. In the first century B.C., cultivated plants included spiked and glutinous millet, winter and spring barley and wheat, soy and lesser beans, hemp, melon, gourd, taro, water darnel, paddy rice, Perilla, sesame, alfalfa, and peas. Fertilizers included human night soil and the excrement of animals such as sheep, goat, pig, ox, horse, fowl, and silkworm, in addition to green manures. The water supply was under control, as canals, ponds, and wells facilitated the change from rain-fed to irrigated agriculture, helping the development of a sophisticated intensification of the farming system.

Intensification of farming, indeed, could greatly reduce some hidden underemployment, especially that of females and minors, who could be helpful at less physically demanding jobs, such as weeding. The practice of multicropping, then, even shortened the length of idle periods. Nevertheless, the long frost season in China, specifically in the north, the heartland of the Han Empire, still put the Han farmer through a prolonged winter period. Accordingly, labor distribution for agricultural work tended to be uneven, with the largest amount of labor required in two peak periods, the growing season and the harvest.

Some nonagricultural work carried out in the dead season by men and/or, in the busy season, women and children would absorb much of the seasonal underemployment. The result could either be of logistic or of market value, for the goods produced could easily be transformed from home consumption to marketable items. Such activities are grouped under one name by some economists as "Z" activities, and conform to the Z activities in the Hymer-Resinick framework.

Family-farm production, therefore, would encourage development of a mercantile agrarianism by which transactions were carried along a treelike structure that seemed to tie the scattered settlements (villages and towns) into a marketing network. Indeed, the geographic area covered by the farmer-trader could be fairly extensive. As a mock contract reveals, transactions were conducted within two circles: the larger with a radius of two hundred kilometers, and the smaller with a radius of fifty kilometers. There are cities, marketing towns, small markets, and side markets, already showing the development of a hierarchical network of trading posts.

Under mercantile agrarianism, some market speculation becomes feasible. In a mid-second-century work, the *Ssu-min-yüeh-ling (Ordinances for the Four Peoples),* the farmer could engage in selling and buying grain several times a year, not for acquiring supplies for consumption, but rather for making a profit. Meanwhile, the production of preserved food, wine, vinegar, medicine, silk and silk fabrics, shoes, and so on, in the farming household were compatible with both home consumption and marketing. There were similar works in ancient times containing calendars of activities, but the *Ssu-min-yüeh-ling* is the first to stress commercial activities. It thus indicates that it was in the Han period that mercantile agrarianism became integrated into the rural economy.

The notion of growing cash crops for profit is included in the *Fan Sheng-chih-shu.* For example, gourds were grown, not only for fodder and to make dippers, but also for making candles. The calculation of such incomes was all done in terms of cash value at the market. Commercialization would link the local economy to an integrated regional and, ultimately, national economic system, which would, in turn, facilitate interregional product specialization. In the time of Ssu-ma Ch'ien, around the first century B.C., there were staple goods, such as chestnuts of Yen and a number of other areas, dates of An-i and Ch'in, oranges of Shu, Han, and Chiang-ling, catalpa of the region between the Huai, Chi, and the Yellow rivers, lacquer in Ch'en and Hsia, mulberry trees and hemp in Ch'i and Lu, and bamboo in the Wei valley. These all appear as products in national markets.

An agrarian, yet mercantile agrarian, society knit the individual farmer into a massive economic network. The situation is, therefore, very different from the common image of an ancient Chinese agrarian society made up of scattered, isolated, self-sufficient cells, which were the farming households.[6] Ironically, the capital accumulated in such mercantile agrarianism could hardly go very far. The commercial urban centers had been stimulated to grow to a remarkable prosperity, as at least twenty-six first-rate urban centers had been formed along major transportation routes in the Western Han period. Yet the merchant-entrepreneur class was nipped in the bud, first by Emperor Wu, who, forging the authority of a monarchy, was jealous of any potential challenge to his sovereign power. And, thereafter, the bureaucracy of the empire was so well established that its members, the bureaucrats and the local leaders, were able to monopolize not only political power but also great wealth. Full development of mercantilization, therefore, never had a chance to start.

The pivot of active transactions hinged on the stability of the political order. When the empire was unified, when roads were good, when travel was safe, interregional specialization would pull individual farmers in the village to participate in a gigantic marketing network. When political conditions deteriorated, as when the empire was divided and the central government failed to rule effectively, and, in more practical terms, when former special export products could not be shipped out and staple goods from outside became too expensive to acquire, the farmer had to change his habits of production and consumption. Local leaders would guide the local area to close in

6. For instance, Etienne Balazs, *Chinese Civilization and Bureaucracy*, pp. 15-16.

on itself. The nonagricultural Z activities would then be shifted from producing market goods back to producing goods and services for their own consumption. There emerged, therefore, a self-sufficient rural community. Since such a small community was often under the dominance of local leaders, it could easily exhibit the characteristics of a manorial system. At the end of the Western Han as well as at the end of the Eastern Han, such communities were numerous. Nevertheless, they were neither closed isolated communities heretofore nor thereafter. Resembling a sea anemone, the frame under the mercantile agrarianism and intensive farming expanded during moments of security and contracted at times of danger.

In summary, agricultural developments during the Han period seem largely in the advancement of intensive farming of the small farm operation as a response to the relative increase of population pressure on arable land, especially within the most well-developed regions. The heavy demand for cash payments, such as the poll tax, and the uneven seasonal distribution of labor under intensive farming is related to nonagricultural activities that, in turn, made possible the development of market-oriented economic behavior. Nevertheless, where the market became inaccessible, the farmer could readjust this behavior by adopting relatively more self-reliant economic patterns.

Part Two: Documents of Han Agriculture

I

GOVERNMENT RESPONSES TO PROBLEMS

1. Land Restored and Granted

The soldiers were all demobilized and returned to their homes. An edict was issued, saying, "The relatives of the [former] feudal lords[1] who are in the Kuan-chung area shall be exempted from service for twelve years; those who have [already] returned home, for half that period. Some people who formerly assembled and took refuge in the mountains and marshes have not enrolled their names on the register. Now the empire has been pacified. Let each of them return to his own prefecture, and let his previous rank, land, and house be restored. The officials shall instruct them clearly through publishing the law, and shall under no circumstances beat or humiliate them. People who because of hunger sold themselves to others as male and female slaves shall all be freed and become commoners. Officers and soldiers who have been pardoned, those who have not committed a crime but are without noble rank, and those whose rank is below that of the *ta-fu*[2] shall all be granted the rank of *ta-fu*. All those with the rank of *ta-fu* or above shall be granted a rank one degree higher. Let all those who have the rank of *ch'i-ta-fu*[3] and upwards be given fiefs. All those whose rank is below that of *ch'i-ta-fu* shall, with their families, be exempted from labor service." 202 B.C.

It also said, "The ranks from *ch'i-ta-fu* and *kung-ch'eng* up are all high ranks. Many of the relatives of the [former] feudal lords and many of those who have been in the army and have returned home possess high ranks. We have several times ordered the officials to give them land and houses first and promptly provide them with whatever they are entitled to request. Some of the holders of ranks have been rulers of their people, have been respected and honored by the emperor, and have been established for a long time; yet concerning their requests the officials have not taken action—this is utterly incomprehensible. In the days past, in the time of Ch'in, the people who held Ca. A.D. 178-79

1. *Chu-hou tzu* means "the descendants of the former feudal lords of pre-Ch'in time."

2. *Ta-fu* was the fifth rank.

3. According to Yen, the *ch'i-ta-fu* was actually the *kung-ta-fu*, so termed because it was the seventh rank.

Fu Tsan comments that according to the Ch'in system only marquises (the twentieth rank) were granted fiefs. The grant of a fief to the *ch'i-ta-fu* and above was a special favor.

the ranks from *kung-ta-fu* up were treated by the prefects [*ling*] and their assistants [*ch'eng*] as equals. Now, we do not think lightly of the ranks; why then should the officials alone do so? Moreover, according to the law those who rendered meritorious service should be given land and houses; yet at present many of the lesser officials who have never served in the army have satisfied themselves while on the other hand those who have rendered meritorious service have received nothing. This is violation of the public interest for the interest of private individuals, [and must be due to] the very unsatisfactory way in which the administrators [*shou*], the commandants [*wei*], and the senior officials [*chang-li*] have issued instructions. Let the officials in conformity to Our wishes be ordered to treat well the holders of high ranks. Moreover, investigations will be made soon; if any are found who have not acted in accordance with Our edict they shall be severely punished." [*HS* 1B:4a-5b; cf. Dubs, *HFHD* 1:103-6]

2. Chia I's Memorial

194-178 B.C. During the reigns of Emperor Hsiao-hui [194-188 B.C.] and Empress Kao [187-180 B.C.] supplies in clothing and food increased. After Emperor Wen [r. 179-157 B.C.] came to the throne, he personally cultivated frugality, intent on comforting the people. As his time was close to the Chan-kuo period, all the people turned their backs on the fundamental [occupations] and hastened to the secondary [ones].[4] Chia I[5] advised the emperor [in 173 B.C.], saying, "The *Kuan-tzu*[6] states: 'only after the granaries are filled will [the people] know the rules of proper behavior.' From ancient times to the present it has never been known that a people who have insufficient food can be kept under control. Men of ancient time said, 'If there is one man who does not cultivate the land, some people will suffer from hunger. If there is one woman who does not weave, some people will suffer from cold.' Production is limited by seasons and if consumption is without restraint, then the material resources will certainly be exhausted. In the administration of the world in ancient times everything was worked out to the last detail; therefore, there were the stored goods to depend on.

4. The fundamental occupation was agriculture, while the secondary occupations included commerce and luxury handicrafts.

5. Chia I, serving at the court around 180 B.C., was responsible for the creation of many new laws. He died ca. 168 B.C. at the age of thirty-three (*SC* 84: 8a-15a; *HS* 48).

6. A book attributed to Kuan Chung, the great statesman of the state of Ch'i. It was a collection of writings mainly by scholars of the Chan-kuo and later periods.

"Now the people turn their backs on the fundamental and rush into the secondary. Those who do nothing but eat are very numerous. This is a great injury to the empire. Licentious and luxurious customs increase day by day; this is a great evil to the empire. Injury and evil are openly practiced, and nothing has been done to stop them. The great mandate is about to fall, and nothing has been done to uphold it. Those who produce are very few, but those who consume idly are extremely numerous. How is it possible that the wealth of the empire will not be exhausted?

"Since Han was established as a dynasty, nearly forty years[7] have passed. The reserves accumulated both by the government and the people are still lamentable. The proper seasons are neglected, and the rain does not fall; people are already looking around [nervously] like wolves.[8] The harvest is bad and there is little income. There has been a petition for permission to sell ranks and children. This has reached Your ear. How can it be possible that Your Majesty is not alarmed at having led the empire into such danger?. . .

"Accumulation of supplies is the very life of the empire. If grain is abundant and wealth excessive, what cannot be accomplished if attempted? Having it, then in attacking, there will be gains; in defending, there will be determination [strength] ; in warfare there will be victories. If the enemy is conquered, and the distant peoples subjugated, who will refuse to come when summoned?

"If now the people are herded and brought back to farming, and are attached to the fundamental, if each one throughout the empire feeds himself by his own labor, and if the people who practice the secondary arts and lead a parasitic life turn back and go into the fields, then the accumulation of supplies will be sufficient and the people will be happy where they are. Thus wealth can be created and the world can be kept in order. Why let the situation be dangerous like this and not act? I personally regret it for Your Majesty."

The emperor was moved by [Chia] I's words, and then he for the first time established the Ceremonial Field,[9] in order to cultivate the

7. The word *ssu*, "four," in the phrase "nearly forty years" must be a copyist's mistake for *san*, "three," since from the rise of Han (202 B.C.) to the year 178 B.C., only twenty-five years had elapsed.

8. According to Li Ch'i, quoted by Yen, the wolf is a coward by nature and looks backward while running away. In the same way, people looking at the sky and seeing no rain about to fall are afraid.

9. "The first time" means in the Han dynasty. The Ceremonial Field was the land that the emperor himself cultivated. The ceremony was designed to show the government's emphasis on agricultural production and to provide the grain for the offerings in the royal ancestral temple (*HS* 4: 9b-10a). Before the emperor

land personally for the encouragement of the people. [*HS* 24A:10-12a; cf. Nancy Lee Swann, *Food and Money in Ancient China,* pp. 152-58]

3. Ch'ao Ts'o's Memorial

178 B.C. Ch'ao Ts'o[10] also advised the emperor [Wen, r. 179-157 B.C.] saying: "When the Sage King is above [the people] and the people do not suffer from cold or hunger, the reason is not that he can cultivate the land to feed them or that he can weave to clothe them, but that he opens up the way towards means and resources. Therefore, although Yao and Yü suffered from floods for nine years and T'ang from droughts for seven years and yet their countries saw no desertions or sickness from starvation is owing to the abundance of their stores and the completeness of their reserves.

"Now, the empire has been unified and the amount of land and the number of people are not less than those of T'ang and Yü. Moreover, there have been no natural calamities of flood or drought for several years. Yet why is it that the stores of provisions have not equaled those [of T'ang and Yü]? It is because there are acreages of land, the people have unused energies, the grain-growing land has not been fully cultivated, the resources of mountains and lakes have not been thoroughly exploited, and the wandering people have not all returned to farming.

"If the people are poor, then depravity follows. Poverty originates in insufficiency; insufficiency originates in not farming. If the people do not engage in farming, they will not become attached to the land. If they do not become attached to the land, then they will leave their native places and regard their families lightly. [Then] the people will be like birds and beasts. Even if there were high city walls, deep moats, strict laws, and severe punishments, it would still be impossible to restrain them. When cold, one does not wait for expensive material for clothing; when hungry, one does not wait for delicacies for food; when one is assailed by hunger and cold, he does not care about honesty or shame. According to human nature, one feels hungry if he does not get a second meal in the day; he suffers from cold if he does not get new clothes in the year. When a man is hungry and he does

---

went to plow the Ceremonial Field he worshiped the god of agriculture and prayed for a good harvest (*Hsü-Han-chih* 4:5b-6a, hereafter cited as *HHC*; see also *LCCS* 14:12a).

The *chi-t'ien,* "Ceremonial Field," according to Hsu Shen, was one thousand mu in extent.

10. Ch'ao Ts'o is an advisor who suggested the emphasis on agricultural production. For his biography, see Burton Watson, *Records of the Grand Historian of China* 1: 527-32.

not get food, and when his skin is cold and he does not get clothes, even his tender mother cannot keep him. How then can a ruler hold his people?

"A wise ruler realizes that this is so. Therefore he urges his people to devote themselves to farming and mulberry culture. He lowers the taxes and enlarges the stores in order to fill the granaries against [the time of] floods and droughts. Therefore, he can win the people over and hold them. . . .

"Pearls, jade, gold, and silver cannot be eaten in case of hunger, nor can they be worn in case of cold. Yet the people value them because of the emperor's use of them. As objects they are very light and small and can be easily concealed. With them at one's disposal, it is possible to travel all around the empire without anxiety about hunger or cold. It is they that cause the courtier lightly to desert his ruler and the people easily to abandon their village; it is through them that bandits receive encouragement and fugitives obtain easy resources.

"Grain and cloth and silk are produced by the land, mature with time, and are gathered together in the markets. All this cannot be accomplished in a day. [Grain or cloth] weighing several piculs *(shih)* cannot be moved by the average man and are not profitable to the dishonest. When they (i.e., grain and cloth) are not obtained for a single day, hunger and cold will arrive. Therefore, a wise ruler values the five grains and belittles gold and jade.

"Now in a peasant family of five members there are not less than two persons who render labor service. All that they can cultivate is not more than one hundred mu. The yield from one hundred mu is not more than one hundred piculs.

"They till the land in spring, hoe in summer, harvest in autumn, and store in winter. [Besides] they have to cut wood for fuel, work in the government buildings, and render labor service. In the spring they cannot escape the wind and dust; in the summer they cannot escape the heat; in the autumn they cannot escape the chilling rain; and in the winter they cannot escape the cold. Throughout the four seasons they do not have a single day of rest. Furthermore, their private expenditures for parting with those who are leaving and for welcoming those who are coming, for consoling with the bereaved, visiting the sick, caring for the orphans, and bringing up the young, all have to be defrayed from that income. Although they are already so industrious and straitened, still they have to suffer from the calamities of flood and drought. Collections are urgent, taxation exorbitant; and the taxes are collected at no fixed date. Orders issued in the morning have to be executed in the evening.[11] Those who have something sell it at

11. The phrase *chao ling erh mu kai,* if literally translated, should be "orders

half price; those who have nothing have to borrow at 100 percent interest. Hence, there are people who have sold their land, their houses, their wives and children in order to pay their debts (i.e., taxes). . . .

"Now, the law despises merchants, but the merchants have become rich and noble; it esteems peasants, but the peasants have become poor and mean. What is esteemed by the people is despised by the ruler; and what is scorned by the officials is honored by the law. The ruler and the people oppose each other, and good and bad are reversed. To hope that the country be rich and the law upheld is impossible.

"The task of the moment is nothing else than to make the people devote themselves to farming. If the people are expected to devote themselves to farming, grain must be made more valuable. The way to make grain more valuable is to allow the people to use grain for gaining rewards and redemption from punishment. If now the people of the empire are called on to offer grain to the emperor and are permitted to obtain rank and to redeem their crimes, then the rich people will have rank, the peasants will have money, and the grain will be dispensed.

"Those who can present grain to receive ranks are all people who possess a surplus. By taking away something from those who have a surplus to provide for the emperor's expenses the levies on the poor people will be reduced. This is what is known as reducing the excessive and supplementing the insufficient. When such an order is issued the people will benefit. This is in accord with the wish of the people and makes three contributions: first, the ruler will have enough for his expenses; second, the people will pay less tax; third, agricultural production will be encouraged.

"Under the present law, whoever has one horse for drawing a carriage or for riding has exemption for three persons from services. Carriages and horses are military equipment of the empire. Therefore their owners are exempted from service. The teachings of Shen-nung say, 'Even though there be a city wall of stone ten *jen* high,[12] a moat of boiling water a hundred paces wide, and a million soldiers, without grain there can be no defense.'

"From this it can be seen that grain is the great resource of a ruler and the fundamental task in civil administration.

"If the people are permitted to present grain and receive ranks, those from the rank of *wu-ta-fu*[13] upward can have only one person exempted from labor service. This is a far greater contribution than

issued in the morning are to be altered in the evening." Yet the *Han Chi* bears a phrase, *chao ling mu te*, which seems the original wording (*HS* 24A: 13b).

12. A *jen* is a length of eight Chinese feet *(ch'ih)*.

13. The ninth rank of twenty honorific titles.

a riding horse. Ranks that are a monopoly of the ruler come from his mouth and are unlimited. Grain that is cultivated by the people is produced by the soil and is inexhaustible. To obtain high rank and redemption for crimes is a great desire of the people. Let the people throughout the empire offer grain for [use at] the frontiers in return for ranks and for redemption of their crimes; in not more than three years the grain at the frontiers will certainly be abundant."

Thereupon Emperor Wen followed [Ch'ao] Ts'o's advice, and decreed that whoever of the people offered six hundred piculs of grain and had it transported to the frontier would receive the rank of *shang-tsao.*[14] If he increased the amount to four thousand piculs he would be made *wu-ta-fu.* For twelve thousand piculs he would be made *ta-shu-chang.*[15] Each was to [receive] a different gradation according to the amount [of grain which he contributed]. [*HS* 24A:12a-15a; cf. Swann, *Food and Money,* pp. 158-69]

4. Tung Chung-shu's Memorial

[Tung Chung-shu] again memorialized: "In ancient times, people were levied no more than one-tenth of the crop as tax, which was an easy to supply requisition. The corvée obligation extended no more than three days each year, which was easy to provide. The people thus had the wealth to nourish parents in order to fulfill the duty of filial piety, to serve the ruler by paying tax, and to support wife and children with utmost affection. Consequently, the people delighted to follow the ruler. *Ca.* **100 B.C.**

"By [the time of] Ch'in, it was not so. The reforms carried out by Shang Yang changed the institutions of the [ancient] rulers. [Hence] the well-field system was abolished and the people were allowed to make transactions [in land]. The wealthy possessed land [so much that the fields] connected with each other, while the poor had not even [enough] land into which to stick an awl. Moreover, [the wealthy] monopolized the profit of streams and marshes and controlled the resources of mountains and forests. . . .

"In the cities there were [some who commanded the same respect as] state rulers. In the neighborhoods there were [some who had the wealth of] feudal lords. How could the lowly people but be held in poverty? Moreover, one had to serve one month of local corvée service each year. Again, one had to serve as a regular conscript one year and frontier garrison duty one year. All the labor requirements totaled thirtyfold those of ancient times. Land tax, poll tax, and salt

14. The second rank.
15. The eighteenth rank.

and iron taxes, altogether were twentyfold those of ancient times. Some, cultivating land belonging to overbearing [rich] people, had to give one-half of the yield as rent. Consequently, the poor wore at all times [garments in quality fit only] to be covering for cattle and horses, and ate food [of the standards for] dog and swine feed. [Conditions] were aggravated by covetous and cruel officials who rashly imposed penalties. The people, desperate, fled to mountains and forests [where they] turned to thievery and robbery. . . . [The same practices survived the Ch'in dynasty, and] when the Han arose they were followed without change. Although it would be difficult to precipitously reinstitute the ancient well-field system, it is proper to [create some institution] somewhat near to the ancient [system]. May limits be set on privately owned land in order to sustain [the poor] in their insufficiency and to block the road towards monopoly." [*HS* 24A:16a-17a; cf. Swann, *Food and Money*, pp. 179-83]

5. Limitation of Land Ownership

7 B.C. When Emperor Ai came to the throne, Shih Tan, who assisted in managing state affairs [as the head of the state], proposed, "None of the ancient sage kings could achieve stability until they established the well-field system. Emperor Wen [of Han], coming to the throne after the empire was exhausted and emptied as a result of military operations during [the chaotic period of] the disintegration of Chou and the disorder of Ch'in, therefore exerted himself to encourage agriculture and sericulture and took the lead in [the practice of] economy and frugality. The people began to have copious abundance; there was no evil of engrossment [of land]. [The government] therefore made no restriction on the quantity of land and slaves owned by any individual. Now, several generations in turn have enjoyed peace. Some wealthy persons, officials as well as commoners, have accumulated a great amount of property, while the poor and the weak, on the contrary, are more and more straitened. Now [it is admitted that] a gentleman, indeed, should honor the practice of traditional methods in his government and consider it a serious matter to make changes, but the reason for making changes is often for the sake of saving [the country] in time of crisis. Accordingly, though it has not been possible to go into details, it is proper to make [certain] general restrictions."

The emperor handed down the proposal [for discussion]. K'ung Kuang, the chancellor, and Ho Wu, the grand minister of works, memorialized, "Let the vassal kings and the marquises own fields [only] within their own fiefs, and let the land in prefectures and border circuits owned by the marquises of the empire who live in Ch'ang-an, and the princesses of the realm, [as well as] the lands owned by *Kuan-nei* marquises, government officials, and the people, in no case be more than three thousand mu [each]. Let vassal kings have [no more than] two hundred slaves; let the marquises and imperial princess have [no more than] one hundred; and let the marquises within the passes, officials, and the people own [no more than] thirty. After a period of three years has been completed [for the law to be put into effect], let those who disobey [these regulations] be subjected to confiscation by the government."

At the time, the price of land, houses, and slaves dropped. [The consort families], Ting and Fu, were in power in the government and Tung Hsien, [the favorite courtier] enjoyed imperial favors. They all [found the regulations] undesirable. An imperial decree instructed postponement of the regulation until a better date. Hence the matter was shelved without putting it into effect. [*HS* 24A:20a-b; cf. Swann, *Food and Money*, pp. 200-204]

6. Nationalization of Land

[Wang Mang] decreed . . . "The Han had lightened land tax on fields to a tax in kind of one-thirtieth of the produce. However, there regularly were commutations of corvée services to conscription taxes that even aged and ill persons had to pay. In addition, powerful commoners encroached upon and oppressed [the poor]; [the powerful] divided their lands, [renting plots to the poor], and plundered the leased land.[16] Hence, in name [the assessment] was one-thirtieth; yet, in fact, it was five-tenths. The rich, being haughty, acted evilly; the poor, being poverty-stricken, acted wickedly. Both have sunk into guiltiness; punishments have been in constant use. A.D. 9

"Now at this time let the term be altered and the land throughout the empire be designated 'king's fields' and slaves be designated

16. The phrase *fen-t'ien-chieh-chia* is an ambiguous one. Ho Ch'ang-ch'ün made a reasonable interpretation of *fen-t'ien* as "share of land" and *chieh-chia* as "forcing the transfer of public land leased by the poor." Cf. Ho Ch'ang-ch'ün, "Ch'in-Han-chien ke-t'i-hsiao-nung-ti hsing-ch'eng ho fa-chan," in *Han-T'ang-chien feng-chien t'u-ti-so-yu-chih hsing-shih yen-chiu*, pp. 229-301. Ho disagrees with the T'ang annotator Yen Shih-ku on the interpretation of *fen-t'ien*. Yen said that the landless poor farmer became a tenant of the rich, while Ho applied his hypothesis that each Han citizen received land from the government as his share and that the poor ones lost the title to the rich. Hence, I prefer Yen's interpretation for *fen-t'ien* and Ho's interpretation for *chieh-chia*; cf. *HS* 24A: 21b and Kuo Mo-jo, *Nu-li-chih-shih-tai*, p. 56.

'private adherents.'[17] Neither [land nor slaves] are to be sold or purchased. Those [families] whose adult males do not number up to eight, yet whose arable land amounts to more than a *ch'ing,* shall divide the surplus land and give it to their relatives of nine generations, and [to the people of their] townships and boroughs."

Violators of the [above] order according to law, could be subject to the death penalty. . . . Three years later [in A.D. 12], Wang Mang, knowing that the people were distressed, promulgated an edict to the effect that those who had an income from the "king's fields" and who had "private adherents" were permitted to buy and sell them without legal restrictions. [*HS* 24A:21a-b; cf. Swann, *Food and Money,* pp. 208-11, and Dubs, *HFHD* 3:477-80]

7. The King's Land

A.D. 9 [Wang] Mang said [in an edict]: "When the ancients established the field cottage and the *ching-[t'ien]* system [based on units of] eight families, and when a husband and a wife had one hundred mu of land, paying one-tenth as tax, the state had a sufficiency, the people were wealthy, and songs of praise came forth. This was the Way of T'ang [Yao] and Yü [Shun] and was what the Three Dynasties followed and practiced.

"The Ch'in acted contrary to the right Way. They exacted heavy taxes in order to serve themselves and exhausted the people's energy in order to gratify their desires to the utmost. They destroyed the institutions of the Sages and abolished the *ching-t'ien* system. Because of this, engrossing arose and greediness originated. The strong measured fields by the thousands while the weak had not even space whereupon to stand an awl.[18] [*HS* 99B:8a-b; cf. Dubs, *HFHD* 3:184-85]

8. Farmers' Hardship

44 B.C. [Kung Yü][19] also said, "The merchants strive for profits. North,

17. The term *private adherent* certainly implies an uplifting of the status of slave to that of retainer, who was attached to another person without losing full freedom.

18. Hu Shih made a translation of the same passage in the *Journal of Royal Asiatic Society*, North China Branch, vol. 59 (1928). Cf. I, 6.

19. Kung Yü (124-44 B.C.) held various offices before he was made grandee remonstrant in 48 B.C. While holding this position he presented a memorial to Emperor Yüan, urging him to reduce expenses. This advice was accepted in part. Later Kung Yü was promoted to the post of grandee illustrious. In 44 B.C. he was again promoted, this time to the office of grand secretary. He died shortly thereafter.

Limited though his tenure was, he presented many memorials, the more impor-

south, east, and west, they everywhere apply their wisdom and craftiness, and are able to dress with fine clothing and enjoy good food. Each year a merchant [makes] 20 percent profit [on his capital], and yet he pays no tax. In the case of the farmer, father and son are exposed in the midst of the field, escaping neither heat nor cold. They pull the weeds and harrow the soil so that they suffer from calluses and blisters on both hands and feet. After having paid a [regular] tax in kind of grain, and also having paid the hay tax, the private demands solicited by authorities of the district and the village become too heavy a burden to bear. Therefore, the people give up the fundamental profession (i.e., farming) and pursue the secondary profession (i.e., commercial activities). The farming population is less than one-half [of the total population]. The poor, though, having been granted land, still sell it cheaply in order to become traders. When their capital is exhausted then they become thieves and robbers. Why? It is in the secondary profession that one makes handsome profits and yet it is money that makes one become bewildered." [*HS* 72:13a-b]

9. Fundamental and Secondary Occupations

The emperor said, "Agriculture is the foundation of the empire. Let a Ceremonial Field be established. We Ourselves will lead in cultivating the land to provide the grain for the offerings in the ancestral temple." [*HS* 4:9b-10a; cf. Dubs, *HFHD* 1:242] **178 B.C.**

10. Encouragement of Agriculture

An edict was issued [by Emperor Wen] saying, "We Ourselves lead the empire in cultivating the land to provide grain for sacrifice, and the empress devotes herself to mulberry culture to prepare the sacrificial robes. Let the rites and ceremonies for these be formulated." [*HS* 4:14b; cf. Dubs, *HFHD* 1:255] **167 B.C.**

11. Ceremonial Field

In the third month the Emperor Wu plowed [the Ceremonial Field] at Chü-ting.[20] [*HS* 6:38a; cf. Dubs, *HFHD* 2:117] **89 B.C.**

---

tant among them dealing with such problems as the lowering of the taxable age, the promotion of agriculture, and the abolition of metal currency. Cf. II, 34; III, 8, 15.

20. Chü-ting was a prefecture in Ch'i commandery. This is a ritualistic practice.

12. Ceremonial Field

86 B.C. The Emperor [Chao] plowed [the Ceremonial Field] in the Amusement Fields of the intendant of [Imperial Palace] parks.[21] [*HS* 7: 2a; cf. Dubs, *HFHD* 2:154]

13. Ceremonial Field

81 B.C. The Emperor [Chao] plowed [the Ceremonial Field] in Shang-lin Park. [*HS* 7:5a; cf. Dubs, *HFHD* 2:160]

14. Imperial Tour of Inspection

A.D. 14 An amnesty was granted to the empire. [Wang] Mang said [in an edict], "I shall perform the rites of a tour of inspection. . . .

"On my eastern tour of inspection I must personally carry a plow and plow the land in every prefecture to encourage the spring activities.[22] On my southern tour of inspection I must personally carry a hoe and weed in every prefecture to encourage the summer ripening.[23] On my western tour of inspection I must personally carry a sickle and reap the crop in every prefecture to encourage the autumn harvest.[24] On my northern tour of inspection I must personally carry a flail and thresh the grain in every prefecture to encourage its storage. When I have completed the rites of the northern tour of inspection, I shall go to the center of the land and dwell in the capital of Lo-yang. If any persons dare to run and make noise, violating the law, they shall be immediately dealt with according to military law." [*HS* 99B: 22a-b; cf. Dubs, *HFHD* 3:333-34]

15. Ceremonial Field

A.D. 61 An edict was issued [by Emperor Ming] saying: "We ourselves plow the Ceremonial Field to pray for good harvests." [*HHS* 2:8b]

21. Ying Shao notes that at this time the emperor was only nine years old and did not perform this rite in person.

22. In ancient times the east was correlated with the spring season (*HS* 21A: 13a). According to K'ung An-kuo, the year begins in the east, and the plowing is the first matter to be attended to; that is what is called *tung-tso*, "the eastern work" (*Shang-shu chu-shu*, SPPY ed., 2:6a, hereafter cited as *SSCS*). Ying Shao and Yen confirm this (*HS* 10: 8a).

23. Literally, "the south ripening," *nan-wei*. The south was correlated with the summer season (cf. *HS* 21A: 21b).

24. Literally, "the west maturity," *hsi-ch'eng*. The west was correlated with the autumn season (*HS* 21A: 21b).

16. Ceremonial Field

The Emperor [Ming] personally plowed the Ceremonial Field. After the ceremony, [the emperor] bestowed food on the spectators in the ceremony. [*HHS* 2:13a] A.D. 69

17. Land Tax Reduced by Half

An edict was issued [by Emperor Wen] stating: "The way to lead the people is to stress the fundamental [occupation]. We Ourselves have led the empire in farming for ten years down to the present. The land under cultivation has, however, not been extended; if for a single year the harvest fails, the people will bear the marks of hunger. This is because the people who are engaged therein are still few, and the officials have not increased their efforts to encourage farming. We have decreed repeatedly that each year people are to be encouraged to cultivate and plant, but achievements have not been forthcoming. This is because the officials do not carry out Our Decrees sincerely and do not encourage people explicitly. Moreover, our farmers do [live] hard lives; yet the officials do not care. How, then, is the encouragement possible? Let the farmers be granted one-half of this year's land tax." [*HS* 4:14a; cf. Dubs, *HFHD* 1:253] 168 B.C.

18. Tax Abolished

An edict was issued, saying, "Agriculture is the foundation of this empire. No task is more urgent. Now [the farmers] are working hard at it, yet they have to pay taxes. This makes no distinction between the fundamental and secondary [occupations].[25] This is not a sound way of encouraging agriculture. Let the tax be abolished." [*HS* 4:14b; cf. Dubs, *HFHD* 1:225] 167 B.C.

19. Inquiry on Food Shortage

An edict [by Emperor Wen] said: "When the [cultivated] land is measured, it has not decreased; then the population is counted, it has not increased. The [arable] land per capita is [actually] greater than in the ancient times, but the food is very deficient. Where does the blame for it lie? Is it that the people who pursue the secondary [occupations] and thereby injure farming are numerous? Is it that those who produce wine and thereby waste the grain are too many? 163 B.C.

25. This statement suggests that, at this time, taxes were not levied on those engaged in commercial activities.

Is it that what is used for feed for the six [kinds of] domestic animals is enormous? We are not able to attain the essential answer. Let it be discussed by the chancellor, the marquises, the officials of 2,000 piculs, and the erudites. If there is anything that is of assistance to Our people, let them assert all their minds and their deepest thoughts on the matter and not conceal anything."[26] [*HS* 4:16b-17a; cf. Dubs, *HFHD* 1:261-62]

20. Promotion of Agriculture and Sericulture

142 B.C. An edict [by Emperor Ching] said, "Carved ornaments and chiseled engravings injure agriculture. Brocade, embroidery, vermilion silk ribbons, and braided ribbons harm women's work.[27] Injury to agriculture is the source of hunger; harm to women's work is the cause of [suffering from] cold. Verily when hunger and cold both come at the same time, there are few who will be able to keep from doing wrong. We Ourselves plow and the empress herself cultivates mulberry culture to lead the empire by furnishing the sacrificial grain and the sacrificial robes from the [imperial] ancestral temples. [We] have not accepted the [yearly] tribute;[28] [we] have reduced [the supplies to] the grand provisioner (*t'ai-kuan*); [we] have lessened labor service and taxes, wishing that the empire should stress agriculture and sericulture and to be full of an accumulation of provisions in order to be prepared for visitations and calamities." [*HS* 5:9a-b; cf. Dubs, *HFHD* 1:328]

21. The Foundation of the Empire

141 B.C. An edict was issued [by Emperor Ching], stating: "Agriculture is the foundation of the empire. As for gold, pearls, and jade, they cannot be eaten in time of hunger, and cannot be worn in time of cold. When used as currency, no one knows their precise value. Recently the harvest has sometimes failed. In our opinion, this is because the people who pursue the secondary [occupations] are numerous, whereas the farmers are few." [*HS* 5:10a; cf. Dubs, *HFHD* 1:331]

22. Diligent Farmers Rewarded

191 B.C. Those among the people who were filially pious, fraternally respect-

26. This is, in fact, an edict issued after consecutive years of poor crops.

27. Women's work was especially raising silkworms, weaving, and sewing. Raising food was men's work; making cloth and clothes was women's work. For further information, see Dubs, *HFHD* 1: 328, n. 9.4.

28. "Tribute" refers to income from commanderies and nobles; ibid., n. 9.6.

ful, and diligent cultivators of the land were recommended and exempted [from taxes and labor service].[29] [*HS* 2:5a; cf. Dubs, *HFHD* 1:182]

23. Quota for Diligent Farmers

An edict [by Emperor Wen] also said: "Both filial piety and fraternal love are [examples of] great submissiveness in the world. Diligent farming is the foundation of livelihood. The thrice venerable are master instructors for the people. Incorrupt officials are models for the people. We indeed appreciate very much the conduct of those gentlemen. Now it is said that there is no such person in a county of ten thousand to meet the standard of recommendation. Does this reflect the reality of human nature? No. It is due to the inefficiency of officials [who fail] to fulfill the responsibility of recommending the virtuous ones. Let internuncios be dispatched to bestow upon the thrice venerable and the filial each five pieces of silk, upon the fraternally respectful and the diligent cultivators each two pieces of silk, and upon the incorrupt officials whose ranks are of two hundred piculs of grain and above each three pieces of silk per each one hundred piculs of grain of salary rank. Furthermore, let [the internuncios] ask the people what they find inconvenient and dissatisfying. Let there be a regular per capita quota for designating the thrice venerable, the filially pious, the fraternally respectful, and the diligent cultivators; each of them is commanded to apply himself with all his mind in guiding the people." [*HS* 4:14a-b; cf. Dubs, *HFHD* 1:254-55] 168 B.C.

24. Defect in Promotion of Farming

An edict was issued [by Emperor Ch'eng] stating: "The 'Great Rules' [of the *Book of Documents*][30] lists, in its eight administrative measures, food supplies as the first. This is the fundamental measure for making every family self-sufficient and for doing away with punishments. The late emperor encouraged the farmers, lightened their land and poll taxes, favored those who were diligent and industrious, and ordered that they be given the same treatment as the filially pious and fraternally respectful.[31] 21 B.C.

29. The recommendation of the people who were diligent in cultivating the fields was at first a measure to encourage farming. Later on, it became one of the ways of selecting candidates for offices in the central government.

30. "Great Rules" (Hung-fan) is a chapter in the *Book of Documents*.

31. The "filially pious and fraternally respectful" were one of the main categories for recommendation of candidates for office.

"Recently the people have become more and more indolent. Those who incline toward the fundamental [occupation] are few and those who go towards the secondary [occupations] are many. What should we do to correct this?" [*HS* 10:8a; cf. Dubs, *HFHD* 2:392]

25. Imperial Land Granted for Cultivation

205 B.C. [It is decreed that] people are permitted to cultivate the land that was formerly the imperial zoos, parks, gardens, and ponds. [*HS* 1A: 30b; cf. Dubs, *HFHD* 1:74]

26. Titles Granted to People

205 B.C. [It is decreed that] people should establish altars for Earth and Millet of the Han to replace those for the Ch'in. Grace was bestowed; the people were granted noble ranks. [*HS* 1A:30b; cf. Dubs, *HFHD* 1:74]

27. Migration to Defend Frontier

178 B.C. [Ch'ao Ts'o memorialized], "Your Majesty has been concerned with the border problem. Officers are dispatched and soldiers are mobilized at stations along the border line. It is indeed your great consideration. Nevertheless, armies from distant places are sent to guard the border for one year after which they are to be replaced [by units]. The soldiers therefore are not able to be acquainted with the capabilities of the Hsiung-nu (Hu). It is better to select permanent settlers who would reside and cultivate there to defend the border. Let there be high walls and deep moats, equipped with catapults and caltrops. Furthermore, let there be fortified cities within [the wall] while forts will be separated from each other by a distance of one hundred and fifty paces.[32] At strategic spots and localities near water supplies, let there be built cities of no less than one thousand households surrounded by fortifications. Dwellings should be completed and implements be provided. Then convicts and those who take the penalty of labor work should be recruited to settle down there. If the number is not sufficient, then recruit slaves and female slaves [who are given by their owners] for payments as fines or as the price for purchasing ranks. If the number still is not sufficient, then recruit ordinary people who are willing to go. They should be granted honorary ranks and their households be given exemption from corvée duty. Winter and summer clothing and food should be provided

32. These forts were distributed along the Great Wall.

until [the settlers] are able to support themselves. People should be allowed to purchase ranks up to that of a *ch'ing*.[33] As to those who have no spouse, the government should purchase one for them, because by nature human individuals are not able to settle down without a companion. The population along the frontier will not stay long in this area of hazard and danger if the benefits are not good. When the Hsiung-nu enter and drive [captives and animals] before them, those who are able to stop those who are being driven away should be granted one half of the value [of the captives and animals]. The government should pay ransoms for the people. When the situation is like this, then towns and villages will come to the aid of each other and in going against the Hsiung-nu they will not fear death. It is not that the people will be trying to establish virtue in the eyes of their superiors, rather they want to provide security for their relatives and benefit by the material incentives." [*HS* 49:14a-15a]

28. Planned Migration to the Frontier

Ch'ao Ts'o again memorialized, "Your subject learned that in antiquity when people were moved to remote regions to fill up the extensive and unoccupied land, [the government] determined the harmony of yin and yang, tasted the flavor of the rivers and springs, examined the quality of the soil, and observed the natural resources of the plants and trees. Afterwards, it built a town and erected a city wall, it created wards, and divided off areas for residences. It opened up roads leading to the fields and rectified borders separating plots. First of all, it built the residences, [each] consisting of one central hall, two bedrooms, an outer quarter, and an inner quarter, and each was provided with implements and things. When the people arrived, they had a dwelling place and when they worked they had what they would use. This is why the people would easily leave their own home regions and be persuaded to emigrate to a new town. It installed physicians and sorcerors to cure illnesses and to take care of worshipping. Men and women had mates [arranged for them]. Only when one is able to take care of the living as well as the dead, when one has [the family] cemetery, when one [can see] what one plants grow and accumulate, and when [one enjoys] a residence which

Ca. 165 B.C.

33. *Ch'ing* is not mentioned as one among the twenty honorific titles of military ranks in the early Han period. Nevertheless, *ch'ing* was customarily regarded as the rank above that of *ta-fu*. This rank, therefore, ought to be one of the upper echelon in the Han hierarchy.

is safe and sturdy, is one willing to take up permanent residence there." [*HS* 49:15b-16a]

29. Migration Permitted

156 B.C. An imperial edict [of Emperor Ching] said: "For the past successive years there have been no good harvests, so that many people are lacking food: early death is cutting short their natural [span of] years. We are very much pained at this [situation]. In some of the commanderies and kingdoms there are stony and narrow [regions] unsuitable for agriculture or sericulture or the feeding and rearing [of domestic animals], [whereas] in others the land is fertile and broad, fodder and grasses are abundant, rivers and springs are advantageous; yet the people have not been allowed to migrate [to these places]. Let it be discussed that the people who wish to migrate to [such] broad and large regions may be permitted to do so." [*HS* 5:2b-3a; cf. Dubs, *HFHD* 1:309-10]

30. Migration for Famine Relief

205 B.C. In Kuan-chung[34] there was a great famine; a *hu*[35] of rice cost ten thousand cash [and] people ate other people. [The King of Han] ordered the people to go to Shu and Han[36] to eat. [*HS* 1A:34b; cf. Dubs, *HFHD* 1:82]

31. Massive Migration

119 B.C. Many of the people east of the mountains who suffered from floods were so destitute that they were starving. Whereupon the Emperor Wu sent messengers to empty the granaries and storehouses of the commanderies and kingdoms to aid the poor. [The food, however] was still not sufficient. Therefore the [local] wealthy were solicited to provide loans, which were still not sufficient to render needed aid.[37] [The government hence] moved more than seven hundred thousand of the poor to the west of the [Han-ku] Pass in order to

34. The capital area of the Ch'in and Han dynasties.

35. *Hu* is a unit of cubic measure.

36. Both are in the present province of Szechwan.

37. Swann renders the passage to mean that the rich were called upon to provide loans. In other words, the loans were directly made between the rich and the poor. The text, however, does not give such an implication. On the contrary, the text following this paragraph tells us that the rich took the opportunity to make profits. It is highly unlikely that they would respond voluntarily to the plea of the government to provide loans.

fill up the region south of So-fang, an area known as Hsin Ch'in.[38] [The migrants] depended upon the government for food and clothing. For several years property was provided as loans.[39] [Imperial] agents were given the responsibility of supervising [the migrants who] were organized into divisions. Agents [of various missions] followed one after another on the road. The cost was hundreds of thousands of cash and the national treasury was greatly depleted. [*HS* 24B:10a-b; cf. Swann, *Food and Money*, pp. 263-65 and Dubs, *HFHD* 2:63-65, n. 15.8]

32. The Emperor's Private Treasury

First century B.C.

In ancient times the famous mountains and great marshes were not given as fiefs to be used for the exclusive benefit of subjects. The profits of the mountains and the seas and the products of the broad marshes were the stored wealth of the empire; all ought to belong to the Privy Treasury.[40] Your Majesty has unselfishly assigned them to the grand minister of agriculture to assist in relieving the people.[41] [*YTL* 1:13a; cf. E. M. Gale, *Discourses on Salt and Iron*, p. 34]

33. The Emperor's Privy Treasury

No date

The Privy Treasury is in charge of the tax from the mountains, marshes, reservoirs, and ponds; [such income] is called palace money.

38. *HS* 6: 15a-b names five commanderies. One of them is a southern commandery covering an extensive area south of the Yangtse River. The other four all lay along the great bend of the Yellow River and coincided with the description of both south of So-fang and Hsin Ch'in.

39. Swann interpreted this phrase as loans with interest. It is more likely that the government granted the migrants open, uncultivated land, houses, and necessary instruments without extracting taxes for five years.

40. The revenue derived from "salt and iron" should be paid into the *shao-fu*, the personal treasury of the emperor, and not into the *ta-ssu-nung*, the Public Treasury, a title established in 104 B.C. (translator's note by E. M. Gale, *Discourses on Salt and Iron*, p. 34, n. 4).

41. The *Yen-t'ieh-lun* was compiled by Huan K'uan and was allegedly the record of a debate in 81 B.C. on the state monopoly of salt and iron. Huan K'uan (73-49 B.C.), living one generation after the great debate took place, probably employed a common style of dialogue to develop his own Confucian ideology. Nevertheless, the debate, which had been participated in by a good number of scholars from various commanderies, could still be remembered as a great event even one generation later. Huan K'uan perhaps did have some firsthand records to develop. Our translation is from the SPPY edition. A partial translation by Esson M. Gale and others is available.

It provides for the emperor's private expenses. [It] is stored separately. What is meant by privy is small. Therefore the treasurer is named privy (*shao,* literally, small, minor) treasurer; his rank is fully two thousand *shih*. Great expenses are paid from the office of the grand minister of agriculture. Small expenses are paid from the Privy Treasury. Hence, it is called the small storehouse.[42]

By privy it implies small. Therefore, the small [treasury] can be called the Privy Treasury. The ruler uses the regular land tax for public expenditures and the taxes from mountains, marshes, reservoirs, and ponds provide for the ruler's personal expenditures. In antiquity [the latter office] was called "small treasury."[43]

The land tax and the hay tax are spent on the regular uses in famine years. The taxes from mountains, marshes, fish, salt, and markets are for the Privy Treasury for private use.[44]

34. Horse Breeding

153-137 B.C. For the first time there were established parks for [breeding] horses in order to extend the use [of horses]. Palaces, residences, rows of halls, carriages, and horses were more than ever increased and improved. [*HS* 24A:15a; cf. Swann, *Food and Money,* p. 172]

35. Confiscated Land Cultivated

117 B.C. The wealth and possessions of people thus acquired[45] was figured by the hundreds of thousands [of cash], male and female slaves were counted by the thousands and ten thousands, while cultivated fields amounted to several hundred *ch'ing* in large prefectures and to more than one *ch'ing* in small prefectures. Dwellings were in the same proportion. As a consequence, traders and merchants, whose fortunes placed them among the well-to-do and above, for the most part were ruined. . . .

The chief commandant of waters and parks *(shui-heng),* the Privy Treasury *(shao-fu),* the Grand Ministry of Agriculture *(ta-ssu-nung),* and the grand keeper of equipages *(t'ai-p'u),* each set up offices for

42. The original source of this passage is not indicated; it is preserved in the Sui encyclopedia, *Pei T'ang Shu Ch'ao*.

43. This is quoted from an annotation to *HHC* 26:3a.

44. Ibid.

45. As a result of the investigation throughout the empire of the amount of property on which the *min* tax was assessed. According to Tsan's commentary, the *min* tax was directed against merchants and artisans and not against income derived from land (*SC* 30: 14b).

agriculture,[46] [officials from which] went frequently to the recently confiscated lands in the commanderies and prefectures and had them cultivated.[47] [*SC* 30:33-34; *HS* 24B:16a-b; cf. Swann, *Food and Money*, pp. 295-99]

36. Commandery Fields

[Huang Hsiang] was transferred to be grand administrator of Wei commandery. Formerly the commandery had possessed fields and lands that were rented to tenants. The annual income therefrom was several thousand *hu* of grain. Hsiang said: "According to the ordinance concerning fields, traders are not permitted to engage in agriculture, and according the the "Wang-chih"[48] officials do not till the land. Those who could store up ice [in the winter for consumption in the summer] and who live on emoluments should not compete with the [common] people for profit." He, hence, granted the [commandery] land to the people to farm and exacted taxes thereon. [*HHS* 80A:11a] A.D. 112

37. Rent Collected by Government Offices as Provisions

The land of the metropolitan area is less than one thousand *li* [in size], yet the expenses for military use are many. Officials from [the three] ducal ministers and the [nine] ministers downward should not request exemption. If their offices have government land then taxes are to be levied according to the salary rank of the offices concerned. Let the rent and taxes be collected by each [of the offices] itself.[49] [*CHHW* 8:4b and *Hou-Han-chih* 28:15a] A.D. 190-195

38. Horse Pastures

An edict was issued, saying, "Let the pastures for raising horses be 140 B.C.

46. The *nung-kuan*, Offices of Agriculture, had under them the cultivated land in the interior confiscated from lawbreakers, particularly from tax-evading merchants. This land seems to have been cultivated either by tenants or by hired laborers under the supervision of the government.

47. Ssu-ma Kuang places this event in the sixth year of the Yüan-shou reign period, 117 B.C. (*Tzu-chih t'ung-chien* 20:2b-3a).

48. Allegedly the regulations of the Chou period.

49. This is a quotation from the *Ch'i-chü-chu*, a daily record of Emperor Hsien, who reigned from A.D. 190 to 219. It is in the form of an edict issued just after Emperor Hsien was enthroned at Ch'ang-an, in A.D. 189.

relinquished and given to the poor." [*HS* 6:2b; cf. Dubs, *HFHD* 2:30]

39. Imperial Park Distributed among the Poor

78 B.C. The Chung-mou Park[50] was abolished and was divided up[51] among the poor. [*HS* 7:8a; cf. Dubs, *HFHD* 2:168]

40. Government Land Lent

69 B.C. Land was lent to the poor people of the commanderies and kingdoms. [*HS* 8:7a; cf. Dubs, *HFHD* 2:215]

41. Government Land Open to the Public

67 B.C. It was also decreed that the imperial water reservoirs and reserved lands[52] that were not visited by the emperor be loaned to the poor people, that the palaces and residences in the commanderies and kingdoms be not repaired or put in order, and that the vagrants who returned home be loaned government land, be loaned seed and food, and be exempted temporarily from poll tax and labor services. [*HS* 8:9a; cf. Dubs, *HFHD* 2:222]

Ca. 67 B.C. Hsien and [Ho] Yü, [Ho] Shan and [Ho] Yün,[53] who saw their power declining day after day, frequently got together, wept, and grumbled among themselves. Shan said: "Now the imperial chancellor[54] is in power and is trusted by the emperor. He has completely changed the laws proclaimed during the time of the grand general [Ho Kuang]. He is distributing government land among the poor people. He is exposing the faults of the grand general." [*HS* 68:15a]

42. Government Lands Distributed

48 B.C. The government lands and the parks in the three metropolitan

50. The imperial park in Chung-mou. Chung-mou is roughly equivalent to modern Chung-mou County, eastern Honan.

51. The word *fu,* which in Han times generally designated the poll tax, is here used in the sense of *pu* or *shih-pu,* "to distribute." See Mao Heng's (second century B.C.) and Cheng Hsüan's commentaries to the *Book of Odes (Mao-shih chu-shu* 18: 3/8a).

52. The early commentators offer a number of different interpretations for the term *ch'ih-yu.* However, the accepted one seems to be that *ch'ih* is a water reservoir and *yu* is an exclusive imperial hunting ground and garden.

53. They were relatives of the deceased powerful grand general, Ho Kuang. Hsien was his wife, Yü his son, and Shan and Yün his grandnephews.

54. This imperial chancellor was Wei Hsiang.

districts and [in the territories] under the jurisdiction of the grand minister of ceremonies,[55] and in the commanderies and kingdoms, which could be spared, were used to give aid and occupations to the poor people. Those whose property was less than a thousand cash were given and loaned seed and food.[56] [*HS* 9:2a; cf. Dubs, *HFHD* 2: 302-3]

43. LAND BELONGING TO THE PRIVY TREASURY LENT TO THE PEOPLE

Again it is decreed, "There are poor crops this year in regions east of the Pass [of Han Ku] ; many people are impoverished. Let the commanderies and the kingdoms that have been severely stricken be exempted from paying annual tax and poll tax. Let the resources of rivers, seacoasts, reservoirs, lakes, gardens, and ponds that belong to the Privy Treasury be used by the poor who also shall pay no land taxes and poll taxes." [*HS* 9:2a-2b; cf. Dubs, *HFHD* 2: 302-3] 48 B.C.

44. PARK LAND LENT

It was decreed that the imperial chariots, carriages, dogs, and horses under the Office of the Yellow Gate, the imperial parks under the office of the chief commandant of waters and parks, the lower park of I-ch'un Palace,[57] the outer ponds of the sharpshooter,[58] and the reserved bird-game grounds, the ponds and fields under the Privy Treasury be abolished and lent to the poor people. [*HS* 9:3a-b; cf. Dubs, *HFHD* 2:306] 47 B.C.

45. LAND DISTRIBUTED TO THE POOR

A decree was issued, saying, "The Five Emperors[59] and the Three Kings[60] put the wise and virtuous into office and employed the able, thus attaining the greatest peace. Yet the country at present is not 43 B.C.

55. The grand minister of ceremonies, *t'ai-ch'ang,* was in charge of the prefectures in which the imperial mausoleums were situated.

56. This order may have been issued at the request of Kung Yü, who asked Emperor Yüan (48-33 B.C.) to reduce the expenses of the imperial household in order to relieve the people's burdens.

57. The I-ch'un Palace was located southeast of Ch'ang-an.

58. The sharpshooter was an official under the Privy Treasury. The *Han-i-chu,* quoted by Ju Shun, says that this official prepared arrows and silk strings for the shooting of wild ducks and geese to be used in sacrifices.

59. They are the five legendary rulers of antiquity. Their names are variously given by different scholars.

60. The founders of the Hsia, the Shang, and the Chou dynasties.

peacefully ruled. Is this because the people are different? The blame lies in our lack of perception and of the means of recognizing the wise and virtuous men. . . . [The people] depart from the rules of proper conduct and equity and collide with the laws and punishments. Is this not indeed lamentable? From this point of view, what blame attaches to the masses? Let a general amnesty be granted to the empire. Let [the pardoned] do their utmost to improve themselves, and let each devote himself to agricultural work. Let those [among the pardoned] who have no land be lent [land], and let seed and food be loaned as in the case of the poor people." [*HS* 9:7a-b; cf., Dubs, *HFHD* 2:318-19]

46. Land Belonging to Empress's Kinsmen Distributed

6 B.C. The grand empress dowager decreed that the land belonging to the Wang clan, imperial relatives by marriage, with the exception of the tomb yards, be distributed among the poor. [*HS* 11:4a; cf. Dubs, *HFHD* 3:26]

47. Private Property Donated for Redistribution

A.D. 2 In the commanderies and kingdoms there was a great drought and [a plague of] locusts; in Ch'ing Province it was especially severe so that its common people became vagrants. The duke who gives tranquillity to the Han [dynasty, Wang Mang], the four coadjutors, the three highest ministers, the high ministers, grandees, officials, and common people, 230 persons [in all], presented their fields and residences [to relieve] the suffering and indigent people. [The land] was distributed among the poor people in accordance with their number. [*HS* 12:5b-6a; cf. Dubs, *HFHD* 3:72-73]

48. New Prefecture Made from Former Park Land

A.D. 2 Hu-ch'ih Park in An-ting [commandery] was abolished in order to create a prefecture named An-min. Government offices, public buildings, marketplaces, and residential blocks were constructed. Poor people were solicited to migrate [into this new prefecture]. During the journey of migration, prefectures en route provided [them] with food until they reached their destination where [the migrants] were granted fields, residences, and implements. Plows, oxen, seeds, and food were loaned [by the government]. [*HS* 12:5a-6a; cf. Dubs, *HFHD* 3:73]

49. Government Land to Rent

81 B.C. At present, [we the literati see] the government expanding many

national parks and enclosures, government fields, ponds, and marshes. The state nominally rents [land to peasants] but the profits go to the powerful families.[61] In the area of the three metropolitan districts, hemmed in by rivers and mountains, the land is constricted and the people numerous. The four quarters of the empire come together there. The supply of grain and fuel falls short. With the government fields being leased out in a rotating manner, mulberries and elms, vegetables and fruit are not planted, and the land is not utilized to full capacity. It is our humble opinion that other than the parks, reservations, ponds, and weirs established by the late emperor,[62] the land should be distributed back to the people, while the government of His Majesty merely levies taxes. Though rent and tax are different terms, they are identical in substance. If this is not done, the efforts of the individual farmer will be exhausted in the southern field and the efforts of the individual woman will be exhausted in working on woven hemp. When fields and land are developed and hemp textiles are produced, both the ruler and the ruled will have plenty. How would there be any deficiency and distress? [*YTL* 13:1b-2a; cf. Gale, *Discourses,* pp. 83-84]

50. GOVERNMENT LAND GRANTED

The commanderies and kingdoms were ordered to grant government land to the poor people, to each according to his degree of poverty. [*HHS* 2:11a-b] A.D. 66

51. LAND FOR THE POOR

[An edict was issued by Emperor Ming saying] "The lowland along the canal[63] shall be distributed to the poor people. Let no power- A.D. 70

61. Gale failed to notice the content that stressed the problem of intervening in the profit-taking by powerful families who seized land that was supposedly given to the farmer.

62. What is involved here is a punctuation problem. One group of scholars tends to separate this sentence into two, thereby making it difficult to understand. The present translation is based on the discussion of Hiranaka Reiji in his "Kandai kōden no 'Ka': Entetsuron Enchi-hen no kisai ni tsuite" in *Wada hakushi koki kinen tōyōshi ronso,* pp. 781 ff. For comparison, cf. Goi Naohiro, "Kandai kōden ni okeru kasaku ni tsuite," *Rekishigaku kenkyū* 220, no. 6 (1958): 11 ff.

63. The Pien Canal, also called the Pien River, branched off from the Yellow River northeast of modern Jung-yang (eastern Honan), flowing northeastward and emptying into the sea in Ch'ien-sheng commandery. It was built in A.D. 69 by combining old and newly dug canals.

ful individuals secure its benefits." [*HHS* 2:13b-14a]

52. LAND GIVEN

A.D. 76 It was decreed that the fields of the reserved land in the Shang-lin Park should be distributed to the poor people. [*HHS* 3:4a]

53. GOVERNMENT LAND GRANTED

A.D. 84 An edict was issued [by Emperor Chang] saying: "The commanderies and kingdoms are ordered to solicit those who have no land and who desire to migrate into other territories that are fertile; the local government agencies are commanded to grant government land [to those people], to assist them in hiring helpers, to lend them seed and provisions, and to rent them implements. Let the land tax be exempted for five years and the poll tax for three years. Later on, those who desire to return to their home places are not to be prohibited from doing so." [*HHS* 3:10b]

54. UNCULTIVATED LAND

A.D. 86 The Emperor [Chang] tilled [the Ceremonial Field] in Huai.[64] An imperial communication was sent to the grand administrators of the commanderies of Wei, Chü-lu, and P'ing-yüan and to the chancellors of the kingdoms of Ch'ang-shan, Ch'ing-ho and Tung-p'ing, stating: "As we were about to offer a sacrifice to Mount Ch'ang,[65] we therefore journeyed to the north. We passed Wei commandery and came through P'ing-yüan. . . . Now there is still considerable fertile land that has not yet been brought under cultivation. Let all of it be distributed to the poor people, and let them be given seed; let them devote themselves to exhausting the potentialities of the land; and let them not be idle." [*HHS* 3:15b-16a]

55. RENT-FREE LAND OPENED TO THE POOR

A.D. 93 An edict was issued [by Emperor Ho] to the responsible officials to reduce the number of horses of the inner and outer stables and of the pastures in Liang Division. The orchards of the detached palaces in the capital region and outside it, in the Shang-lin and the Kuang-

64. Huai, capital city of Ho-nei commandery, was located west of the modern city of Wu-chih in northern Honan.

65. Mount Heng was called Mount Ch'ang at the time because Heng was the personal name of Emperor Wen of Western Han and therefore taboo.

ch'eng Parks,[66] were all to be lent to the poor people, who were to be permitted to pick and hunt freely, and were not to have any rent demanded of them. [*HHS* 4:6a-b]

56. Rent Not Levied on Mountains, and so Forth

Messengers are dispatched to circulate in the commanderies and kingdoms to give food and loans to those who could not support themselves because they were suffering from misfortune and calamity. They are permitted to fish and collect firewood from mountains, forests, ponds, and swamps without being levied rent. [*HHS* 4:11a] A.D. 99

57. Park and Public Land Leased

The hunting grounds of Kuang-ch'eng Park and the government land in those commanderies and kingdoms that had suffered from calamities were leased to the poor people.[67] [*HHS* 5:3a] A.D. 107

58. Park Land Leased or Given

It was decreed that Hung Lake[68] was to be lent to the poor people. [*HHS* 5:6b] A.D. 109

It was decreed that the cultivable land of the Shang-lin and Kuang-ch'eng Parks be distributed among the poor. [*HHS* 5:6b] A.D. 109

59. Corrupt Officials Warned

An edict was issued, saying, "Now for some years there have not been good harvests and the food of the common people has been quite scanty; where does the fault lie? Perhaps dishonest and hypocritical officials take presents and bribes, stealing by fraud or by force the people's [property], encroaching upon and consuming the common people. An assistant prefect is a ranking official; [for him] to pervert the laws and rob with the robbers is utterly unspeakable. 142 B.C.

66. The Shang-lin Park of Eastern Han was situated just west of the imperial capital Lo-yang. It is to be distinguished from the park of the same name of Western Han. Kuang-ch'eng Park was situated west of Ju-chou of T'ang (modern Lin-ju County, southeast of Lo-yang).

67. A passage in *HHS* 32:11b says that Empress Dowager Teng distributed all of the public land to the poor people. Judging from the context, this was the relief measure recorded in the above passage.

68. Hung Lake was located 20 *li* east of the imperial capital.

"Let it be ordered that two-thousand-picul [officials][69] should each look after his own charge. As to those who do not occupy themselves with their official duties or who govern unintelligently, the imperial chancellor should report it and beg [the throne to order punishment] for their crimes. Let this be published and told to the empire and cause Our will to be clearly known."[*HS* 5:9a-9b; cf. Dubs, *HFHD* 1:329]

60. Punishment of Officials Neglecting Agriculture

141 B.C. An edict was issued, stating: "Let the commanderies and kingdoms be ordered to stress the promotion of farming and mulberry culture and to increase the amount sown and planted in order to obtain the materials for food and clothing. Officials who mobilize the people or commute their labor services for money in order to hire laborers to search for gold, pearls, and jade shall be sentenced for having taken illicit gain and committing theft. The two-thousand-picul [officials] who permit such actions shall be punished with the same punishment [as their guilty subordinates]." [*HS* 5:10a-b; cf. Dubs, *HFHD* 1:332]

61. Evil Officials Investigated

34 B.C. It was also decreed: "It is spring; husbandry and mulberry culture begin; and it is the time when the people exert themselves to the utmost. Therefore, in this month we encourage the farmers and exhort the people in order not to let them miss the proper season. But evil officials, in reconsidering law cases involving petty crimes and in summoning witnesses in such cases, take up matters of no urgency and interfere with the people [in their work]. They cause them to miss the task of the moment and [thereby] to lose the work of a whole year. Let the ministers carefully investigate and reprimand them." [*HS* 9:12a-b; cf. Dubs, *HFHD* 2:334]

62. Promotion of Agriculture

21 B.C. An edict was issued, stating: "Now it is the time for the spring activities. Let the officials of two thousand piculs be ordered to exhort and encourage farming and mulberry planting, crossing in and out of the boundaries of the fields, and giving consolation and encouragement." [*HS* 10:8a; cf. Dubs, *HFHD* 2:392-93]

69. Officials of the two-thousand-picul rank include grand administrators of commanderies and chancellors of kingdoms.

63. Agricultural Officials Created

Thirteen circuit assistants under the grand minister of agriculture [were set up]. Each had charge of one division[70] for the encouragement of agriculture and mulberry culture.[71] [*HS* 12:4b; cf. Dubs, *HFHD* 3:70] A.D. 1

64. Able Official Rewarded

During the reign of Emperor P'ing (r. A.D. 1-5) there was a serious plague of locusts all over China. [Although] more than twenty prefectures in the Ho-nan [commandery] all suffered this calamity, [the locusts] did not enter the territory of Mi prefecture, [the magistrate of which was Chou Mou]. . . . At this time Wang Mang was in power. [He] created six circuit assistants under the grand minister of agriculture, [whose duties were to] encourage agriculture and mulberry culture. [He] promoted [Chou] Mou to be a circuit assistant in the imperial capital. [*HHS* 25:2a] A.D. 1-5

65. Able Official Rewarded

[Ch'en Li] was transferred to the post of grand administrator of T'ien-shui commandery. He was the foremost in the empire in encouraging the people [to devote themselves] to agriculture and mulberry culture. He was rewarded with forty catties of gold. [*HS* 95:7a] *Ca.* 32-9 B.C

66. Officials Instructed to Observe Agricultural Seasons

An edict was issued, saying: "We offer sacrifice to Heaven. . . . Spring is the beginning of the year. If at the beginning good order is obtained, then the three seasons[72] will produce results. Recently there have been untimely floods and immoderate droughts; the people on the frontiers have but little food. When the administration errs above, the people suffer from its faults. Let the responsible officials exert themselves to follow the auras of the seasons, and encourage A.D. 60

70. The circuit *(pu)* and the division *(chou)* were identical administrative units.

71. Throughout Western Han edicts encouraging farming and mulberry culture were repeatedly issued. The establishment of the thirteen circuit assistants to facilitate the promotion of agricultural production was an important step in this direction.

72. Spring, summer, and autumn.

and supervise farming and mulberry culture. Let them exterminate caterpillars, locusts, and grubs. Let them be careful about punishments, circumspect about inflicting penalties, and perspicacious about arguments without adequate evidence. They shall work diligently day and night to comply with Our wishes." [*HHS* 2:7a-b]

67. Amnesty Proclaimed to Celebrate Good Crops

A.D. 67 An edict was issued saying: "In the past year the five grains ripened abundantly; and now both silkworms and the wheat have yielded a good harvest. Let a general amnesty be granted to the empire.

"Just now summer is at its prime, the time when [the grains] are growing. Let the past evils be eliminated in order to reward agricultural efforts. Let the people exert and devote themselves to mulberry culture and crops in preparation against calamities. Let the officials respect their duties and not be indolent." [*HHS* 2:11b]

68. Efficient Official Rewarded

A.D. 30 [Li Chung][73] in the sixth year [of Chien-Wu] was transferred to be grand administrator of Tan-yang commandery. At this time all within the seas had just been pacified. . . . The cultivated fields increased greatly. In three years the wanderers who settled there numbered more than fifty thousand persons.

In the fourteenth year [A.D. 38] the three highest ministers[74] reported that [he was rated] as the first in the empire in administration. [*HHS* 21:4a-b]

69. Land Reclaimed

A.D. 31 In the seventh year [of Chien-wu, Tu Shih][75] was promoted to the post of grand administrator of Nan-yang commandery. . . .

He also repaired and constructed embanked reservoirs and enlarged the amount of arable land. Every home in the commandery was well provided for. The people of his time compared him to Shao Hsin-ch'en.[76] Thus the people of Nan-yang coined the following saying:

73. Li Chung was one of the meritorious officers under Emperor Kuang-wu. He died in A.D. 43.

74. The *ta-ssu-ma*, the *ta-ssu-t'u*, and the *ta-ssu-k'ung*.

75. Tu Shih was renowned as an honest and just administrator. Cf. V, 20.

76. Shao Hsin-ch'en is another grand administrator of the same commandery; cf. IV, 33.

"Previously there was Father Shao; then there is Mother Tu." [*HHS* 31:3a-b]

70. Rules for Farming

[T'ung Hui] was promoted to the post of prefect of Pu-ch'i [prefecture].[77] Whenever the officials or people violated the law, he always showed them the proper way regarding the matter. If the officials had been competent in their duties or the people had performed good actions, he offered all of them the courtesy of a feast to encourage and stimulate them. For all cultivating, weaving, sowing, and harvesting he laid out rules. Peace and calm prevailed throughout the jurisdiction. For a number of successive years there were no prisoners in the prisons. Wanderers from the neighboring prefectures who came to him and who moved there to settle down numbered more than twenty thousand households. [*HHS* 76:16b] *Ca.* A.D. 178-79

77. T'ung Hui was promoted to this position in 178. Pu-ch'i prefecture was near the present Chi-mo, Shantung.

II

LAND AS WEALTH

1. Freedom of Trade in the Early Han

Ca. 200 B.C. When Han arose, all within the seas was united. Barriers and bridges were opened; restrictions were relaxed regarding [the use of resources of] mountains and marshes. Therefore, rich traders and great merchants roved around the empire. No commodity was not in circulation so that each one was able to obtain what he desired. [*SC* 129:16-17; cf. Nancy Lee Swann, *Food and Money in Ancient China,* p. 437 and Burton Watson, *Records of the Grand Historian of China* 2:484]

2. Regional Specialization and Interregional Dependency

Ca. 200 B.C. Now the region west of the mountain[1] is rich in timber, bamboo, the *ku* tree,[2] hemp fiber, yaktails, jade, and [other precious] stones. The region east of the mountain has plenty of fish, salt, lacquer, silk, musical [instruments] and cosmetics.[3] The region south of the Yangtse River produces cedar, catalpa, ginger, cinnamon, gold, tin, crude lead,[4] cinnabar, rhinoceros horns, tortoise shells, pearls, [elephant] tusks, and hides. The region north of Lung-men and Chieh-shih[5] has plenty of horses, cattle, sheep, felts, fur garments, tendons, and horns. Copper and iron [deposits] are available over a space of a thousand [square] *li,* as chess pieces scattered [on a board]. These are the things the Chinese like and customarily use for bedding, clothing, beverage, and food, and that people provide for the living and prepare for the dead. Therefore, it takes the farmers to feed [the people], the woodsmen[6] to bring out [products from mountains and marshes], the

1. Mount Hua, at the eastern corner of today's Shensi Province, marked the border between the eastern and western regions.

2. *Ku* is a tree of the mulberry family. Its bark was later used for pulp.

3. Watson, misled by the commentator Ch'en Tzu-lung, translated the term *sheng-she* as "singers and beautiful women."

4. Cf. Swann, *Food and Money,* p. 311, n. 675, for identification of this term.

5. Lung-men is the present province of Shansi, and Chieh-shih is in Hopei. The region referred to here is the area along the Great Wall as well as the steppe beyond it.

6. *Yü,* the woodsmen, actually meant the watchmen for the forests. The text here, however, implies all the professions involved in bringing out products from mountains and marshes. Cf. Swann, *Food and Money,* p. 421.

craftsmen to manufacture, and the merchants to make [these articles] circulate. [*SC* 129:4-5; cf. Swann, *Food and Money*, pp. 420-21 and Watson, *Grand Historian* 2:477]

3. Large-Scale Enterprises

Therefore, it is said that [if one has] on dry land 50 horses, or 166 head of cattle, or 250 sheep; on marsh land, 250 swine; in ponds or reservoirs, 1,000 piculs of fish; on hilly land, 1,000 pieces of timber; [or that if one possesses] in An-i 1,000 date trees; in Yen and Ch'in, 1,000 chestnut trees; in Shu, Han, and Chiang-ling, 1,000 orange trees; in the region north of Huai River, south of Ch'ang-shan between the Yellow and Chi rivers, 1,000 catalpa trees; in Ch'en and Hsia, 1,000 mu of lacquer trees; in Ch'i and Lu, 1,000 mu of mulberry trees and hemp; along the Wei River, 1,000 mu of bamboo groves; or that [if one owns] in the suburbs of a large city of 10,000 households in a famous kingdom, 1,000 mu of land [which yield] one *chung* per mu,[7] or 1,000 mu of gardenia or madder plants[8] for yellow and red dyes, or 1,000 rows[9] of ginger and leeks, the owner [of any of these above-mentioned items] has a wealth equivalent to that of an enfeoffed lord [who has the revenue collected from] 1,000 households [of tax-paying subjects]. [*SC* 129:31-32; cf. Swann, *Food and Money*, pp. 432-33; Watson, *Grand Historian* 2:492-93] *Ca.* 200 B.C.

4. Discrimination Against the Merchants

Because the currency of the Ch'in was heavy and difficult for practical use, it was ordered that the people could cast a new denomination of coins.[10] Each unit of gold was [fixed at] one catty.[11] With laws simplified and prohibitions reduced, people who were pursuing profit by unscrupulous means accumulated and hoarded a surplus of commodities; because market goods were hoarded, the prices rose 200 B.C.

7. One *chung* is roughly 3.616 U.S. bushels (Swann, *Food and Money*, p. 360). One *chung* of grain per mu was an extremely high yield in the Han period.

8. The seed of the gardenia produces a yellow dye and the petals of madder yield a red pigment.

9. According to the commentator Hsü Kuang, 1,000 rows is 25 mu. Yet another commentator, Liu Hsi, suggested that the large *hsi* was 50 mu (*SC* 129:32).

10. According to *HS* 24B:3a, the new coin was called elm-pod, a comment on its light weight.

11. The annotator Tsan said that the weight of one unit of gold in Ch'in was one *i*, which consisted of 20 taels, and the Han unit was one catty, or 16 taels. The change of the unit of gold, therefore, means a devaluation of 20 percent since the coin value was regulated as ten thousand cash for one unit of gold. Cf. *SC* 30: 4, and Swann, *Food and Money*, p. 364.

by leaps and bounds. The grain price went to ten thousand cash per picul, and a horse [cost] one hundred units of gold. After peace had been restored to the empire, Kao-tsu issued an edict forbidding merchants to wear silk or to ride in carriages, and increased the [business] taxes to hamper and humiliate them. During the reigns of Emperor Hui (194-188 B.C.) and Empress Lü (187-180 B.C.),[12] because the empire was only recently recovered [from the civil wars], the laws pertaining to the merchants were again relaxed. The sons and grandsons of the merchants who traded in the market,[13] however, were still banned from serving in the government either as an official or as a staff member. [*SC* 30:34; cf. Watson, *Grand Historian* 2:79-80]

5. Salt and Iron Monopolies

117 B.C. The minister of agriculture[14] presented to the throne a proposal made by K'ung Chin and [Tung-kuo] Hsien-yang,[15] the assistants in charge of salt and iron controls, that said, "The mountains and the seas[16] are storehouses of heaven and earth. The [control of] resources thereof ought to belong to the Privy Treasury.[17] Your Majesty, being unselfish, has transferred the authority to the Ministry of Agriculture to supplement revenues from [regular taxes]. It is proposed that [the government] summon people who would furnish their own capital and undertake salt-evaporation through the use of government-owned equipment. To them the government shall furnish evaporating pans on a rental basis. The parasitic and irregular people[18] would control the resources of mountains and seas in order to gain wealth. They have belabored and exploited the poor people. Deliberations

12. Emperor Hui was the eldest son of Kao-tsu. After Hui passed away, his mother, the wife of Kao-tsu, ruled as empress.

13. The term *shih-ching*, "market-wells" means simply "marketplaces," a survival of the ancient custom when fairs were held around public wells. Cf. Swann, *Food and Money*, p. 231, n. 395.

14. The minister of agriculture was actually the state treasurer in charge of finance.

15. K'ung Chin and Tung-kuo Hsien-yang were wealthy iron and salt merchants, respectively. Cf. *HS* 24B: 12a.

16. Mountains and seas refer here to natural resources such as iron and salt.

17. The Privy Treasury, *shao-fu*, was in charge of the private revenues of the imperial house; see I, 33.

18. Gale translated this phrase in a very similar passage in the *YTL* as "Ne'er-do-wells and upstarts," where he based himself on the translation of *SC* by Edouard Chavannes (E. M. Gale, *Discourses on Salt and Iron*, pp. 34-35, n. 1). The *YTL* commentator, however, said that *ch'i-min* means vicious people.

regarding their efforts to nullify state policy have been numerous." [*HS* 24B:12b-13a; *SC* 30:22-23; cf. Swann, *Food and Money*, pp. 275-77, and Watson, *Grand Historian* 2:90]

6. Rich Persons as Potential Challengers to Political Authority

In the past the great families, aggressive and powerful, were able to control the profits from the resources in the mountains and the seas. They mined iron ore and, using bellows, smelted it; they also manufactured salt by evaporating sea water. One such family sometimes gathered a host of more than a thousand men, for the most part vagrants who had drifted far away from their native villages, abandoned their own family cemeteries, and adhered to a great family. Gathering in the midst of deep mountain areas and extensive marshes, they carry out their illicit occupations and bring to completion the power of an [armed] band. Great, indeed, is their willingness to look lightly upon doing evil.[19] [*YTL* 1:13b; cf. Gale, *Discourses*, p. 35] 81 B.C.

7. Preventing the Emergence of Powerful Groups

The imperial secretary said: "When otters appear in the water, the fish in the pond are disturbed. With powerful groups in the state, the common people will be ruined. Thus there cannot be luxuriant herbage beneath a flourishing forest; there cannot be good shoots between large clumps of earth. The principle of governing a state consists of removing the noxious and hoeing out the powerful. Only then will the people enjoy equal treatment and find satisfaction under their roofs. The commandant of justice, Chang [T'ang][20] codified the laws and statutes, published them to give a common standard to the empire, executed the evil and the crafty, and exterminated the ones who [were profiting] by land engrossment [by means of seizing land from the weak]. As a consequence, the strong could not take advantage of the weak, and the many could not ill-treat the few, the grandee[21] has busied himself with statistical calculations to increase the state revenue. The various profits from salt and iron are monopolized in order to suppress rich traders and big merchants." [*YTL* 3:3a; cf. Gale, *Discourses*, p. 87] 81 B.C.

19. This passage is preceded by a paragraph very similar to the one in II, 5. Both are obviously from the same source.

20. Chang T'ang who served as commandant of justice during 126-120 B.C.

21. The word *ke* is superfluous.

8. Property Taxes

120 B.C. [The courtiers proposed]: "At an earlier date[22] there had been levied a tax on small [animal-drawn] carts and on strings of cash possessed by tradespeople proportionately. We propose that a tax be levied as was done in the past. The various tradesmen and others who engage in secondary occupations,[23] those who make interest on loans, those who sell and buy, those who manage storages for commodities in towns, and those who in any entrepreneurial role make profits, even though they are not on the market registries, shall each, according to his possessions, make his calculations for a report. The tax rate, in terms of strings of cash, is at the rate of one *suan*[24] on each two thousand cash. For those among the craftsmen who have paid tax [or rent], as well as founders of metal, the rate should be one *suan* for each four thousand cash. Those who have not the privileges of government offices, and those who are neither local elders,[25] nor cavalrymen of the northern frontier regions, [should pay a tax of] one *suan* on each small cart. [The tax for] boats of five *chang*[26] or more is one *suan* on each. One who fails to report his evaluation or makes a false report should be sent to a frontier post for one year garrison duty; meanwhile, his cash property should be confiscated. Those who are able to accuse [another of violating these regulations] should be given half [of the confiscated property].

"In order to benefit the farmers, neither the merchants who are listed in the market registers nor their household members should be permitted to register title to land. Anyone daring to violate this ordinance should have his land and slaves confiscated."[27] [*SC* 30: 24-25; *HS* 24B:13a-b; cf. Watson, *Grand Historian* 2:91-92, and Swann, *Food and Money*, pp. 279-82]

9. Accusations of Wealthy Persons

117 B.C. Yang K'o's[28] accusations [of violations] of the property tax

22. The date is 130 B.C. (*HS* 6:7a).

23. By "secondary occupations" is meant commerce and industry.

24. One *suan* is 120 cash; cf. Swann, *Food and Money*, p. 281, n. 531.

25. The local elder, *san-lao*, was appointed local petty officer whose responsibility was "giving instruction" to the people. Cf. *HS* 19A: 30a.

26. One *chang* is equivalent to ten *ch'ih* or 7′ 6.94″, Swann, *Food and Money*, p. 362.

27. In the *Han-shu* version of the same edict, the possessions to be confiscated are "fields" and "money." Cf. *HS* 24B: 13b (Swann, *Food and Money*, pp. 282-83). There is no way to determine which of the two is the original one.

28. The person who initiated the widespread accusations.

law spread throughout the empire. The families of medium wealth and above practically all suffered arraignment. Tu Chou[29] tried them and very few of the charges were reversed. Hence, [the throne] dispatched separately the supervisors of the imperial secretary and the commandant of justice; they went with their respective staffs[30] to commanderies and kingdoms to hear the cases of accusations [of false reports] of property. The wealth and possessions of the people thus acquired was figured by the hundreds of thousands, male and female slaves were counted by the thousands and ten thousands, while cultivated fields amounted to several hundred *ch'ing* in large prefectures and to more than one hundred *ch'ing* in small prefectures. Dwellings [confiscated] were in the same proportion. As a consequence, traders and merchants whose fortunes placed them among the well-to-do and above were, for the most part, ruined.[31] [*HS* 24B:16a-b; *SC* 30:32-33; cf. Swann, *Food and Money*, pp. 294-96, and Watson, *Grand Historian* 2:97]

10. Confiscated Land Managed by Government

In the beginning, the grand minister of agriculture *(ta-ssu-nung)* administered the salt and iron offices [as well as] legal coinage and currency, [which became] too much. [So] the office of commandant of waters and parks *(shui-heng tu-wei)* was established to be in charge of salt and iron. At the time of Yang K'o's accusation[32] (in 117 B.C.), the wealth and resources [under the control of the superintendent of the imperial] park Shang-lin [park] were in large quantity. Then [it was] instructed that the commandant of water and parks be placed in charge of the Shang-lin. Shang-lin was then full. . . . [It was] instructed that [the resources acquired as a consequence of] the property tax were to be handled by several offices. [Hence] there were offices of agriculture[33] settled [within the jurisdictions of] those of the commandant of waters and parks, the Privy Treasury *(shao-fu)*, the grand keeper of equipages *(t'ai-p'u)*, and the grand 115 B.C.

29. Tu Chou was then a palace assistant *(chung-ch'eng)* who served as the prosecutor.

30. It seems that each team dispatched from either of these two judicial organs was headed by one senior official who was assisted by a staff so that the hearings could be determined right on the spot rather than being sent back to the capital.

31. Cf. I, 35.

32. Cf. II, 9.

33. These offices were branches of the central government offices and reported directly to their own main office in the capital.

minister of agriculture. Such offices were often placed in the commanderies and prefecture to [manage the] cultivation of the confiscated fields. [*HS* 24B:16b; cf. Swann, *Food and Money,* pp. 297-99]

11. Trade Taken Over by Government Offices

110 B.C. In the first year of the Yüan-feng reign period (110 B.C.) . . . Sang Hung-yang,[34] then the chief commandant for grain *(chih-su-tu-wei),*[35] took charge of the Ministry of Agriculture and completely supplanted [K'ung] Chin in controlling salt and iron production throughout the empire. Hung-yang considered that various government agencies each in trading [in the markets] competed with each other, and that the merchandise therefore [rose] by leaps and bounds, and that when some of the imperial poll taxes were transported they did not cover shipping costs. Therefore he proposed: "Several tens of regional assistants should be appointed within the Ministry of Agriculture; let them be divided so that each region has charge of commanderies and kingdoms in each of which there would be Offices of Equalization and Transportation[36] and of salt and iron. Let it be instructed that the distant places should present as poll tax those [local] goods that heretofore merchants had shipped and sold and [such goods] will then pour in. An Office of Standardization of Prices *(p'ing-chun)* would be established in the capital to receive what is shipped from throughout the empire. Let the manufacturing offices[37] be ordered to manufacture carts and related equipment, all of which shall be turned over to the Ministry of Agriculture. The various offices of the Ministry of Agriculture will control the money and merchandise of the empire. When prices are high, they will sell; when low, they will buy. In this manner rich traders and great merchants will lose the means by which they gain great profits. Thus will [the people] return to the fundamental profession,[38] and [prices of] numerous items of merchandise will not rise. Therefore, the way to suppress [the price] of merchandise of the empire is termed the *p'ing-chun* system [standardization of prices]." The emperor [regarded the idea] right and granted approval. [*HS* 24B:19b-20a; *SC* 30:41-43; cf. Swann, *Food and Money,* pp. 314-16]

34. Sang Hung-yang, the son of a merchant, was one of the main economic advisors at the court of Emperor Wu.

35. This office was subordinate to the minister of agriculture. The latter was in charge of financial affairs.

36. Sang Hung-yang had experimented with the idea of using governmental resources to trade in the market in 120 B.C. (*HS* 24B: 14a).

37. The government-operated manufacturies.

38. Agriculture.

12. Registration of Trade and Tax on Profits

All those who gathered things of any kind, [such as] birds, beasts, fish, turtles, and insects from mountains, forests, rivers, and marshes; those who engaged in husbandry; the housewives who engaged in sericulture, spinning, weaving, mending, and sewing; craftsmen, artisans, physicians, sorcerers, diviners, invokers, and other professions of special techniques; traders and merchants who sat in stores and shops in the markets, in [residential] areas, and at inns; all these were required to register their [assets] at the imperial offices of their locales. [In each case, the local officials must] subtract the capital and divide the profit into eleven parts, one of which was to be presented [to the government] as a tributary tax. One who dared refuse to register or who registered untruthfully should have what he gained entirely confiscated and should work as a convict laborer for one year for the government. A.D. 10

The several masters of markets *(ssu-shih)* regularly in the middle month of each of the four seasons took the actual price [as standard] of the things in their charges and fixed the price into [three grades], high, medium, and low. Each [market] determined its own standard without being restrained by other places. In transactions of the five grains, textiles, clothes, and silk that were common for daily use, if there was any that was not sold, the [market] agency should investigate its real value and purchase thereof by paying the cost so that the seller would not lose money. Whenever the price of the merchandise was higher by one cash[39] than the standard price, [the market agency] should sell [that particular article] to the people at the standard price. If the price [of certain articles] were normal, people should be allowed to trade by their own agreement. This was for the sake of preventing hoarding. [*HS* 24B:24a-b; cf. Swann, *Food and Money*, pp. 339-41]

13. Government Control of Trade

[It is decreed] . . . "Now salt is the leader of foods and delicious viands. Wine is the senior of the numerous medicines and the delight of felicitous gatherings. Iron is the fundamental in farming. Famous mountains and great marshes are depositories of rich natural resources. The credit and loan system [operated] by the Market Control Office is looked to by the people to provide them with low-interest or no- A.D. 17

39. The real implication of "one cash" here is not clear since the unit prices of merchandise are not the same.

interest loans in times of want.[40] Cash coins[41] that are cast from copper are circulated for the people to use as [transaction] media. These six[42] are not what the common people are able to do in their own homes; they must look into the market to acquire them. Even though the prices demanded are several times more expensive [than they are worth], [the people] have no alternative but to pay them. Wealthy people and rich merchants, therefore, [are able to] exploit the poor and the weak. The Ancient Sages knew that this was the case. Hence, they controlled these [six activities]."

For each of these [six] controls, there was formulated a series of articles in law of preventive and prohibitory measures. Those who violated [the regulations] were punished, even including death. The vicious officials and the treacherous persons all encroached on the common folk, who were not able to have a stable livelihood. [*HS* 24B:25b-26a; cf. Swann, *Food and Money*, pp. 347-49]

14. Merchant Investment in Land

Ca. A.D. 26-30 [Huan T'an memorialized[43] . . .] "The way of ruling a state should be to promote the fundamental occupation and suppress the secondary ones. Therefore, the preceding emperors forbade individuals from having two occupations and excluded the merchants from serving as officials. By such means, they [the earlier emperors] suppressed land engrossment and cultivated moral conscience. Now many rich merchants and wealthy businessmen are [profited by] leasing land [for rent] and lending money [for interest].[44] Youths of medium rank families, being the guarantors, serve them as diligently as if being their subjects and servants. The incomes and rents [the rich collect]

40. In A.D. 10, market control was instituted to regulate prices. From revenues from such maneuvering as well as tributary tax, the officials made low-interest or contingent loans without interest available to the people. Cf. II, 12. In Swann's translation these two related programs are separated as though independent from each other. Cf. Swann, *Food and Money*, p. 348.

41. The text gives "iron" *(t'ieh)*, which, according to a Ch'ing annotator, should be "cash" *(ch'ien)*; *HS* 24B:26a and Swann, *Food and Money*, p. 348, n. 719. Swann mistakenly translates *pu*, a denomination of copper coins, as cloth money, ibid.

42. The six categories consist of salt, liquor, iron, natural resources in mountains and marshes, the credit system, and coinage.

43. Huan T'an presented this memorial to Emperor Kuang-wu as a part of a general discussion of state affairs shortly after the latter was enthroned.

44. One version contained no mention of "land," yet the same passage quoted in the *Tung-kuan Han-chi* includes both land and money as does the text here (*HHS* 28A: 3a).

are comparable to the income drawn by a lord from his fief. Therefore, the people admire and imitate [their practice] of making a living without farming. There are hence extravagant expenses for sensual pleasures. Now the merchants should be ordered to investigate one another and to report [to the officials]: if there is anyone who has gains not by his own efforts, then all the accumulations should be bestowed on the informer. If this is done, then the merchant will work alone, only by himself, and will not dare to loan money to others [to invest]. When his business is scanty and his [economic] strength is weak, he is bound to return to till the land. When the fields are well taken care of, the harvest of grain will be abundant and the potential of land will be well used." [*HHS* 28A:3a-b]

15. LAND ACQUISITION BY THE WEALTHY AND THE POWERFUL

The king of Chao then made [Chao] Kuo replace Lien P'o.[45] . . . 257 B.C.
At the time [Chao] Kuo was about to go, his mother memorialized to the king: "Kuo should not be made to command." The king replied, "Why?" [She] answered, "Formerly when I was with his father who was commanding [the troops], the persons to whom he in person served meals numbered by tens and those whom he befriended numbered hundreds. The gifts bestowed [on him] by Your Majesty and the nobles were all given to the soldiers and officers. The moment he received an assignment, he was concerned no more with the affairs at home. Now as soon as Kuo was assigned to command, he sat on a seat facing the east[46] and held court; and the army officers dared not to look directly at him. The gold and the silk that were bestowed by Your Majesty were stored at home; then he daily looked for good fields and residences to buy and bought them. How does Your Majesty view him in comparison to his father? The father and the son are not alike. May Your Majesty not send him." [*SC* 81:16-18]

16. LAND ACQUISITION THROUGH GRANT OF THE RULER

Then Wang Chien commanded an army of 600,000 men. Ch'in 229 B.C.

45. Chao Kuo, the son of Chao She, who was the best strategist of Chao, was assigned the mission to command the army of Chao at Ch'ang-p'ing where the Chao troops and the invading Ch'in troops had a confrontation. The campaign ended as a complete defeat of Chao in 257 B.C.

46. The seat facing the east was the seat of honor.

Shih-huang[47] in person saw him off at Pa-shang.[48] On departure, Wang Chien requested many good fields, residences, gardens, and ponds. Shih-huang said, "General, proceed. Why should you be worried about poverty?" Wang Chien said, "When someone is a general of yours, Great King, even if he has merit, he cannot be enfeoffed. Therefore taking this opportunity when Your Majesty has trust in me, I venture to request gardens and ponds that can be the property of my descendants." Shih-huang laughed. Wang Chien, reaching the pass, sent messengers five times to request fine fields.[49] [*SC* 73:16]

17. Tax Shelter under the Powerful

Ca. 238 B.C. Collection of taxes and mobilization of manpower are for preparation against emergencies as well as for the full reserve of the granary. Nevertheless, the conscripts who flee from service by means of hiding and subordinating [themselves] to [the protection] of powerful households[50] to evade tax and labor corvée are numbered by tens of thousands.[51] [*HFT* 17:13a-b]

18. Kuan Fu's Wealth

133 B.C. General Kuan Fu[52] was a native of Ying-yin. His father, Chang

47. In 229 B.C., Ch'in Shih-huang-ti (or the First Emperor of Ch'in) had not proclaimed the imperial title.

48. The junction of the River Pa and the highway leading from the Ch'in capital to the east.

49. In this military campaign, Wang Chien totally defeated the Ch'u army (*SC* 73:18).

50. The implication of this phrase is better appreciated if it is read together with another paragraph of the *HFT*: "People make their decisions by taking advantage of safety and benefit as well as avoiding hazards and poverty. Now in battle, one either is killed by the enemy when one advances or is executed when one retreats. It is hazardous indeed. Abandoning one's own private affairs and toiling along on horseback, one has one's own house impoverished, yet is not much regarded by the ruler. One indeed becomes poor. These are hazards and poverty. How can people not avoid them? Therefore [they] serve private [powerful] households to have [tax and corvée] exemptions. . . . Therefore there are few [under the] government and many retained by the private" (*HFT* 19:8a).

51. Han Fei began to write after he failed to win the confidence of his ruler, King An of Han, who was enthroned in 238 B.C. He then was dispatched to Ch'in five years later (in 233 B.C.) as an envoy of Han; he died in Ch'in shortly after going there. Therefore most of his works were written in the period 238-233 B.C.

52. Kuan Fu himself had served as grand administrator and had held other ranking jobs in the government. For a related passage, see II, 21.

Meng, had been a favored retainer of Kuan Ying, the Marquis of Ying-yin. [Chang Meng] thus was recommended [eventually to the position of] two-thousand piculs. . . . [Kuan Fu] associated with no one but local leaders, magnates, and big bullies. His wealth amounted to several tens of millions [of cash]. Every day he entertained several tens to one hundred guests. [He owned] ponds, fields, and gardens. His family and retainers exercised power and made profits; they were unchallenged in Ying-ch'uan. The children of Ying-ch'uan therefore made up a song:

While the waters of the Ying run bright,
It means the Kuans are still all right.
But when the waters run polluted,
We'll know they've all been executed.

In the spring of the fourth year of Yüan-kuang, T'ien Fen[53] reported to the emperor that Kuan Fu's family in Ying-ch'uan were behaving with much disregard for the law and that people suffered thereby. He [thus] requested an investigation. . . . The chancellor also reported:[54] Kuan Fu associated with evil and lawless men, encroaching on the small people, amassing a fortune of millions, bossing the affairs in Ying-ch'uan. [*SC* 107:12, 14, 18, 23; cf. Watson, *Grand Historian* 2:118-19, 121, 125]

19. Population Moved to the Capital

Third to the first century B.C.

When Han arose, the capital was established at Ch'ang-an. The T'ien [family] of Ch'i; the Chao, the Chü, and the Ching [families] of Ch'u,[55] and the households of the meritorious officials were moved to the Ch'ang-ling mausoleum.[56] Later, in successive reigns, the families of two-thousand-picul officials, of wealthy persons with much property, and of the locally powerful persons who engrossed lands were all moved to the mausoleums in order to strengthen the trunk and to weaken the branches; it was not merely for the sake of serving the mausoleums.[57] [*HS* 28B2:51b]

53. T'ien Fen, the chancellor, was a political foe of Tou Ying, a former chancellor and a patron of Kuan Fu.

54. This passage is a part of the comment of Han An-kuo, the grandee secretary, who cited the accusation made by T'ien Fen at a court debate between Tou Ying and T'ien Fen.

55. These are the kinsmen of the royal houses of Ch'i and Ch'u.

56. Ch'ang-ling was the mausoleum of Emperor Kao. A town was established at each of the imperial mausoleums.

57. Such a forced migration was an old measure that appeared as early as

20. Hsiao Ho Seizes Land

195 B.C. In the autumn of the twelfth year of Han, Ch'ing Pu revolted. The emperor in person commanded the troops to attack him. From time to time [the emperor] sent messengers to ask what the chancellor (i.e., Hsiao Ho) was doing. . . . [One of Hsiao's] retainers advised him, "It will not be long before your family is terminated. Your position is that of chancellor; your merit is regarded as the foremost. Can there be anything added to it? It has been more than ten years since you first entered the Pass[58] and won the hearts of the people. They are all attached to you. Yet you are still laboring for popularity. [The reason] the emperor frequently sends regards to you is that he is afraid that you may shake the area within the Pass.[59] Why don't you purchase a lot of farmland on credit and lower the credit rate in order to tarnish yourself? Then the emperor will feel eased." Then the chancellor followed his suggestion. The emperor was very pleased. When he returned after the campaign against [Ch'ing] Pu was over, the people blocked the road to submit accusations that the chancellor forcibly bought at reduced prices the people's lands and houses at a total cost of tens of millions of cash. After the emperor arrived, the chancellor was granted an audience. The emperor laughed, "So, chancellor makes profit from the people!" He then handed over the submitted accusations to the chancellor and said, "You personally apologize to the people."

[Hsiao] Ho always purchased lands and houses in out-of-the-way places. He did not build residences with elaborate walls or roofs. He said, "If my descendants are worthy, they will follow my example of frugal living. If they are not worthy, [the property] will not be taken away by [more] powerful families." [*SC* 53:10-11, 13; Watson, *Grand Historian* 1:130-31, 132-33]

21. T'ien Fen's Greed for Land

135 B.C. Once [T'ien Fen] asked the emperor for some land of government artisans to add to his residence lot. The emperor angrily replied, "And

---

in the Ch'in dynasty. Emperor Kao of Han also ordered such a move when Ch'ang-an was made the capital. The most explicit mention of the motivation for such migration is an order by Emperor Wu of 127 B.C. (*HS* 6:10b; 64A:19b). Cf. Ch'ü T'ung-tsu, *Han Social Structure*, pp. 410-11, 419, 196-98.

58. "The Pass" is the Han-ku Pass that separated the capital district from the eastern part of the empire.

59. This is a subtle way of saying that Hsiao Ho might revolt in the capital district.

then take the arsenal!" . . . [T'ien] built for himself the best residence in the capital and possessed the most fertile fields. . . . Some time later [T'ien] Fen sent Chi Fu[60] to ask [Tou] Ying for [the latter's] farmland south of the city. [Tou] Ying was furious and said, "Although this old servant has been abandoned and although the general is honored, how can he use his power to seize my possessions." He would not allow it. When [Kuan] Fu[61] heard of the incident he cursed [Chi] Fu furiously. [Chi] Fu, feeling sorry to see a schism develop between the two men,[62] offered fabricated and ameliorative excuses to [T'ien] Fen, saying, "Marquis Wei-chi[63] is old and soon will die. Be patient and wait." Later on, [T'ien] Fen heard that [Tou] Ying and [Kuan] Fu in fact had been angered at his request and refused to yield, whereupon he too became furious. "Once in the past the son of Marquis Wei-ch'i committed homicide. It is I who saved his life. In serving Wei-ch'i, there is nothing I have not done for him. Why is he so fond of several *ch'ing* of land? And why is Kuan Fu involved in this matter? I will not venture to make a further request for the land." [*HS* 52:5a-b and 8b; cf. Watson, *Grand Historian* 2: 115, 120-21]

22. Wang Li's Occupation of Public Land

At the time the maternal uncle of the Emperor, [Wang] Li, Marquis of Hung-yang,[64] sent a retainer to Li Shang, grand administrator of Nan-yang commandery, with whose assistance [the retainer] seized several hundred *ch'ing* of cultivated and grass lands. Much of it was shore and marsh land that the people had leased from the privy treasurer and generally it was reclaimed and developed. [Wang Li] submitted a memorial stating that he was willing to put [the land into the hands of] the government. It was decreed that the commandery should assess the land and grant payment for it; the money [paid] *Ca. 16 B.C.*

60. Chi Fu was a courtier who kept on good terms with both Tou and T'ien.

61. Kuan Fu, a military man with great wealth, was a protégé of Tou Ying. His biography is contained in the same chapter (*HS* 52:6a ff). Cf. II, 18.

62. Tou and T'ien.

63. Tou Ying was ennobled as Marquis Wei-ch'i.

64. Wang Li was one of the younger brothers of Wang Shang who was then the grand minister of war and general of the guard *(ta-ssu-ma, wei-chiang-chün)*. As the head of the imperial consort family, Wang Shang was the de facto prime minister. Wang Li was next in line to succeed Wang Shang for the same position. He was by-passed, however, because of this incident cited here and his younger brother Wang Keng was made the grand minister after Wang Shang died.

was in excess [of the market value] by over one hundred million cash.[65] [*HS* 77:10b-11a]

23. Empress Kuo's Father

Ca. First century B.C.

The Empress of Kuang-wu, née Kuo Sheng-t'ung, was a native of Kao prefecture of Chen-ting [kingdom].[66] Her family was renowned in the commandery. Her father, [Kuo] Ch'ang, yielded his [inherited] fields, houses, and [other] properties valued at millions to his half-brother. His countrymen therefore regarded him as righteous. He served in the local government [67] as a *kung-ts'ao*[68] and married the daughter of King Kung of Chen-ting,[69] who therefore was called Princess Kuo. [The princess] bore the empress and a son [Kuo] K'uang. [*HHS* 10A:4a]

24. Land Seized by Consort Family Member

Ca. A.D. 80

In the second year of the Chien-ch'u period (A.D. 77), a younger sister [of Tou Hsien] was made the empress. . . . [Hsien and] his brothers enjoyed great favor, attending both the court and the palace. . . . Hsien, depending upon the influence of the Inner Palace, requested and took the garden and fields of Princess Ch'ing-sui[70] at a cheap price. The princess, frightened by the pressure, did not dare to bargain [with Hsien]. Later, Emperor Su-tsung [r. 76-88] passed by the garden and pointing to it made inquiry of Hsien, who muttered, not able to give an explanation. Later on, the situation was disclosed to the emperor who, greatly enraged, summoned Hsien

65. This incident occurred when Sun Pao served as director of justice *(ssu-chih);* he prosecuted the misdeed of Wang Li. Sun's tenure in this position was one or two years after the middle of the Hung-chia reign period (20-17 B.C.) when Sun was still inspector of I-Chou and at least three years before the second year of the Yuan-yen period (11 B.C.) when Sun was appointed to the position of mayor of the capital. Therefore the date of this incident should be sometime between 17 B.C. and 14 B.C. (*HS* 77:10b-11a, 19B:47).

66. The present-day Kao-ch'eng County of Hopei Province.

67. The text refers to the commandery government. Since Chen-ting was a kingdom, the local administration should be that of the state chancellory of Chen-ting kingdom.

68. *Kung-ts'ao,* "merit evaluator," was a ranking staff member of the local administration. He was chief of the personnel section.

69. King Kung of Chen-ting is Liu Pu, a seventh-generation descendant of Emperor Ching.

70. Princess Ch'ing-sui was a daughter of Emperor Ming, hence, a sister of the reigning emperor.

and rebuked him, "Recall deeply the time when we passed by the princess's fields and garden that you took. How does that differ from the incident when Chao Kao indicated that the deer was a horse![71] For a long time I have thought that you frightened people. Formerly, during the Yung-p'ing period (58-75), Yin Tang, Yin Po, and Teng Tieh[72] were regularly ordered to investigate each other. Therefore the powerful consort families dared not violate the law. The imperial decrees, nevertheless, still sternly warned the maternal uncles on their land and residences. Now even an honored imperial princess was crookedly deprived [of her property]. How much more the case for a small man! The state will abandon Hsien as if he were a solitary chick or a rotten rat." [*HHS* 23:11b-12a]

25. Fertile Land in the Capital District Possessed by Consort Families

[Ma] Fang and his brothers[73] were honored and prosperous.[74] *Ca.*
Each possessed more than one thousand slaves and property by the A.D. 79
millions. They all purchased fertile land in the capital district. They also constructed residences and observation towers; the connected pavilions overlooked the thoroughfares and completely filled lanes and streets. [*HHS* 24:20b]

26. Imperial Consort Family's Extensive Realty

[Liang Chi] also appropriated forests and parks; the prohibitions A.D. 147
[against entering them] were the same as those issued by a royal family. His holdings reached to Hung-nung in the west, Jung-yang in the east, Lu-yang in the south,[75] and the Yellow River and Ch'i

71. Chao Kao, a powerful eunuch at the court of Emperor Erh-shih of Ch'in (209-207 B.C.), alleged that a deer was a horse to test whether the other courtiers would go along with him on this false identification or would disclose the truth to the emperor (*SC* 6:80).

72. All three were consort family members.

73. Ma Fang was a son of Ma Yüan and a brother of the empress of Emperor Ming. Though members of a consort family, none of the Ma males was made chancellor or grand general. The empress consciously restrained her brothers from getting into power positions or even receiving feudal titles. Cf. *HHS* 10A: 8b-11b.

74. In A.D. 79, Ma Fang and his brother Ma Kuang were both enfeoffed by Emperor Chang. Four years later (83) the Ma brothers lost the favor of the throne after being criticized for having indulged in extravagant living.

75. These three places are all located in the vicinity of the Eastern Han capital of Lo-yang. Hung-nung is near today's Ling-pao; Jung-yang remains the same

River[76] on the north. His holdings contained mountains and marshes, and distantly included hills and wildernesses, with perimeter boundaries of almost one thousand *li.*[77] . . . [In 159, Liang Chi was executed.] The confiscated property [of Liang Chi] was auctioned by the government. The total value amounted to more than three billions, which was returned to the imperial treasury and the taxes of the whole empire were reduced by one half. His parks and gardens were redistributed to the poor people. [*HHS* 34:13a-16b]

27. Eunuch's Behavior

A.D. 147-172 Hou Lan . . . was enfeoffed as Marquis of Kao-hsiang as a reward for his alleged merit in participating in the deliberations regarding executing Liang Chi.[78] Tuan Kuei, a lesser attendant of the Yellow Gate,[79] had his home in the Chi-yin commandery. He and Lan both established landed properties. . . . Their servants, retainers, and guests encroached upon the people and robbed travelers. . . . In the second year of Chien-ning (A.D. 169) Hou Lan returned home to mourn for his mother and constructed a tomb on a large scale. Chang Chien, a commandery investigator of transgressions, took the occasion to charge Lan with being greedy and extravagant and having requested and seized altogether 381 residential sites and 118 *ch'ing* of land. [*HHS* 78:11b-12a]

28. Land Granted to Half-Sister of the Emperor [Wu]

140 B.C. A daughter named Su, whom the empress dowager had borne to Chin Wang-sun when she was still a commoner, was living among the common people.[80] The matter was concealed. Shortly after Emperor

name; Lu-yang is today's Lu-shan.

76. Ch'i is a tributary of the Yellow River flowing through today's Hui-hsien, Lin-hsien, and Ch'i-hsien.

77. According to another Han record, such mountainous lands and wilderness areas were marked with huge banners bearing warning signs, "No Trespassing." Cf. quotation by Hui Tung, *HHS* 34:13a.

78. In A.D. 159, a group of eunuchs staged a coup d'état to terminate the influence of Liang Chi, the power holder of an imperial consort house. Cf. II, 26.

79. The Yellow Gate was the inner gate of the Han palace. The title of lesser attendant of the Yellow Gate, *hsiao huang-men,* indicates a eunuch of junior rank.

80. Empress Dowager, née Wang, first married Chin Wang-sun before her own mother presented her to the then crown prince, who later ascended the throne

Wu came to the throne, Han Yen[81] told him of it. The emperor said, "Why did not you speak of this earlier?" . . . He granted to his elder half-sister ten million cash, three hundred slaves, one hundred *ch'ing* of government land, and a first-class residence. [*HS* 97A:10a-b]

29. Land Bestowed to Wet Nurse

When Emperor Wu was still young, the mother of the Marquis of Tung-wu had nurtured him. When the emperor grew up, he bestowed upon her the title of grand wet nurse. She regularly came to court twice a month. . . . Once the mother nurse presented a petition that said that in a certain place there was government land and she would like to be granted the lease. The emperor said, "Does my wet nurse wish to have it? Let it be granted to her."[82] [*SC* 126:13-14] 140-87 B.C.

30. Government Land

Between the double railing of the rear tower of the Chien-chang Palace there appeared an animal that resembled an elk. This was reported. Emperor Wu went and saw it. He questioned those of various attending ministers who were practical men and those who were knowledgeable in Classics and [various] techniques. But none of them was able to answer. He summoned Tung-fang Shuo to look at it. Shuo said, "Your servant knows it. If Your servant be granted fine wine and good food for a large meal, then Your servant will tell [what it is]." It was decreed, "Approved." *Ca.* 122 B.C.

[After he had eaten] he also said, "In a certain place there are several *ch'ing* of government land, fish ponds, rushes, and reeds. If Your Majesty grants them to Your servant, then Your servant, Shuo, will then speak." It was decreed, "Approved."[83] [*SC* 126:21]

31. Land Granted

[The Emperor] exceptionally conferred upon [Pu] Shih[84] the 119 B.C.

---

and was known as Emperor Ching. Emperor Wu was the only boy among the four children born to Emperor Ching by Empress Wang. Cf. *HS* 97A:8b-9a.

81. Han Yen, the grandson of a Han nobleman, was Emperor Wu's schoolmate.

82. This passage is one part of the supplement added by Ch'u Shao-sun to the original text of Ssu-ma Ch'ien's work. No exact date is given.

83. This passage is taken from the supplement added to the *Shih-chi* by Ch'u Shao-sun.

84. Pu Shih's biography, *HS* 58:8b-10b.

position of "gentleman of the palace" *(chung-lang)*, bestowed upon him the rank of *tso-shu-chang*,[85] and granted him ten *ch'ing* of land. [*HS* 24B:13b; cf. Swann, *Food and Money*, p. 284]

32. Prince's Private Land

A.D. 83 Su-tsung[86] gave back to [K'ang][87] all the territory that had been taken from him [previously].[88] Thereupon, K'ang built up a huge fortune and erected palaces and houses on a large scale. His male and female slaves numbered fourteen hundred; the horses in his stable, twelve hundred; and his private land, eight hundred *ch'ing*. [*HHS* 42:8a]

33. Land Tax Owed by a Marquisate

A.D. 20 [Liu Hsiu] on behalf of his uncle, the former Marquis of Ch'ung-ling,[89] went to the office of the grand minister of war *(ta-ssu-ma)* to place a suit for alleged unpaid land tax due before the date of *jen-yin* of the twelfth month, first year of the Ti-huang period. The total amounted to twenty-six thousand *hu* of grain and several tens of thousands of hay tax money.[90] [*Tung-kuan-chi*, as quoted in commentary on *HHS* 1A:4b]

34. Good Income of a Bureaucrat

48-44 B.C. [Kung] Yü[91] memorialized: "Your subject Yü was aged and poor. My whole property was valued at less than ten thousand cash.[92] My

85. *Tso-shu-chang*, the tenth grade of honorary rank.

86. Emperor Chang (r. 76-88 A.D.).

87. K'ang, one of the sons of Emperor Kuang-wu (r. A.D. 25-57), was enfeoffed at Chi-nan.

88. In the reign of Emperor Ming, K'ang was punished as he was suspected of plotting rebellion. Five prefectures were stripped from his domain. Emperor Chang, the nephew of K'ang, restored the lost territories to K'ang.

89. The text of *HHS* gives "Formerly, Emperor Kuang-wu, for the household of the Marquis of Ch'ung-ling, placed a suit for the unpaid land tax at the office of Yen Yu." Yen Yu, a ranking official at the court of Wang Mang, was *ta-ssu-ma*, grand minister of war, at the time of the incident. In A.D. 23, Yen Yu was one of the commanding generals to lead an army to suppress the uprising at Nan-yang in which Liu Hsiu participated.

90. The quotation was provided by Li Hsien, the early T'ang commentator of the *HHS*.

91. On Kung Yü, see I, 8, n. 19.

92. The Han taxation system made property worth ten thousand cash the

wife and children were not well fed, with unhusked grain and beans, and not well clothed, with coarse materials and poor fabrics. I possessed 130 mu of land.[93] Your Majesty unexpectedly summoned me to court. Your subject sold 100 mu of land for the provision of a carriage and horses. After my arrival I was then appointed to the position of grandee remonstrant with the rank of eight hundred piculs; I received a monthly salary of 9,200 cash. The daily meal was provided by the imperial kitchen; furthermore, gifts were bestowed, including clothes and cloths of the four seasons, wines, meats, as well as fruits. The grace of Your Majesty was indeed abundant. Once, when I was sick, the palace physician was dispatched to cure me. Because of the benediction of Your Majesty, I survived [the sickness]. I was then appointed imperial court grandee with the position of two thousand piculs; I received a monthly salary of 12,000 cash.[94] The more the emolument, the wealthier the family is, and the more honorable I become. This, indeed, is beyond what Your foolish subject of humble origin deserves."[95] [*HS* 72:12a-b]

35. Income from a Marquisate

Formerly, K'uang Heng[96] was enfeoffed with a domain in Lo-an district of T'ung prefecture [of Lin-huai commandery]. The lands of the district were demarcated at 3,100 *ch'ing*. The southern boundary was Min Path. In the first year of the Ch'u-yüan period (48 B.C.) the commandery map mistakenly showed P'ing-ling Path instead of Min Path [as the southern boundary of Lo-an district]. Some ten years later, K'uang Heng was enfeoffed in Lin-huai commandery, 36 B.C.

lowest bracket on the property tax scale. Yoshida Torao, *Ryōkan sozei no kenkyū*, pp. 48-49.

93. Kung Yü's statement may be a little exaggerated here. A farmer who had 130 mu of land should have been able to make both ends meet; see Ch'ü, *Han Social Structure*, p. 110. Li Ya-nung cited another case in which the value of 100 mu of land and one residence came to the equivalent of 100,000 cash (Li, *Hsin-jan-chai shih-lun-chi*, p. 964).

94. The exact way of making salary payment is not quite clear. Utsunomiya suggested a ratio of 30 percent cash and 70 percent grain instead of 50 percent each as given in *Hsü-Han-chih*. Cf. Utsunomiya, *Kandai shakai keizaishi kenkyū*, p. 203 ff.

95. This document is the first half of a memorial in which Kung Yü requested the throne to release him from government service. The emperor, instead of granting the resignation, promoted him twice and eventually appointed him to the post of grandee imperial secretary.

96. K'uang Heng rose from a poor hired hand to chancellor. For his biography, see Ch'ü, *Han Social Structure*, pp. 356-57. K'uang was ennobled as Marquis of Lo-an in 36 B.C.

whereupon P'ing-ling Path was taken as the boundary of his fief—which gave him an extra 400 *ch'ing*. In the first year of the Chien-shih period (32 B.C.), the commandery resettled the boundaries of the marquisate. The annual report and the revised map were presented to the office of the chancellor. . . .[97] [Two of K'uang's subordinates in the] Chancellory questioned the commandery, "According to the old map, the southern border of Lo-an district is P'ing-ling Path. Why do you not follow the old [map] and instead take Min Path to be the border?" The commandery immediately returned the 400 *ch'ing* of land to Lo-an marquisate. Heng dispatched a clerk to T'ung prefecture to collect and pay to Heng's family over one thousand piculs of tax grain from the returned land.[98] [*HS* 81:10a-b]

36. Lo P'ou and His Protectors

1 B.C. At the time of Emperors Ch'eng and Ai (31-1 B.C.), Lo P'ou of Ch'eng-tu had a wealth that amounted to one hundred million. At the beginning, [Lo] P'ou came to the capital to engage in trade, bringing along with him millions in cash. He [then] managed the finances of the Shih family of P'ing-ling.[99] Because of his effort, the Shih family [made] a fortune at the level of the Ju and the Chü families. The Shih family confidentially entrusted [Lo P'ou] with a large sum of money and assigned him to trade in the Pa-Shu region.[100] In a few years, [Lo P'ou] made more than ten millions. [Lo] P'ou gave half of that amount as a gift to Marquis Ch'ü-yang and Marquis Ting-ling[101] whose influence he relied upon to make loans in the provinces[102]

97. K'uang Heng was then the chancellor. One of his staff members suggested to him that the modification could be contested. Heng neither approved the proposal nor stopped it. As a consequence of the staff member's action, Heng regained the 400 *ch'ing* of land. Heng was later impeached for this action.

98. The grain which Heng's domain retrieved from the commandery equals the revenue of three years from the area of 400 *ch'ing* of land. Reduced to the year/*ch'ing* unit, the whole 3,100 *ch'ing* of land of the marquisate ought to yield an annual income of

$$3100 \times 1000/400 \times 3 = 2580 \text{ piculs}$$

In addition to this, K'uang Heng also received about this amount of annual salary as chancellor.

99. P'ing-ling was one of the tomb counties in the vicinity of the capital.

100. The present-day province is Szechwan.

101. Marquis Ch'ü-yang was Wang Keng and Marquis Ting-ling was Ch'un-yü Ch'ang. Both were powerful courtiers then.

102. Literally, "the commanderies and the kingdoms," which means all areas in the empire.

and people did not dare default. He also took the profits from a salt-well monopoly. In one year he doubled his capital. This is [the way] he multiplied his wealth. [*HS* 91:9a; cf. Swann, *Food and Money*, pp. 453-54]

37. Ning Ch'eng as Landlord

Ning Ch'eng[103] said, "How can an official who is not able to advance to the rank of two thousand piculs or a merchant who is not able to make ten millions be compared to others?" Therefore he bought on credit some one thousand *ch'ing* of slope land and leased [the land] to poor people,[104] and several thousand families labored [on the land] and served him. After a few years a general amnesty was issued [absolving him from his former offenses]. By that time, he had already made a fortune valued at several thousand [catties of] gold. [*SC* 122:9; cf. Watson, *Grand Historian* 2:424] *Ca.* 140 B.C.

38. Officials-Merchants Unity

Ho Wu[105] and his brothers, all five, were on the staff of the commandery government. [Officials of] the commandery and the prefecture looked up to them in reverence and fear. The household of Hsien,[106] [Ho] Wu's younger brother, had their name on the market register. They regularly did not pay their taxes and the prefectural [administration] on several occasions failed its rating examination. The Market Sheriff Ch'iu Shang arrested and humiliated members of Hsien's household. Hsien, enraged, intended to attack [Ch'iu Shang] *Ca.* 43 B.C.

103. Ning Ch'eng, one of the most harsh of officials, served Emperors Ching and Wu to establish absolute imperial authority. When Ning Ch'eng was the mayor of the capital district, he was restrained by influential persons in the court and was punished to serve as a conscript. He nevertheless managed to escape from the capital district and returned to his home county where he purchased land.

104. The *HS* text gives no indication of his purchasing the land. It can be interpreted to mean, as the commentator Yen Shih-ku suggests, that Ning Ch'eng rented the land from other sources, then leased the land to the poor (*HS* 90:4a). If the *HS* version is correct, Ning Ch'eng could have rented either privately owned or government-owned land.

105. Ho Wu began his career of government service in 43 B.C. after studying at the Imperial University. He enjoyed a good reputation at court, yet Wang Mang identified him as one of the old political figures to be eliminated. He was eventually accused by Wang Mang of plotting against the government; Wu committed suicide in A.D. 3.

106. Hsien later reached the position of grand administrator of Ying-ch'uan commandery.

via some administrative fault. Wu said, "With [the status of] our family, [yet our] taxes and corvée [are not rendered] first among the many [as an example]. Wouldn't it be fitting to have a just official [who behaves as such]?" [*HS* 86:1b]

39. Strict Land Survey

A.D. 39 At that time [the register of] cultivated land in the empire for the most part did not correspond to reality. Moreover, the number of the people and their ages both were either increased or reduced. . . . An edict was issued to the divisions and commanderies to check these matters. Yet most of the inspectors and grand administrators were not equitable in reporting the results. Some treated the powerful individuals with great kindness and exploited the poor and the weak. The people lamented and grumbled, blocking the roads and wailing.

At this time each of the commanderies sent up a commissioner to report on the matter. The emperor saw a note on the documents of the official from Ch'en-liu commandery. He looked at it. It read: "[The situation] in Ying-ch'uan and Hung-nung can be investigated, but in Ho-nan and Nan-yang it cannot be investigated." The emperor asked the official about its source and significance. The official did not want to comply and falsely stated that he had obtained it in Ch'ang-shou street. The emperor became angry. Hsien-tsung,[107] who was at this time the Duke of Tung-hai, and twelve years old and who was behind the hangings, said, "At the order of the commandery government the official naturally wanted to find out the amounts of cultivated land and to compare them." The emperor said, "Even if it is so, why should he say that Ho-nan and Nan-yang cannot be investigated?" He replied, "In Ho-nan, where the imperial capital is located, there are many officials who are close to the throne. In Nan-yang, which is the emperor's native place,[108] there are many close relatives. Their lands and houses exceed the norms and cannot be taken as standard [for comparison with Ch'en-liu]."

The emperor ordered the general of the palace-gentlemen-as-rapid as-tigers to question the official. Then the official admitted that it was as Hsien-tsung had stated. The emperor thereupon sent internuncios *(yeh-che)* to investigate the reality of the situation and he learned fully the villainous activities of the commandery officials. [*HHS* 22:8a-b]

107. Hsien-tsung is the posthumous title of Emperor Ming (r. 58-75), the son of Emperor Kuang-wu.

108. Emperor Kuang-wu's birthplace, Ts'ai-yang (in modern Tsao-yang County, Hupei), belonged to Nan-yang commandery.

In the next year, Liu Lung was found guilty [of false land registration], was summoned to court, and was imprisoned. Of his colleagues more than ten were put to death. Because Lung was a meritorious minister, the emperor by special favor exempted him from punishment, but made him a commoner.[109] [*HHS* 22:8b] A.D. 40

The grand administrator of Ho-nan, Chang Chi, and more than ten administrators who were found guilty of making false registrations of land[110] were all imprisoned and died. [*HHS* 1B:12b] A.D. 40

[Wang] Yüan was first made prefect of Shang-ts'ai prefecture. [Later] he was transferred to the position of chancellor of Tung-p'ing kingdom. Found guilty of making false registrations of the cultivated land, he was imprisoned and died. [*HHS* 13:12a] A.D. 40

[Pao Yung][111] was sent out to the post of chancellor of Tung-hai kingdom. Found guilty of making false registrations of land, he was recalled. [*HHS* 29:7a] A.D. 40

Later on [Li Chang][112] was found guilty of making false registrations of the people's land and was summoned to the court. [*HHS* 77:4b] A.D. 40

[Mou Ch'ang][113] soon was promoted to the post of grand administrator of Ho-nei commandery. Found guilty of making false registrations of the cultivated land, he was dismissed. [*HHS* 79A:9a] A.D. 40

40. Land Opened

[Chang K'an][114] then opened up more than eight thousand *ch'ing* A.D. 40-48

109. This change of status was temporary. Four years later, in A.D. 44, Liu Lung was appointed to a much higher post in charge of the affairs of the grand minister of war (*HS* 1B:21a).

110. According to the *Tung-kuan Han-chi,* some administrators, for the purpose of making a good record, even assessed the house land and the villages as taxable land. The commoners, being distressed, howled in congregations beside the roads.

111. Pao Yung, the shepherd of Yang division, became colonel director of retainers in A.D. 35. Four years later he was demoted to the post of chancellor of Tung-hai kingdom.

112. At this time Li Chang was a grand administrator of Lang-ya.

113. Mou Ch'ang was a scholar who gave special attention to the study of the *Book of Documents*. In A.D. 26 he was appointed erudite and then grand administrator.

114. Chang K'an was then the grand administrator of Yü-yang commandery

of paddy fields in Hu-nu[115] and encouraged the people to cultivate them. They thereby became extremely rich. The people made a song that says:

Mulberry trees have no useless branches,
The ears of wheat are double.
Under the administration of Sire Chang
Our happiness is more than we can bear. [*HHS* 31:7a-b]

41. Land Record

A.D. 76-81 [Ch'in P'eng][116] opened up several thousand *ch'ing* of rice fields. During every agricultural season he personally measured the acreage [of the cultivated land], distinguished the fertile from the barren soil, and classified it according to three grades. [For each grade] he made separate registers that were kept in the district and prefectural governments. Thereupon corrupt officials became constrained and were no longer able to practice their trickery.

P'eng then presented a memorial saying that it would be advisable to have the whole empire adopt the same system. By an edict the regulations he had set up were conveyed to the Three Administrations,[117] and were also transmitted to the divisions and commanderies. [*HHS* 76:7b]

42. Land Reclaimed

A.D. 76-87 [Yang Jen][118] was appointed prefect *(ling)* of Shih-fang prefecture.[119] . . . More than a thousand *ch'ing* of land were opened up. [*HHS* 79B:5a]

---

in the northern part of the present province of Hopei. He remained at his post for eight years.

115. Hu-nu was located in the southwestern part of the modern Mi-yun County, Hopei. The water was drawn from the Po River, which flows through the territory in a southerly direction.

116. Ch'in P'eng, a native of Fu-feng commandery, was appointed grand administrator of Shan-yang commandery in A.D. 76. He was a benevolent and good administrator, and the people liked him. He remained at this post for six years.

117. The *san-fu*, the Three Administrations, were the three highest offices in the central government.

118. Yang Jen, a native of Lang-chung prefecture of Pa commandery was appointed to this post when Emperor Chang ascended the throne.

119. Roughly equivalent to modern Shih-fang County, central Szechwan.

43. Reclamation of Land

[Ts'ui Yüan][120] was promoted to the post of prefect of Chi prefecture.[121] During his term of office he several times presented proposals that were advantageous and suitable. He opened up several hundred *ch'ing* of rice fields for the people. [*HHS* 52:13b] A.D. 135-141

44. Call to Restore the Well-Field System

With the adverse change of the well-field system, powerful men multiplied their wealth [through trade]. They spread restaurants and inns throughout provinces and commanderies; their fields are connected together in regions and kingdoms. Even though one of them does not have the commission of a half seal with blue silk cord,[122] he unlawfully wears garments decorated with three stars and dragon designs.[123] Even though he does not hold the position of chief of a pentad of registered households,[124] he commands services [like those] of a famous domain of one thousand families.[125] His prestige and his pleasures surpass those of an enfeoffed lord; his power equals that of a [commandery] administrator or a prefect. In pursuit of wealth he indulges himself, and when he commits a crime, he is not prosecuted. Assassins and dare-to-die bravos sacrifice their lives for him. It has come to the point that the weak and the ignorant have ragged blankets [in bed] and torn curtains [in their houses]. They die in places away from home with their bodies uncared for. Oppressed and wronged, distressed and needy, they dare not care for themselves. Although this is in part due to the looseness of the law, it is the limitless division of the land that has caused such a state of affairs. Now if one desires to uphold the principles of Great Peace, to lay the foundation of the highest transformation, to equalize the abundance and dearth of the wealth of the people, and to rectify the extravagance and frugality in customs, there is, in fact, no alternative but the well-field system. The change [of the well-field system] has produced this ruination and the system ought to be restored.[126] [*HHS* 49:15b-16a] A.D. 208-220

120. Ts'ui Yüan was a famous scholar.

121. In Ho-nei commandery.

122. The cord was attached to an official seal as a handle; the blue cord was used by officials of the lowest grade.

123. The ornament on the dress of ranking officials and nobles.

124. In Ch'in and Han, five or ten households were organized into units for mutual aid and legal responsibilities.

125. This is equivalent to the fief of a noble.

126. This is taken from Chung-ch'ang T'ung's *Ch'ang-yen*. The author, born

45. The Wish of a Scholar

A.D. 208-220

[Chung-ch'ang T'ung][127] always considered that those who wandered after emperors and kings wished to establish themselves and to make a name for themselves. But reputations do not endure and the life of a man is easily obliterated. Freely roaming and leisurely relaxing, one could entertain himself. It was his desire to divine for a residence that was pure and spacious where he could take pleasure in following his desires. In discussing [his ideals] he said, "Let there be fertile land and a dwelling place that has a large residence, mountains to the rear and overlooking a river, ringed by canals and ponds, surrounded by bamboo groves and woods. In front, there will be built the threshing ground and the vegetable plot; in the rear there will be planted fruit trees and a garden. Boats and carriages will be sufficient to spare us the trouble of walking and wading; servants will be sufficient to give rest to the labors of our four limbs. For nourishing the parents there will be a variety of delicious repasts; for the wife and children there will be no labor that distresses the body." [*HHS* 49:11a]

46. Hsün Yüeh's Compromise

Ca. A.D. 190-210

The feudal lords do not have a monopoly on fiefs. The privately owned land of the wealthy exceeds the limitations and their wealth surpasses that of the feudal lords. This is self-enfeoffment. The grandees do not monopolize land; people buy it and sell it of their own volition. This is monopolizing the land. Someone has said, "Should the well-field system be restored?" I answer, "No. Monopolizing the land was an ancient [practice]. The well-field system is not a modern [practice]." "If such is the case, then what is to be done?" I say, "Let the land be cultivated, but not owned, while awaiting proper institutionalization; this may be done."[128] [Hsün Yüeh, *Shen-chien* 2:7b-8a]

---

in 180, died at forty-one years of age. *Ch'ang-yen* is a book of thirty-four chapters, though only a small portion is now preserved in the *HHS*. For Chung-ch'ang T'ung's life, see Etienne Balazs, *Chinese Civilization and Bureaucracy*, pp. 213 ff.

127. For Chung-ch'ang T'ung, see II, 44, n. 126.

128. Hsün Yüeh was a Confucian scholar who served in the Han court when Ts'ao Ts'ao had already become the real master of North China in the third century. Hsün wrote the book, *Shen-chien*, to comment on current problems.

III

THE FARMER'S LIVELIHOOD

1. Ordinances for the Four Peoples: Ssu-Min-Yüeh-Ling

A.D. Mid-second century

A.1. The first day of the first month is called the New Year. [The head of the household] personally leads his wife and children reverently to offer sacrifice to ancestors and [deceased] parents. Three days before [the ceremony], the head of the household and those who have assignments should have observed [a period of] purification.[1] On the day of the ceremony, wine is served to cause the spirits to descend. After the ceremony is over, the whole family, including the old and the young, sit according to seniority in front of the ancestors. The sons, their wives, the grandchildren, and the great-grandchildren each presents pepper [blossom] wine to the head of the household,[2] to toast to health and longevity. They do so happily. One should call upon and send seasonal greetings to the lord, the teachers, the former superiors, the kinsmen, the seniors in the clan, the friends of the father, friends, relatives, and elders in the community.[3]

A.2. In this month, select a good day to have the capping ceremony for young adolescents.[4]

A.3. The hundred kinds of plants sprout. The dormant insects emerge. On the first *ting* day of the month,[5] the guardian deity of

1. This ritual should consist of several days of fasting and abstinence.

2. According to the note, possibly by the author Ts'ui, wine scented with cypress leaves and pepper blossoms is helpful to the aged. The toasting, according to the note, should start from the youngest.

3. The "lord" refers to the local administrators or the princes and marquises. "Former superiors" refers to officials under whom one had served or by whom one was recommended.

4. According to the original note, a youth is capped at nineteen years (or eighteen years in the Western counting of age).

5. *Ting* is one of a series of ten symbols that are used to designate the order of days. In combination with another series of twelve symbols, there are sixty days in a cycle. This cycle goes on independently from the months. Each month is divided into three sections of ten days. The first *ting* day is the day with a *ting* designation during the first ten days of the month. In the following sentence, a *hai* day is also referred to, meaning the day of a *hai* designation falling in the first twelve days of the month.

the travelers is worshipped at the gateside so that the *yang* atmosphere (i.e., aura) is ushered in and the stagnant atmosphere is expelled; [thereby] one prays for blessings and good luck.[6] Also, on the first *hai* day of the month, the agricultural deity[7] and the ancestors are worshipped to pray for a good harvest for the year.

A.4. On the *Ch'u*[8] day of the first ten days of the month, or on the fifteenth day of the month, one prepares medicinal pastes, pills, homemade remedies and a kind of powdered medicine.[9]

A.5. Farming activities are not started yet so that the adolescents and those older are sent to the school of upper grades,[10] where the five classics are taught. The other historical materials are not to be studied unless there is a good faculty and a good tradition of teaching.[11] When the water on the ink stone is not frozen, the young children are sent to grade school to learn reading, writing, and arithmetic. The girl workers are urged to weave cloth.

A.6. From the beginning to the end of the month, the trees that can be transplanted are bamboo, lacquer tree, the tung tree, the *Catalpa kaempferi,* pine, cypress, and other trees. Only fruit trees are not transplanted after the fifteenth day of the month.

A.7. On the day of *Yü-shui,*[12] as the vapor of the earth starts

6. Since in the winter one is largely confined at home, going out of doors in the spring bears the significance of the beginning of activities throughout the whole year.

7. Literally, "former farmer," which means the "inventor of agriculture."

8. *Ch'u* is one of another series of day designators, the "Twelve Spirits of Chien and Ch'u" *(chien-ch'u shih-erh-shen).* Each of the twelve was correlated with one of the twelve "earthly branches" of the duodecimal cycle and used for prognosticatory purposes. See Morohashi Tetsuji, *Dai Kan-Wa jiten,* IV.9574.118.

9. Two of the given names of medicine are general names of paste and ready-to-use-remedies. The other two seem to be specific kinds of pills and powders that, however, are not identified.

10. The text gives senior school here in contrast to the minor school, which is translated "grade school" in the following sentence. From the text it is clear that in the former, the curriculum is mainly the study of classics while the latter offers the three R's. According to the original note, adolescents include youths between fifteen and twenty years old and young children include those between ten and fourteen years old.

11. The text here is somewhat ambiguous. The translation is based upon the revision suggested by Shih Sheng-han.

12. *Yü-shui* is one of the twenty-four days that serve as demarcations of twenty-four periods of climatic changes in a solar year. *Yü-shui,* literally the "rain water," usually falls between the sixteenth and the eighteenth of February.

to ascend, the planted peg is covered by the upswelled earth,[13] and the roots of the last year's crop can be uprooted, and one must hasten to remove the grass from fields of superior [rich] soil and black earth.

A.8. [From now] until the second month, it is time to plant spring wheat and small peas. [It is] also time to plant melons, gourds, mustard greens, mallow, leeks, large and small green onions, water-peppers, thyme, alfalfa *(Medicago denticulata),* a variety of garlic, and taro.[14] [It is] also time to transplant leeks and mustard greens and to put manure on the soil.

A.9. On the first *hsin* day of the month, clean up the dry leaves in the leek furrows.

A.10. This month and throughout the second month, trees can be pruned.

A.11. The [head of the household] orders the one in charge of food in person to brew spring wine reverently and cleanly for provisions for the offering at the summer sacrifices. One can make various sauces. In the first ten days of the month, the soybean is fried; in the second ten days, soybean is boiled. The chopped soybean is made for the *mo-tu* paste,[15] which, by the conjunction of the sixth and seventh month, is to be used for preserving melons [pickles], as well as for making fish sauces, meat sauce, and basic soy sauce.

A.12. From this month to the late summer, one cannot fell bamboos and trees [because the timber] will have worms and insects.

A.13. White dogs are collected, as well as the liver and the blood thereof, which can be used to concoct remedies.[16]

B.1. In the second month, on the day of offering sacrifice at the shrine of the Grand Altar (i.e., the local patron god), one presents

13. A peg is planted in the soil during the winter. When the ice is thawing, the moisture then makes the loose loess soil push up. The peg of two *ch'ih* (feet) in length, as buried, leaves two *ts'un* (inches) above the ground. The pushed-up soil, therefore, would cover the exposed two *ts'un* of the peg. It indicates that the soil contains sufficient moisture for cultivation.

14. According to the original note, the summer green onion is the small onion; the winter green onion, the large onion. Today, the root *yü* [taro] is identified as *Colocasia antiquorum*.

15. It is not clear what this *mo-tu* is. It seems similar to our soybean sauce, which is fermented salty soybean.

16. In Chinese medicine, the blood and the meat of white dogs are preferred to that of dogs of other colors.

leeks and eggs to ancestors and [deceased] parents. Beforehand, purification, food preparations, cleaning, and washing, all are according to those of the formal sacrificial ceremonies. In the evening, one also sets out grave registers and the food and sacrificial vessels. In the morning, the offerings are presented in the tomb. If on the day[17] there is an order from the ruler or other case of emergency, a good day for offering sacrifice at the cemetery should be selected by divination.

B.2. In this month, a date can be selected for a wedding.

B.3. On fine days, one practices archery to be prepared for the unexpected.

B.4. When ice in the shadows has thawed completely, one can turn over [the grass] in fields of good soil, soft soil, and small plots on riverbanks. One can plant spiked millet,[18] soybean, female hemp, and sesame.

B.5. On the day of the spring equinox, thunder will be first heard. Five days before and after, husband and wife should sleep separately.[19]

B.6. Activities for sericulture are not yet started. The tailor is ordered to wash winter clothes and to remove the wadding and [thereby] make lined garments. The surplus cloth is made into autumn clothing.

B.7. In this month elm pods are fully developed. They should be collected when they are still green and dried for storage. If the color changes to white and they are about to fall, the pod can be harvested and made into elm sauce.

B.8. From this month through the third month one can bury tree branches.[20] One can plant foxglove *(Rehmannia glutinosa,* Li.), collect peach blossoms, madder *(Rubia cordifolia),* rootstocks of *kua-lo (Trichosanthes japonica),* and *t'u-kua (Trichosanthes cucumeroides);* in the hilly areas, collect *wu-t'ou (Aconitum carmichaeli* Debx.), *t'ien-hsiung (Aconitum fischeri* Rich.), and *t'ien-men-tung*

17. On that particular day, the second day of the second month. The text gives two corrupted characters, which are noted by Shih Sheng-han.

18. Here spiked millet is given as a special variety called *chih-ho,* which is to be planted in early spring.

19. The original note says, "Those who do not observe the taboo will bear handicapped children."

20. According to the original note, the branches will take root and, after two years, can be transplanted.

(Asparagus lucidus).[21] One can sell millet, panicled millet, soybeans, hemp, and wheat seed and collect charcoal.

B.9. When the swallows build their nests, then one plasters the wall.[22]

C.1. On the third day of the month, one can plant melons. On this day and the first day of *Ch'u* one collects artemisia leaves, *wu-chiu* moss *(Stenoloma chusaum,* L. ching), the fringed pink *(Dianthus superbus),* and willow catkins.[23]

C.2. On the day of *Ch'ing-ming,*[24] the women [who take care] of the silkworms are ordered to prepare the silkworm chamber by plastering the cracks and holes in the walls and by preparing the frameworks and trays for silkworms and the baskets for mulberry leaves.

C.3. Ten days after *Ch'ing-ming,* fresh ginger is sealed [in mud] ; after the day of *Li-hsia,*[25] as soon as sprouts emerge, it can be planted.

C.4. By the time of *Ku-yü,*[26] when the silkworms are all hatched, one joins together wives and children to diligently carry out the work. Do not allow them to devote themselves to other tasks and thereby bring disorder to the primary occupation. Those who disobey such an order should be punished without hesitation.

C.5. In this month, the apricot blossoms are in full bloom. One can remove the grass in the fields of sandy soil, white soil, and light soil. When it rains, it [is time] to plant [nonglutinous] *keng* rice, early ripening spiked millet, hemp,[27] broadbeans, and sesame, and to transplant small green onions. When the constellation Shen does

21. Foxglove and *kua-lo* are raw materials for medicine; peach blossoms and madder are for dye pigment; and *t'u-kua* is for starch. The three following plants are medicinal herbs.

22. The original text is ambiguous. Translation is based upon the correction of the corrupted characters made by Shih Sheng-han.

23. The day of *Ch'u,* cf. n. 8. These four plants are for medicinal use. Willow catkins, according to the original note, are for external use on infected wounds.

24. *Ch'ing-ming* is one of the demarcation days of the twenty-four sections of the calendar. It falls usually one or two days preceding or following the third or fourth of April; cf. n. 12.

25. Beginning of summer; one month after *Ch'ing-ming.*

26. *Ku-yü* falls fifteen days after *Ch'ing-ming*, cf. n. 4.

27. The female hemp does not yield tissue fibers. It is planted for use in torches and candles, according to the original note.

not appear at the south horizon in the evening,[28] and when the mulberry turns red, one can plant soybeans. This is the best time. When the elm pods fall, one can plant indigo.

C.6. In this month, the winter grain is about exhausted, yet the mulberry and the wheat are not ripe. One should coincide with the *yang* element and extend grace[29] by relieving the poor and those in need; devote attention first to the nine grades of relatives, starting with the closest ones. Neither hold back money without being sympathetic toward the poverty-stricken, nor be motivated for reputation by exhausting one's own property for the sake of imitating the wealthy. The expenditures should be within the range of income in order to reach an appropriate balance.

C.7. Agricultural activities are still leisurely. One can drain the drainages and irrigation canals, repair walls and houses while waiting for the [monsoon] rain to come. Doors and windows are fixed, and security facilities are made ready to guard against bandits and thieves during spring famines.

C.8. From this month until the summer solstice, the warming vapors will flourish; the heat of the sun will dry and heat [things]; hence, it is advantageous to use lacquer and oil, and to make various sun-dried medicines.[30]

C.9. [It is time] to sell millet and to buy cloth.

D.1. In the fourth month, after the day of *Li-hsia,* the silkworm eats voraciously. One can plant ginger and make sauce from the roe of bonito fish.[31]

D.2. Silkworms are ascending the hay bundles;[32] the timely rain

28. Evening stars were always important signs for the ancient people to determine time. In the third month, the Shen constellation does disappear below the southern horizon.

29. *Yang* is the element of growing, benevolence, and dominance. When the *yang* element is gaining strength, one should behave accordingly. Also note this time of the year is a period of scarcity as the crops of the last year have been consumed and the new crop is not ready yet.

30. E.g., Chinese medicated pad, which requires a thorough drying of the medicine between two pieces of skin sheets.

31. By the time of Tu T'ai-ch'ing, the identity of this particular fish was already uncertain. Our identification is only tentative.

32. The silkworm, after hatching, grows gradually toward the stage of making a cocoon. At this stage it ascends a hay bundle standing on a large tray and then wraps itself with silk in the midst of the hay.

falls. One can plant millet, spiked millet[33] —which is the best time—as well as soybeans, lesser beans, and sesame.

D.3. On the fourth day of this month, one can brew vinegar.

D.4. The silk cocoons appear in the hay bundles. One hastens to the reeling. The cocoons are opened and the weaving looms are readied; one is careful of the way to keep them [the filaments] unbroken. Gathered are seeds of rape, turnips, seeds of mustard, of parsley *(Petroselinum hortense* hoffm.*)*, of winter hibiscus, and of *lang-tang (Scopolia japonica* max.*)*.[34] When the cuckoo is calling, one can harvest small garlic. When the grasses grow luxuriant, they can be burned for ashes.

D.6. In this month, one can make date cake to treat guests. One can buy oats,[35] and barley, as well as silk wadding. One can transplant small green onions.

E.1. In the fifth month, after the day of *Mang-chung,*[36] the *yang* force starts to decrease and the influence of the *yin* is about to emerge. The weather begins to be warm. Insects and worms are plentiful. One should relax the bows and crossbows with parts of horn and remove the strings. One should stretch the bamboo and wooden bows and relax the string on them. Using ashes, one should store felt, fur, and furry things, as well as arrow feathers. Raincoats are hung on bamboo poles. Do not fold them to store them.

E.2. On the fifth day of this month, one can make vinegar, compound pills of *huang-lien (Coptis chinensis)* for diarrhea and cholera, collect the herb *hsi-erh,* catch toads for making external medicine, and collect crickets [for obstetrics medicine].[37]

E.3. On the day of the summer solstice, wheat and fish are presented to ancestors and [deceased] parents. When that day dawns, the sacrifice is offered. One day before [the ceremony], [one follows] the procedures for food preparation, purification, and cleansing, all

33. The character *ho,* which is translated "spiked millet," seems to have been entered here by mistake, according to Shih Sheng-han's note.

34. The last three species are collected for medical uses.

35. For identifying *k'uang* as oat, Amano Motonosuke gives a good discussion on the prevalent theories. Amano, *Chūgoku nōgyōshi kenkyū,* pp. 60-62.

36. *Mang-chung* usually falls about the fourth of June, cf. n. 12.

37. The character *nien* is too ambiguous to render any sense here; and it is left untranslated. *Hsi-erh* is not identified. The toad is used externally for skin infections. The cricket is used for curing childbirth delivery complications, according to the original note.

of which are similar to [the ceremony for] presenting leeks and eggs.

E.4. When the timely rain falls, one can plant sesame. Five days before and after the summer solstice, one can plant late ripening spiked millet[38] and male hemp; two days before and after [that date], one can plant millet.

E.5. In this month, one can transplant rice and indigo. Until twenty days after the summer solstice, one can turn over the wheatfield and cut the hay stock. After wheat is harvested, plenty of cakes should be made as provisions for travelers. The rainy season is about to come; rice, unhusked grain, firewood, and charcoal should be stored to prepare for the road being impassable from water and mud.

E.6. In this month, the *yang* and the *yin* forces are contending [for dominance],[39] and the energy of the blood is dissipated. During the five days before and after the summer solstice [the husband and the wife should] sleep separately. The *yin* force enters [the body] and becomes blocked in the stomach, which cannot digest oily foods; [hence, for] ten days before until ten days after the summer solstice one should go light on rich flavors and spices and not have many fatty, rich meals. [By the time of] *Li-ch'iu*,[40] do not eat boiled pastries and pastries made by soaking meal.[41]

E.7. In this month, one can prepare elm sauce and minced meat sauce. One sells beans, lesser beans, sesame; buys oats, barley, and wheat; and collects cloth rags and clothes. After the summer solstice, one can buy wheat husks, which after drying, are tightly sealed in urns in order to have no worms. In the winter, they can be fed to horses.

F.1. In the first part of high summer in the sixth month,[42] wheat and melon are presented to ancestors and [deceased] parents. Purification, food preparation, cleansing, and washing, all are the same as [the procedure for the ceremony of] presenting wheat and fish.

38. This unidentified variety of spiked millet is planted almost three months after the early one.

39. The summer solstice, around June 21, is regarded as the day on which the *yang* reaches its fullness.

40. *Li-ch'iu* is the beginning of autumn and usually falls around August 6, cf. n. 12.

41. According to the original note, food made from flour dough is hard to digest; it is the fermented flour food that is easy to dissolve.

42. High summer, the thirty days between mid-July and mid-August, has three divisions: *ch'u-fu*, *chung-fu*, and *san-fu*.

F.2. In this month, one is urged to weed with hoes. Do not miss the time! The wheatfield is to be turned over. The female workers are ordered to weave white and light silk.

F.3. On the sixth day of this month, one can plant mallow *(Malva verticillata)*. After the midpart of high summer,[43] one can plant winter hibiscus; one can plant turnips, winter indigo, and small garlic, and transplant large green onions. One can burn [wood for] ashes for dying green, purple, and other colors.

F.4. After the day of *Ta-shu,* and the midpart of high summer is over,[44] one can store gourds, lay up melons, and gather mustard seed, until the seventh month.

F.5. On the twentieth day of the month, one can hull and sift wheat and grind it. On the twenty-eighth day, one soaks it and keeps it in the bedroom. By the seventh day of the seventh month, one ought to make yeast leaven.[45] One must personally and reverently [carry out these steps] to provide the resources for sacrificial [liquor] offerings throughout the whole year. The quantity [of the yeast] is not fixed but is determined by the financial condition of the house.

F.6. One can sell soybeans, buy oats and wheat, and collect plain white silk.

G.1. On the fourth day of the seventh month, one orders the preparation of the brewing chamber, one readies mats (i.e., screens for walls), and frameworks, and gathers clean artemisia leaves.[46] On the sixth day, the five grains and the mill are all ready.

G.2. On the seventh day, one then makes yeast. On this day, one can compound medicinal pills and *shu-ch'i* pills, expose books and clothes to the sunlight, prepare dried food, and pick *hsi-erh.*[47]

G.3. In this month, one can plant rape, turnips, mustard, alfalfa, large and small green onions, small garlic, and shallots, and transplant

43. In late July.

44. *Ta-shu* usually falls around July 22; cf. n. 12.

45. According to the original note, the period of fermentation takes at least six or seven days and as much as ten days.

46. Artemisia is burned to smoke the insects out of a room, a practice commonly observed in China during summertime. Here this grass is burned to keep the brewery clean from insects and worms.

47. *Shu-ch'i,* another name for *ch'ang-shan,* is a fever medicine. On *hsi-erh,* see n. 37.

leeks, store leek flowers, cut hay, turn over wheatfields,[48] and collect cypress seeds.

G.4. The day of *Ch'u-shu*[49] is close to the autumn season.[50] Old clothing is washed; the new is tailored. Heavy clothing is made to prepare for the beginning of cold weather.

G.5. One can sell lesser beans and beans, buy wheat, and collect raw silk and white silk.

H.1. In the eighth month, a good day after the day of *Pai-lu*[51] is selected by divination to offer sacrifices to the honored deities who are regularly worshipped through the year. During the seven days before [the ceremony], the whole family shall not go to houses where there is mourning or a new birth. The young, the old, and those who have assignments, all should observe purification. One sets forth the sacrificial registers, and devotes oneself with exceptional care to cleaning and washing. In this month, sacrifices are offered to the Grand Altar. On the day of worship, millet and pigs are presented to ancestors and [deceased] parents. On the morning of the day, one worships at the tombs with the same [ceremony as when] presenting wheat and fish.

H.2. When the summer heat is receding, the small children are sent to school as they are in the first month. The cool breeze heralds the cold. One should hurry to select various kinds of plain, white silk to be dyed in colors. One can tear apart the floss and make ready the wadding. New clothes are made; the old, washed. Reed sandals are purchased when they are cheap and good in anticipation of need to prepare for wintry bitter cold.

H.3. On the eighth day of this month, one can collect the herbs, *chü-ch'ien-tzu (Plantago major* L.), *wu-t'ou, t'ien-hsiung,* and *wang-pu-liu-hsing (Vaccaria pyramidata medic.).*[52]

H.4. In this month, one can marry [a wife].

H.5. One can cut open the gourds for use,[53] dry foxglove, make

48. In one version, this step of turning over the grass in wheatfields is work carried out in the fifth and sixth months, not in the seventh month.

49. *Ch'u-shu* usually falls around August 23; cf. n. 12.

50. Here the "autumn season" *(ch'iu-chieh)* does not refer only to the season of autumn; it means the demarcation day of *Pai-lu,* the mid-autumn period, which falls around September 8.

51. *Pai-lu,* cf. n. 50.

52. All these herbs are for medicinal uses.

53. According to the original note, the white inner tissue of the gourd is good for hog food and the tissue close to the shell is good for candles.

mo-tu preserve paste, cut weeds and hay [for fodder], collect leek flowers and make leek sauce. One can dry mallow, collect leaves of beans, and plant large and small garlic and mustard. Whenever barley and wheat are planted, the inferior fields should be planted on the day of *Pai-lu*; the ordinary fields, on the day of the autumn equinox; and the fertile, ten days later. Only oats [can be planted] either early or late; there is no regularity.

H.6. As it is cool and dry, bows with parts of horn may be set up. The structure [for assembling the bow] is repaired and corrected and the string is tied on; then it is time to practice archery. Bamboo and wooden bows are to be relaxed.[54]

H.7. Sell wheat seed, buy millet.

I.1. In the ninth month, the pounding field is to be cleaned, the [walls of] the granary, plastered; the storage pit, repaired. The five weapons are repaired and military exercises and archery are practiced to guard against bandits in the time of coldness and poverty. Relatives and kinsmen are comforted: the widows, the orphans, the aged, the sick, and those who are not able to take care of themselves. Divide the wealth and distribute what one has an abundance of to relieve them from the cold.

I.2. On the ninth day of this month, one can pick chrysanthemum flowers and collect *chih* oranges (*Citrus trifoliata* L.).

I.3. The ninth month [is the time] to make preserved and dried mallow, to store fresh ginger and wild ginger. If it is still warm in that year, all [these works] should be delayed to the tenth month.

J.1. The tenth, walls are reconstructed and thickened,[55] north windows are blocked, and holes are plastered up. One should hurry to store up the harvest, which ought not to be left in the open field! One can collect turnips and store up melons.

J.2. On the first *hsin* day[56] of the month, the cook is ordered to steep the yeast leaven. When the yeast is dampened one brews

54. The complex bow, which consists of horn, bone, and other materials, is not practical in humid summer days. In the autumn, the bow is assembled once again since the weather is dry and cool. Setting up the bow, one needs to correct the positions of the parts on a frame.

55. Walls in North China are constructed of tamped earth. After the rainy period of spring and summer, walls require reinforcement by filling holes and by tamping more earth for solidity.

56. *Hsin* is the eighth of the ten-day series.

winter wine. [The master] must [prepare this] reverently in person to provide the wine for use in the sacrifices at the winter solstice, in the *la* sacrifice,[57] New Year's Day, for the guardian deity for travelers, and in the presentation of leeks and eggs [in the second month]. In this month one makes dried and salted meat for use in the *la* ceremony.

J.3. When agricultural activities are over, one orders the adolescents and those older to enter the senior school, the same as in the first month.

J.4. The five grains are harvested; the household collects its stores. Then, according with ordinances of the seasons, one orders that funerals [be carried out]. If there is a poor clansman who has been dead for a long time and for whom burial could not be afforded, one should then organize the clan members to jointly give assistance [for the burial]. The responsibility is determined by the criteria of relationship and wealth. The share should be fairly decided so that there is no excessive burden. Pay heed to putting first one's own exaction in order to lead others to follow.

J.5. In this month, one can transplant large green onions. Before freezing begins, let jelly food be prepared and sweetcakes be made. One can split hemp; one hastens to spin hempen thread. One makes white shoes[58] and grass sandals. One sells clothes, silk, and waddings. One purchases spiked millet, beans and lesser beans, and hempseed. One gathers *kua-lo* gourds.

K.1. In the eleventh month, on the winter solstice, one presents millet and lamb; first, sacrifices are offered to the water deity at the well, and then to ancestors and [deceased] parents. Purification, food preparation, cleaning, and washing—all are similar to [the procedure for the ceremony of] offering millet and pigs. Toasting the seniors, and preparing name cards, calling upon, and offering greetings to the lord, the teachers, and the elders are similar to what was done in the first month.

K.2. In this month, the *yin* and the *yang* elements are contending for dominance, and the energy of the blood is dispersed. For five days before and five days after the winter solstice, the husband and the wife should sleep separately.

57. *La* sacrifice was held in the twelfth month as the last sacrificial ceremony before the New Year festival, cf. paragraph L. 1.

58. The words "white shoes" are dubious. We may speculate that the term means plain, undecorated.

K.3. Water on the ink stone freezes; one orders the young children to read the texts of the *Classic of Filial Piety* and the [Confucian] *Analects,* and to enter the lower school.

K.4. Gradually, one should prepare millet, spiked millet, and rice food,[59] which are all provisions for the sacrifices in the *la* ceremony. One can brew [fermented] meat sauce, and cut bamboo and timber. White dogs are purchased and raised for the sacrifice to ancestors and [deceased] parents.

K.5. Buy *keng* rice, spiked millet, beans,[60] lesser beans, and hempseed. [Addendum] On the day of the winter solstice, one measures one *sheng* of each of the five kinds of grain, which are put in small urns and buried in the shadow of the northern wall. Fifty days after the winter solstice, [the grains] are taken out and measured. The one that has gained most will be the most suitable for the coming year.

L.1. On the day of the *la* sacrifice in the twelfth month, one offers rice and wild geese. Five days before the date, pigs are butchered. Three days before the date, sheep are butchered. Two days before the date, there is purification, food preparation, cleaning, and washing—then the *la* sacrifice is offered to ancestors and to the five deities. The day after the *la* sacrifice is called "Minor New Year"; wine is offered and the deities descend. Toasting the seniors, preparing name cards, and offering greetings to the lord, the teachers, and the elders are similar to that on the New Year's Day. On the next day, once again sacrifices are offered; it is called the *cheng* sacrifice. Three days later, the sacrificial ceremonies in the house are concluded. Then, to promote goodwill, clansmen, relatives, and guests are invited and entertained. Agricultural activities are stopped and labor is brought to a halt; the workers have a rest; grace must be downward extended.

L.2. In this month, a multitude of deities incessantly move around. A *ta-cha*[61] ceremony is held; one then offers sacrifices to

59. In the text, three varieties of millet are given. Yet the combination of these four grains is general usage for all cereal food. The translation, therefore, is not intended to make distinctions.

60. The text includes in the list an item, *mi,* "husked rice." Shih Sheng-han suggests that it is a corrupted form of *ta,* "large." Therefore, the item has been translated "bean."

61. *Ta-cha* is a ritual in which all the miscellaneous deities are thanked and worshipped.

[deceased] lords,[62] teachers, kinsmen, and friends to esteem the principle of lasting relationships.

L.3. Then there is rematching of farming teams and selecting implements. Plow oxen are raised; those good for field work are chosen, awaiting the beginning of farming.

L.4. Store the pig jaw bone[63] and burn the bamboo fan used during the sacrifices of the *la* ceremony.[64] At the East Gate, the head of a white rooster is crushed [as a talisman]. Find oxen gall to make pediatric medicine.[65] [*Ssu-min-yüeh-ling*]

2. Fan Chung's Large Landholdings

Ca. First century B.C.

[The Fan] was a prominent clan in its local area. [Fan Hung's] father was named Chung, styled Chün-yün. [The Fans] for generations had been good in farming and skilled in trade. [Fan] Chung was warm and generous. For three generations members of a common ancestry lived together to share the property. [Chung's] children and grandchildren paid him respect in the morning and in the evening [twice daily], as if the household were a court. Chung managed the property without wasting anything. The servants and the slaves were assigned tasks according to their individual aptitudes. Therefore, the master and the subordinates cooperated to double the profit annually. They

62. "Lords" means the local officials and those under whom one had served or been recommended.

63. According to the original note, it is used for making external medicine for infections.

64. According to the original note, drinking a solution containing the ashes of a burned fan is effective in curing the wound made by a thorn in the flesh, and burning it at the corners of the melon field drives away insects.

65. This translation of the full text of Ts'ui Shih's *Ssu-min-yüeh-ling* is based upon the reconstruction of the text by Shih Sheng-han. The original work, written in the mid-second century, has long been lost. Paragraphs are cited in works written later, such as Chia Ssu-hsieh's *Ch'i-min-yao-shu* (mid-sixth century). Four reconstructed versions had been published before Shih Sheng-han's effort. Shih's reconstruction is based upon the paragraphs quoted by Tu T'ai-ch'ing's *Yü-chu-pao-tien* (of the late sixth century). The strength of the Tu edition is its clearer distinction between the texts and the notes, although notes made by the early and later commentators, namely by Chia and Tu, are blended. Shih Sheng-han, using an early hand-copied edition, makes a better reconstruction than the one previously published. The translation presented covers only the text and omits the notes unless they are relevant. The edition I use is *Ssu-min-yüeh-ling chiao-chu*, published in 1965 by Chung-hua Press, Peking. Another reconstruction, based upon virtually the same sources, is by Moriya Mitsuo in his *Chūgoku kosaijiki no kenkyū*, pp. 9 ff. and pp. 272 ff. See also Patricia B. Ebrey, "Estate and Family Management in the Later Han," *Journal of the Economic and Social History of*

reclaimed more than three hundred *ch'ing* of arable land. They constructed residences equipped with terraces and high halls. Reservoirs and canals irrigated the fields.[66] Fish were in the ponds; animals, in the pastures. Every needed item was available [at home].[67] Once, when there was the need to make furniture, Chung started by planting catalpa and lacquer trees. His contemporaries laughed at him. Yet as the time went by, all these [accumulated items] became useful. Those who previously had laughed at him turned to him for provision. His property had a worth of billions. The clan members were provided with assistance and relief, and grace was extended to the neighborhood. . . . He was elected as the *san-lao* (thrice venerable) of the local district. He died in his eighties. In his will, it was instructed that the debt records, which amounted to several millions, be destroyed and burned. The debtors, after hearing the news, strived to return the money. Chung's sons, however, followed the instruction and declined to accept it.[68] [*HHS* 32:1a-1b]

3. Wealth of the Powerful

The powerful persons possess several hundred houses, which are connected one after another, and extensive fertile land stretching all over in the countryside. Retainers form groups by the thousands, and those who are attached to [the powerful persons] are counted by the tens of thousands. Carriages and ships are dispatched to make trade in all directions. To disperse and to store up [commodities, their agents] fill the cities. Precious jade and valuable things are too numerous to be kept in a large room; horses, cattle, sheep, and pigs are too many to be enclosed in a valley.[69] [*HHS* 49:13b] A.D. 208-220

the Orient 17 (1974): 173-205.

66. According to the *Shui-ching-chu*, the Fan reservoir was ten *li* in length and five *li* in width. It also recorded the saying that "Water being abundant [in the Fan reservoir], the lower land is indeed fertile."

67. The *Hsü-Han-shu* (quoted in the commentary of *HHS*) described the wealth of the Fan house by pointing out that "within the household, it was as well-stocked as a market: weapons, bows, utensils, furnishings. The value [of the goods[reached one million. . . . Their wealth was comparable to that of an enfeoffed aristocrat."

68. Cf. VI, 23 for some of the activities of Fan Chung's son, Hung. This document, with some variations, is in T'ung-tsu Ch'ü, *Han Social Structure*, pp. 286-87.

69. This is an excerpt from the chapter on "Order and Disorder" in the book called the *Ch'ang-yen* written by Chung-ch'ang T'ung (see II, 44). In this paragraph, it is clear that a wealthy person was engaged not only in agriculture, but also in commercial activities. Meanwhile, such a wealthy person had the service of both retainers and "attached persons," who probably were tenants and even ordinary persons who were dominated by the power of the wealthy one.

4. The Subordinated Population

A.D. Second century

The dominative households[70] accumulate fortunes with values of millions, and their houses and lands are comparable to those of an enfeoffed lord. They use bribes to distort the administrators' behavior and hire assassins to intimidate the common folk. They arrogantly murder the innocent; yet [the powerful person who is responsible for the murder] boasts that no child [of his] has died in the market (i.e., has suffered capital punishment). The style of life for the living and the funerals for the dead match those of a ruler. Therefore, lower households have a rugged path to tread, nay, even no space to stand! The father and the son, lowering their heads, serve the rich as if they were slaves. They lead the whole household, wife and children, to render services. Hence the rich, who have a surplus, become prosperous day by day, while the poor, who are short of reliable resources, become impoverished year after year. For generations, they have been almost as captives, yet they are still in need of clothing and food. Alive, they toil all their lives; dead, they worry about having their bodies exposed. Any slightly poor crop causes them to wander in the roadside ditches and to sell their own wives and children.[71] [*Ch'üan-Hou-Han-wen* 46:10b, hereafter cited as *CHHW*]

5. Household Register (1)

Ca. Mid-first century B.C.

Li Chung, age 30, order of *kung-ch'eng,*[72] of Kuang-ch'ang hamlet, Lu-te prefecture, fortress commander.

2 nonadult,[73] male slaves, valued at 30,000	5 working horses, valued at 20,000	1 residence, 10,000
1 adult female slave, valued at 20,000	2 ox-drawn carts, valued at 4,000	5 *ch'ing* of land, 50,000
2 light, horse vehicles, valued at 10,000	2 yoke-oxen, valued at 6,000	Total assessed value 150,000[74]

[Michael Loewe, *Records of Han Administration* 1:72]

70. Here "dominative households" refers to the landlords and "lower households" indicates the subordinated population, including both tenants and other already subjected people.

71. This is an excerpt from the *Cheng-lun* (Treatise on Politics), written by Ts'ui Shih, the author of the *Ssu-min-yüeh-ling.*

72. *Kung-ch'eng* was an honorific title, eighth on the twenty-rank scale.

73. The original text has *hsiao-nu,* "minor slave," which means below fourteen years old. See Hiranaka Reiji, "Kyoen Kan-kan to Kandai no zaisanzei," *Ritsumeikan daigaku jinbun kagaku kenkyūjo kiyō*, no. 1 (1953).

74. This tablet is from Chung-kuo K'e-hsueh-yuan, K'ao-ku yen-chiu-so, comp.,

6. Household Register (2)

Hsü Tsung, age 50, order of *kung-ch'eng*, of Hsi-tao hamlet, Chü-yen prefecture, commanding officer of San-wu section.[75] *Ca.* Mid-first century B.C.

1 residence, valued at 3,000	1 wife
50 mu of land, valued at 5,000	2 sons
2 yoke-oxen, valued at 5,000	2 daughters
	2 male *t'ung-ch'an*[76]
	2 female *t'ung-ch'an*[77]

[Loewe, *Han Administration,* p. 71]

7. A Mock Contract between a Slave and His Master

Wang Tzu-yüan of Shu commandery came to Chien on personal affairs and stayed at the house of the widow, Yang Hui. Yang Hui had a slave named Pien Liao. Tzu-yüan asked the slave to buy wine [for him]. Pien Liao, dragging a large stick, ascended the tomb of the husband [of Yang Hui] and said, "When the master bought Pien Liao he only instructed [Pien Liao] to guard the tomb and not to buy wine for other men." Tzu-yüan, greatly enraged, said, "Do you want to sell this slave?" Hui said, "This slave is very offensive. No one wants him." Tzu-yüan then fixed a contract to buy the slave. The slave again said: "The work you desire me to take on should all be specified in the contract. What is not specified in the contract, Pien Liao cannot take on." Tzu-yüan said, "Yes." 59 B.C.

The text of the contract reads: "On the fifteenth day of the first month of the third year of the Shen-chüeh reign period,[78] Wang Tzu-yüan, a male from Chih-chung, purchases from Yang Hui, a female of An-chih district of Ch'eng-tu, the bearded slave, Pien Liao, who

---

Chü-yen Han-chien chia-pien, 181 B. The translation here is largely based upon Loewe, *Han Administration.* For detailed studies see Ch'en P'an, "Yu Han-chien chung-chih chün-li ming-chi shuo-ch'i," *Ta-lu tsa-chih,* vol. 2, no. 8 (1951): Hiranaka, "Kyoen Kan-kan to Kandai"; Loewe, *Han Administration,* p. 145, n. 75; Ho Ch'ang-ch'un, "Ch'in-Han chien ke-t'i-hsiao-nung-ti hsing-ch'eng ho fa-chan," pp. 147 ff.

75. Hsu commanded a section on the fortified defensive construction against the nomads.

76. *T'ung-ch'an* implies siblings and the children of siblings. Ho Ch'ang-ch'un, "Ch'in-Han-chien ke-t'i-hsiao-nung-ti," p. 181, n. 15; p. 186.

77. See III, 5, n. 74. This tablet if from Lao Kan, *Chü-yen Han-chien k'ao-shih: Shih-wen-ch'ih-pu,* p. 455.

78. This is in the reign of Emperor Hsüan; the third year is 59 B.C.

belongs to the household of her deceased husband. The price is 15,000 cash.[79] The slave should do various tasks without arguing.

"[He should] sweep the floor in the morning when he arises, wash the dishes after eating, and as routine work: bore mortars, make brooms, cut bamboo poles, make containers by chiseling wood, clean the drainage, mend fences, hoe the garden, repair field paths, stop leakages in the field banks, make large flails, twist bamboo to make rakes, scrape and repair well pulleys. He should not mount a horse nor ride in a cart, nor sit with outstretched legs, nor utter complaints. When he gets up from his bed,[80] he should go fishing, cut the hay into bundles, weave mats, and split hemp tissues. He should draw water for gruel and help in making *tsu-mo* drink.[81] He should weave sandals, make bird nets to trap birds and catch crows, weave nets to catch fish, shoot wild geese with stringed arrows and shoot wild ducks with pellets, ascend the mountain to shoot deer, dive into the water to capture turtles, in the pond keep fish, in the backyard raise geese by the hundreds, chase away owls and other [evil] birds, use a rod to herd the swine, grow ginger,[82] raise sheep, rear small pigs and colts, clean and plaster halls and walls [of pens, stables, and barns], feed the horses and cattle, rise before dawn as well as feed the animals at midnight.

"In the second month on the vernal equinox repair dikes, cultivate areas on and around the field banks, trim mulberry trees, peel bark from palm trees, plant melons, make dippers from gourds, transplant the vegetables and onions, burn brush, dig trenches, pile up the ridges [between furrows], break up lumps in the field. [He should] light the fuel early at noon, pound grain in the mortar when the rooster crows, tame horses and mules, put on horse-shoes. . . .[83]

"When there are guests [he should] grasp the pot to buy wine, draw water to make the meal, wash dishes, set the table, pick garlic in the garden, cut up vegetables, chop dried meat, slice pork, bake taro, cook fish, roast turtle, and make tea. He will clean up the left-

79. The slave price seems quite consistent with information given in other sources. Cf. III, 5.

80. This line, according to the rhyme arrangement reconstructed by Utsunomiya, should fall in the place preceding the sentence, "take an ax into the mountains."

81. According to Wilbur, this is a fine brew, made from the skimmings of boiled butter; Wilbur, *Slavery in China,* p. 391.

82. Ginger certainly is not adequate feed for animals. It is included here simply for the sake of rhyme between *chiang* (ginger) and *yang* (sheep).

83. The text here is too ambiguous to make sense; cf. Wilbur, *Slavery in China,* p. 391, n. 18.

overs and store them, close the door, shut the small windows, feed the pigs, and unleash the dog.

"[The slave] should not quarrel with the neighbors. The slave should only eat beans and drink water and should not be drunken. If he desires to drink good wine he can only wet his lips and his mouth. He should not turn over the cup or invert the dipper. He should not go out in the morning and return at night. He should not make friends or find female company.

"The trees behind the house should be cut to build boats with which [the slave should] sail upstream to Ch'ang-chou and downstream to Chien-chu, working for cash under [the supervision of] the staff of the local administrations.[84] He is to search for plaster and trade in cordage made of palm tissue.[85] At Mien-t'ing, [he should] purchase mats. Traveling between Hsin-tu and Lo-hsien, he should find cosmetics for women and sell them at small market fairs. On his return trip, he should carry hemp tissue and sell it along side routes, from whence he brings dogs and trades geese. At Wu-yang he should buy tea and carry away lotus from the Yang family pond. When he travels to markets and villages, he should maintain careful guard against tricks and theft. When he enters the markets he should not squat, lie on the ground, speak evilly or revile others. He should amply make knives and bows that he brings into I-chou to trade for cattle and sheep. The slave should teach himself, being clever and witty, and should not pretend to be dull and stupid.

"After awaking from sleep, he should take an ax into the mountains [to cut wood] for making wheel rims and cart shafts; from the remaining [wood] he should make small tables, wooden sandals, and feeding plates for dogs and pigs. He should burn firewood to make charcoal; pile rocks to make bank walls; repair the houses and put roofs on the buildings; and cut wooden tablets for use in writing. In the evening [when he] returns he ought to bring back home two or three bundles of dried firewood.

"In the fourth month he should transplant [in the field]. In the ninth month he should harvest.[86] In the tenth month he should collect

84. The sentence that involves "the staff of local administrations" seems out of place in the context. In several versions, this sentence and the following sentence are both missing; cf. Utsunomiya Kiyoyoshi, *Kandai shakai keizaishi kenkyū*, p. 270, and Wilbur, *Slavery in China*, p. 391, n. 21.

85. This sentence is ambiguous because of the uncertainty of the readings. I take *o*, "evil" to be *o*, "plaster for building walls."

86. The text given here has "fifth month" instead of "ninth month"; cf. Utsunomiya, *Kandai shakai keizaishi kenkyū*, p. 270; and Wilbur, *Slavery in China*, p. 392, n. 29.

beans and store hemp, from which he should make rope and cord. On rainy days when he has nothing to do he should weave screens and mats. He should plant peach, plum, pear, persimmon, and mulberry trees. There should be one tree each three *chang* and eight *ch'ih* between rows. [He should] plant fruit trees according to their species in a checkerboard arrangement. When the fruits ripen for picking, he should not sit and taste any. When the dog barks he should rise to wake the neighbors, close the doors, close the windows, ascend to the upper floor to beat the drum, carry a shield and grasp a spear, and patrol three times around the settlement. [He should] work hard and promptly and not play around. When he is old and his strength is exhausted [he still should] plant the weaving grass and weave the matting. When he takes a rest after work he should pound one *shih* of rice. At midnight when he has nothing to do he should wash clothes until they are white. If he has secretly made private savings, the master can spend it to entertain guests. The slave should not be allowed to have vicious and selfish activities; everything should be reported [to the master]. If the slave does not comply with the instructions, he should be flogged one hundred strokes."

After the contract was read, the slave was utterly speechless. In reckless agitation, he knocked his head repeatedly, and struck himself with both hands. Tears fell from his eyes and snivel from his nose hung down a foot. He said, "If things will be as Master Wang has said, I'd rather return to the yellow earth,[87] and let the earthworms penetrate my forehead. If I had known earlier what would happen, I would have bought wine for Master Wang and dared not to offend indeed."[88] [*Ch'üan-Ch'ien-Han-wen* 42:11b-12b; cf. C. Martin Wilbur, *Slavery in China during the Former Han Dynasty*, pp. 383-92]

8. Agricultural Productivity

48 B.C. [Kung Yü][89] again memorialized: "In ancient times metal coins

87. "Yellow earth" is a common expression in the Han period referring to the world of the dead.

88. The translation is based upon the studies made by Utsunomiya, *Kandai shakai keizaishi kenkyū*, and Wilbur, *Slavery in China during the Former Han Dynasty*, pp. 383-92. The original text is a jocular piece of verse in rhyme by Wang Pao, a poet in the first century B.C. The mock contract gives practically all the activities, agricultural as well as nonagricultural, on the farm of a small landlord. The labors listed are beyond the possible capacity of a single human being, naturally. The text is now available in four versions contained in encyclopedic or anthological works. Details of textual comparison are given by Utsunomiya and Wilbur. Only the major problems are brought into the notes given above, however.

89. On Kung Yü, see I, 8, n. 19; see also II, 34 and III, 15.

were not used as currency. People were devoted to farming. Therefore, if there was one man who did not cultivate the land, some would necessarily suffer from hunger. Now [under the rule of] the House of Han the Office for Coinage and the Office for Iron both have set up officials and have conscripts and convict laborers dig in the mountains to extract copper and iron. A year's work utilizes more than a hundred thousand men. As a farmer of medium means can feed seven persons,[90] the result is that there are seven hundred thousand people who are constantly suffering from hunger. [*HS* 72:13a-b]

9. Livelihood of a Farmer

Ca. Fourth century B.C.

Today[91] if a man supporting a family of five persons cultivated 100 mu of land, each year from each mu he would harvest 1½ piculs [of grain], making [a total of] 150 *shih* of unhusked grain.[92] Out of this would be taken 1/10 for taxes-in-kind, [that is], 15 *shih*, and there would remain 135 *shih*. For food, each person each month should consume 1½ *shih*, and five persons for an entire year should consume 90 *shih*. Remaining is 45 piculs, which would be worth 1,350 cash [at the rate of] 30 cash for each *shih*. Deducting expenses of 300 cash for offerings at community [Earth] shrines, offering of new [crops to deities], and spring and autumn sacrifices,[93] there would remain 1,050 [cash]. For clothing, usually each person should expend 300 cash and five persons for an entire year should consume 1,500. [The balance is] a deficit of 450 [cash]. Not included here are expenses in times of affliction, illness, death, burial, and government levies [other than the land tax-in-kind]. That is why farmers are

90. This figure probably only refers to the grain consumption of seven persons. Suppose the farmer has a five-person family to support, he can have 28.5 percent of his production as surplus after he has fed his own dependents. This "surplus," however, should be converted to cash for taxes, as well as payment for clothing and other consuming items that he does not produce on his own farm.

91. This passage is attributed to a Chan-kuo statesman, Li K'uei, who was the minister of a ruler of Wei, 424-387 B.C. On the identification of Li K'uei, see Swann, *Food and Money*, p. 136, n. 100.

92. There is considerable differentiation among the standards of measurement in various Chan-kuo states. This *shih* unit given here should be a unit of capacity in the state of Wei. We have difficulty in determining its equivalent, especially as on one of the ancient Wei measures, the reading of the inscription is dubious. Ma Ch'eng-yüan, "Shang Yang fang-sheng ho Chan-kuo liang-chih," *Wen-wu*, 1972, no. 6, pp. 18-19. There is little possibility of establishing the capacity in terms compatible with Han measurements.

93. It is not impossible that the new crop offerings, as well as the spring and autumn sacrifices, were all held at the community Earth shrines, which served as the center of religious activities in the village communities.

constantly in want, and have no incentive to engage in farming. Meanwhile the grain price is excessively high. Therefore those who are good in planning price control at a constant level ought to observe that there are three grades of harvest: the best, the better, and the good. In a best harvest, [the product] would be four times [as much as the ordinary harvest]; and [after deducting for food] then there would be a surplus of 400 *shih.* In a better harvest, [the product] would be three times [as much], with a surplus of 300 *shih.* In a good harvest, [the product] would be doubled, with a surplus of 100 *shih.* [In a year of] a slight famine, the harvest is 100 *shih;* [in that of] a medium famine, 70 *shih;* and [in that of] a great famine, 30 *shih.* Hence, in a year of best harvest the government then would purchase three [quarters of the surplus], leaving one [for the people]. In a year of better harvest [the government] then would purchase two [thirds], while in the year of good harvest, [the government] would purchase one [half of the surplus]. This causes the people to have just a sufficiency for all needs; and when the price returns to its ordinary level, [government purchase] would be stopped.[94] [*HS* 24A:7b-8a; cf. Nancy Lee Swann, *Food and Money in Ancient China,* pp. 140-43]

10. Temporary Military Colonies

60 B.C. [Chao Ch'ung-kuo][95] then submitted a memorial on the establishment of military colonies,[96] "Your subject has heard that the armed forces are for the purpose of manifesting virtue and eliminating harm. Therefore, if undertakings are competent externally then good fortune is produced internally. One indeed ought to be very cautious. Your subject has under his command officers, soldiers, horses, and cattle, which totally consume monthly provisions [as follows]: 199,630 *hu*

94. Calculating the remainders of the surpluses, the sufficiency level of a farming household is given as 200 *shih.*

Please note the following calculation, given in *shih* of grain:

	Product	Government purchase	People's share
Best year	600	300	300
Better year	450	200	250
Good year	300	50	250
Mediocre year	150	0	150

95. Chao Ch'ung-kuo, a senior warrior at that time, was commanding the expedition against the Ch'iang tribes in the present provinces of Kansu and Chinghai.

96. It is by Chao's suggestion that military colonies were organized to sustain a prolonged military operation.

of grain, 1,693 *hu* of salt, 250,286 *shih* of fodder. . . . [I have] estimated that there are more than 200,000 mu of land, including the fields formerly owned by the Ch'iang, the government fields, as well as the land that is not cultivated by the people, in the area between Lin-ch'iang and Hao-meng to the east. . . . Your subject previously ordered soldiers to fell trees in the mountains: timber, large as well as small, amounting to some sixty thousand pieces in total, were stored at the riverside. I would withdraw the cavalry units, yet retain the conscripts,[97] the mercenaries, as well as the infantry from Huai-yang and Ju-nan,[98] and the private retainers of the officers,[99] for a combined force of 10,281 persons in total. They would consume 27,363 *hu* of grain and 308 *hu* of salt monthly. Let them be stationed at the strategic locations. When the ice thaws and transportation by waterway becomes possible, let [the soldiers] repair the local guard posts,[100] drain the canals and ditches, construct some seventy bridges to the west of Huang Pass so that [the army can] reach the vicinity of the Hsien River.[101] At the time of farming, each soldier is to be assigned twenty mu of land. By the fourth month, when the grass grows, [Chinese] cavalry units from the commanderies and the barbarian cavalry units from the subordinate tribes and states will be dispatched. Each group shall consist of 1,000 strong cavaliers supplemented by a 20 percent complement of spare horses to patrol the pasture to protect the farming units. The crops should be shipped to Chin-ch'eng commandery to increase the stores and to save the great cost [of transportation of provisions] so that with the grain shipped by the Grand Ministry of Agriculture,[102] the total can reach an amount sufficient to provide 10,000 persons with one year's provision. I hereby respectfully submit lists of field localities and [the necessary] implements. May Your Majesty grant approval." [*HS* 69:10b-11a]

11. SALARY OF A GOVERNMENT STAFF MEMBER

A senior staff member . . . draws a monthly emolument of twenty *Ca.* A.D. 150

97. In Han times, the convict could volunteer to serve at the frontier as a substitute penalty.

98. These two regions traditionally were assigned to organize infantry units from their draftees.

99. In Han foreign wars, it was not uncommon for ambitious persons to join the expeditions voluntarily, supported by their own means, in order to win merits that eventually could gain for them positions in the military service. Cf. *HS*, 61:10b.

100. The original text refers to the buildings for the local village administration. Such a building, called a *t'ing*, served both as post station and guard station.

101. This particular river is not identified in the Han sources.

102. The Ministry of Agriculture, in fact, was the State Treasury in the Han government structure.

hu of millet and 2,000 cash. . . . Now [he spends] 1,000 cash as monthly wage to a retainer, 500 cash for fodder, oil, and meat, and 500 cash for fuel, salt and vegetables. Two persons consume six *hu* of millet [each month]. The remainder is barely enough for horse fodder.[103] How is he able to provide himself with winter and summer clothing and blankets, sacrifices at seasonal ceremonies, and entertainment for friends! How can he afford to welcome parents and family [to live with him]? [*Cheng-lun,* as cited in *CHHW* 46:9a-9b].

12. Taxation According to Acreage

Ca. 81 B.C. The Secretary:[104] In antiquity, it was regulated that 100 paces of field formed a mu, which the people farmed according to the well-field system[105] in which one-tenth was presented [to the lord], following the principle that the "lord's interest cometh first, private concerns, second." Such was the fundamental duty of citizens and subjects. The late emperor,[106] taking pity upon the hardships and sufferings of the people and their insufficiency in food and clothing, promulgated new regulations whereby 240 paces of field formed one mu and the tax was levied at the rate of one-thirtieth. But the idle people, not working strenuously on their farms, cause hunger and cold to fall upon themselves. This indeed is what they deserve. They want to sow without having plowed, and to reap without having planted. Why lay the blame for this on the salt and iron monopoly?

The Literati:[107] The tithe collected for the lord came [from

103. There seem to be certain exaggerations in this document. Taxes, clothing, and housing are not included in the expenditures. The cost of maintaining horses is given under "fodder," together with oil and meat. Yet it is mentioned again as the main expense after the monthly provisions are deducted.

104. The secretary is Sang Hung-yang, who was then the major promoter of state control of commerce and monopolies on salt and iron. He held the position of grandee secretary.

105. The well-field system was an idealized land system alleged to have been in practice in antiquity. The first mention of such a system is made by Mencius, who suggested that eight families work on a ticktacktoe-patterned complex in which each household worked one plot of 100 mu of land and eight families jointly worked on the central plot wherefrom the crops went to the lord as tax. James Legge, trans., *The Chinese Classics* 2:245.

106. Emperor Wu, r. 140-87 B.C.

107. These are the "worthies" and "literati" who took part in the debate. They were specially selected and recommended from the commanderies to serve in the court both as spokesmen of their own home areas and as candidates for government positions, after having served a period as court attendants. Cf. Gale, *Discourses,* p. 1, n. 4. It must be noted that they often exaggerated the hardship of the populace to decry the complaisant attitude held by the administration.

nothing] but the labor of the people. The government thus shared with the people in the good as well as the bad crops. It would not get more when the people had less, nor would it get less when the people had more. Hence it is said that the tithe was the most proper and just measure for the whole empire. But now, though the farmers are taxed merely one-thirtieth, the rate is based upon acreage. Therefore, in good years when the grain lies about in abundance, the actual exaction would be small [in proportion], while in bad years with famine rampant, the full stipulated amount still would be demanded. In addition, there are poll taxes and corvée duties. The taxation burden actually would be one-half of the labor of one person. The farmers are forced not only to yield all of their produce, but are often forced to borrow in order to fulfill the required amount. People are overtaken with hunger and cold, in spite of their strenuous farming and intensive labor. . . . [*YTL* 3:5a-b; cf. E. H. Gale, *Discourses on Salt and Iron,* pp. 94-95]

13. Chung-ch'ang T'ung's Estimation of Productivity

Now taking into consideration the rate of fertility and estimating the produce of grain, we can assume that three *hu* of grain can be produced from each mu. From each *hu,* it is not too much to take one *tou* [as tax]. Within one single year there will be reserves for several years.[108] Even if there are unnecessary operations or expenses for luxury or grace bestowed upon favorite persons [in the court], [the reserves] still will not be exhausted. Deviating from the ancient way, [the Han government] has created a light tax. Then whenever there is alarm [or disturbance] in any corner of the empire or natural disaster in one region, the accounts become short in less than three years. Hence, one has to witness that the soldiers have only vegetables and the starved fall on the road. How is a ruler to rule in such a manner? To tax one-twentieth is called the barbarian way.[109] How much more if the tax is one-thirtieth?[110] [*HHS* 49:18a-b] *Ca.* A.D. 200

108. One *tou* is one-tenth of a *hu,* the measure of grain. "Reserves for several years" obviously is in the context of comparison against the storage of grain exacted according to the rate of one-thirtieth.

109. The term *barbarian way* is a reference to a saying of Mencius, who, advocating the one-tenth tax rate, mentioned that only the barbarians in the northeastern area levied a tax at the rate of one-twentieth; see Legge, *Chinese Classics* 2:441-42.

110. This is one section from Chung-ch'ang T'ung's political essay contained in his biography in the *Hou-Han-shu.* On his life, see II, 44, n. 126.

14. First Poll Tax in Han

203 B.C. In the eighth month, for the first time a poll tax was levied.[111] [*HS* 1A:39b; cf. Dubs, *HFHD* 1:93]

15. Poll Tax Levied at Age Seven

44 B.C. [Kung] Yü again considered that in antiquity people were not subjected to poll taxes, which began when Emperor Wu[112] sent expeditions against the four barbarians. Then heavy taxes were imposed upon the people. People who have children are [required] to pay the *k'ou-ch'ien* tax as soon as the child has reached three years old. Thus the people are so much burdened that they kill the children at the moment of birth. How lamentable it is! May it be ordered that the *k'ou-ch'ien* not be levied until the child reaches seven years old when he loses his teeth, and that the *suan-fu* is not levied until he reaches twenty. . . . The emperor sent down the memorials [of Kung Yü] to be reviewed [by the court]. It was ordered that thereafter the *k'ou-ch'ien* was not to be levied until the child reached the age of seven.[113] [*HS* 72:13a-15b]

16. Poll Tax Paid in Grain

79 B.C. In the sixth month [of the second year of Yüan-feng] a general amnesty was granted. A decree stated: "We feel sorry that the people do not have sufficient supplies. In the previous years, the transportation of three million *shih* of grain was canceled.[114] Horses for the

111. According to the comment here by the annotator Ju Shun, who quoted the now lost work, *Han-i,* people between fifteen and fifty-six years old should each pay one unit of *suan,* which was 120 cash, to maintain the government arsenals, chariots, and horses. The same information is also available in *HS* 2:6a (Dubs, *HFHD* 1:183-84), in the passage in which it was decreed that females who failed to marry between fifteen and thirty years of age should pay five units. The annotator Ying Shao mentioned that merchants and slaves had to pay two units. The tax from the latter, of course, was paid by their masters.

112. *Suan-fu,* the adult poll tax, actually had been in effect since the beginning of the Han dynasty. Although the date of origin of the *k'ou-ch'ien,* the poll tax for children, is not definite, it had been in existence by the reign of Emperor Wu. Kung Yü probably refers to the increase of the latter by three cash in the time of Emperor Wu; *Han-i* quoted in note, *HHS* 1B:16b-17a.

113. It seems that only one part of the recommendation in the memorial was granted, namely, the lower limit of the *k'ou-ch'ien* taxable age being raised from three to seven, while the first taxable age for the *suan-fu* remained at fifteen.

114. Since transportation of grain to the capital region and the frontiers was

imperial chariots and in the pasture-parks are somewhat reduced to supplement the horses needed on the frontier and in the three capital district post stations. Let the surtax for horses[115] be not levied in this year. [And let people] in the three capital districts and the areas under the management of the grand minister of ceremonies *(t'ai-ch'ang),*[116] be granted [the option of] paying their poll tax with beans and millet." [*HS* 7:7b; cf. Dubs, *HFHD* 2:166-67]

In the sixth year of the Yüan-feng reign, it was decreed: "It is the low price of grain that hurts the farmer. Now in the three capital districts and the areas under the management of the grand minister of ceremonies, the price of grain is low. Let it be ordered that beans and millet be paid as tax payment this year. [*HS* 7:10a; cf. Dubs, *HFHD* 2:173] 75 B.C.

17. Labor Duties and Substitute Fees

[It is granted that] the labor duties fee payments due in the past three years are not to be collected. 77 B.C.

(The commentator Ju Shun[117] said: "There were three kinds of *keng*[118] duties: The *tsu-keng,* the *chien-keng,* and the *kuo-keng.* In antiquity, there was no regular standing army. Everyone had to serve in rotation. Service once a month each year is called the *tsu-keng.* The poor desired to have the wage, which was paid by those whose turn came, in the sum of 2,000 cash each month—this is called the *chien-keng.* In the empire, every [male] had to serve three days at the frontier as a garrison soldier; it is called *keng,* which in the legal

done by corvée labor, decrease of the shipment did mean the reduction of corvée and perhaps even reduction of some taxes in kind.

115. This particular surtax on horses and other domestic animals was initiated in the reign of Emperor Wu; see *HS* 96B:38b (Dubs, *HFHD* 2:170, n. 8. 7). It is sometimes confused with the practice of raising horses in the countryside, especially at the frontiers, as the government supplied those who were willing to raise horses with a colt or even a mare; cf. *SC* 30:39 and Yoshida Torao, *Ryōkan sozei no kenkyū,* pp. 63-65. Yet this particular practice had been abolished in 82 B.C., three years prior to the horse tax discussed here; see *HS* 7:4a.

116. *T'ai-ch'ang,* the grand minister of ceremonies, was in charge of all the religious activities in the imperial household. The prefectures established around the imperial mausoleums, therefore, were under the jurisdiction of the *t'ai-ch'ang* office.

117. Ju Shun (A.D. 189-266) is a *Han-shu* annotator who frequently quotes sources that are no longer available. Thus his comments are often useful and contain valuable information.

118. *Keng,* literally, means "rotation."

code is named garrison duty. Even the son of a chancellor had to go. Yet it is impossible for every person to go to the frontier for three days. Moreover, those who did go to the frontier for three days were not to return immediately. They therefore stayed for a whole year as one term of *keng.* Those who did not go paid 300 cash to the government, which then paid the garrison guard—this is the so-called *kuo-keng.* The legal code said that the *tsu-* and *chien-keng* were for the residents. Those who were residents in the prefecture should serve five months until replaced. Later it was modified to serve one month and take a rest for eleven months. In the 'Treatise on Food and Money,' it is said: 'One had to serve one month of local corvée service each year. Again, one had to serve as a regular conscript one year and on frontier garrison duty one year. All the labor requirements totaled thirtyfold those of ancient times.'[119] This was the practice that the Han inherited from the Ch'in dynasty. Later it was changed. Only those who were punished took the garrison duty at the frontier for one year. By payments due, it means the unpaid *keng-ch'ien,* or duty fees.") [*HS* 7:8b-9a; cf. Dubs, *HFHD* 2:170, 176-77]

18. Age Limits of Military Services

205 B.C. In the fifth month, the king of Han[120] was stationed at Yung-yang. Hsiao Ho[121] mobilized from the area of Kuan-chung the aged, the weak, and the young below twenty years of age who were not registered yet, and sent all of them to join the army [of the king

119. This passage is quoted directly from the memorial of Tung Chung-shu. The different ways of punctuating at the point after "regular conscript," however, create different interpretations to mean either (a) one served one year as a regular conscript, and another of frontier garrison duty, or (b) being a regular conscript, one served one year at the frontier and one year doing labor corvée. See Lao Kan, "Han-tai ping-chih chi Han-chien-chung-ti ping-chih," *Bulletin of the Institute of History and Philology* 10:23-55, and Hamaguchi Shigekuni, "Senkō to kakō," *Tōyōgakuhō* 19, no. 3:370-93, and the *Supplement,* vol. 20, no. 2. Also *Shin Kan Zui To shi no kenkyū* 1:459-95; Yoneda Kenjirō, "Kandai yōeki nissū ni kansuru ichi shiron," *Tōhōgakuhō* 27 (1967):189-212; Lien-sheng Yang, "Notes on Dr. Swann's *Food and Money in Ancient China,*" in *Studies in Chinese Institutional History,* pp. 113-16, and A.F.P. Hulsewe, *Remnants of Han Law,* p. 363, n. 97. The reading of interpretation (a) is preferred here because of consideration of grammatical structure. The phrase "labor corvée" of reading (b) has to belong to the next sentence as its subject to be related to the phrase "totaled thirtyfold those of ancient times."

120. This is the title Liu Pang held before he took up the imperial title.

121. Hsiao Ho was the chancellor who stayed at the home base of present-day provinces of Shensi and Szechwan. The mobilized conscripts were from Shensi.

of Han]. (. . .The commentator Ju Shun said: ". . .The *Han-i-chu:* Men at twenty-three years old became regular conscripts—for one year they are guards and for one year they are infantrymen and cavalrymen who learn chariot driving, archery, horse riding, and battle formations. It also says, at age fifty-six, when one is weak and old, one is then entitled to retire home as a common civilian in one's own home district. Here the aged, the weak, and the youths below twenty years of age who were not registered yet were all mobilized. The weak are those below twenty-three years of age. The old are those older than fifty-six.")[122] [*HS* 1A:34a; cf. Dubs, *HFHD* 1:80-81]

19. Defect of the Ever-Normal Granary System

The emperor intended to establish the *ch'ang-p'ing* or "ever-normal" A.D. 68
granary.[123] Most of the ministers who discussed this affair felt it convenient. [Liu] Pang answered [to the emperor]: "The ever-normal granary has had a reputation of benefiting the people, yet it actually has exploited them. The local powerful ones take advantage of the situation to make deceitful profit. The common people have no way to enjoy the normalcy." [*HHS* 39:8b-9a]

122. This particular quotation cited by Ju Shun (A.D. 189-266) is largely consistent with another quotation from the *Han Kuan-i,* quoted by the annotator Liu Chao as a footnote to the term *t'ing* in the "Treatise of Offices" in the *HHS* 28:8b. Also, see Yang, "Notes on Dr. Swann's *Food and Money in Ancient China,*" pp. 112-13.

123. This price controlling, state-owned granary system was established in 57-54 B.C. (*HH* 24A:19b).

IV

FARMING RESOURCES

1. Winter Wheat

Ca. 100 B.C. Tung Chung-shu memorialized:[1] "In the *Ch'un-ch'iu,* no grain [other than millet and wheat] was recorded. [It is the] failure of millet[2] and wheat that was recorded. Hence it is perceivable that the Sage[3] regarded millet and wheat the most important among the five grains. Now people in the region of Kuan-chung[4] customarily do not like to plant wheat. It indeed means annually to miss what was regarded as important in the *Ch'un-chiu* and to lose subsistence for the people's livelihood. May Your Majesty decree that the grand minister of agriculture urge the people of the Kuan-chung region to plant more [winter] wheat[5] so they will not miss the season." [*HS* 24A: 16a; cf. Nancy Lee Swann, *Food and Money in Ancient China,* pp. 178-79]

2. Wheat and Barley in a Ditty by Children

A.D. 146-167 During the reign of Emperor Huan, a children's ditty said:

The wheat turns green;
The barley withers.
Who harvests the crop?

1. This passage comes right after the observation by the historian that the people preferred business to farming, and preceding another memorial submitted by Tung Chung-shu on reduction of taxes and limitation of individual landholdings. Such a particular passage, rather technical, sandwiched between discussions on general policy, does appear somewhat out of place. I suspect that this passage was a separate memorial, which Pan Ku grouped together with other memorials simply because it was also a suggestion of Tung.

2. The word *ho* could be a general term for any cereal plant as well as a name specifically for spiked millet in Han times. Since in the Ch'un-ch'iu period (8th-5th centuries B.C.), common millet and spiked millet were both important, the annalist perhaps referred to both. Tung Chung-shu, however, being in the Han period, might mean only spiked millet by the term *ho*.

3. Confucius, the alleged author of the *Ch'un-ch'iu*.

4. Including the metropolitan area of the Han capital and its surrounding land, the southern part of the present province of Shensi.

5. The name *Su-mai,* according to Yen Shih-ku's annotation, means wheat that grows in winter.

Daughter-in-law and mother.
Where is the man now?
In the West, fighting the barbarians.[6]

[*T'ai-p'ing yü-lan,* Chung-hua reprint, 838:5b, hereafter cited as *TPYL*]

3. Field Record at a Military Settlement

A.D. Third century

1. Officer Chang Ch'ien is currently commanding twenty-one soldiers. There are 200 mu of barley; 20 mu have been harvested. There are 90 mu in lowland; 70 mu have been irrigated. There are 37 mu of wheat; 29 mu have been cut. There are 185 mu of millet; 20 mu have been irrigated and 50 mu have been hoed. There are 90 mu of millet of inferior grade; 70 mu have been irrigated.[7]

2. Officer Liang Hsiang is currently commanding twenty-six soldiers. There are 66 mu of barley; 50 mu have been harvested. There are 80 mu in lowland; 70 mu have been irrigated. There are 63 mu of wheat; 50 mu have been irrigated. There are 175 mu of millet; 50 mu have been hoed, and 50 mu irrigated. There are 80 mu of millet of inferior grade; 70 mu have been irrigated.[8] [Lo Chen-yü, *Liu-sha chui-chien k'ao-shih,* tablet no. 31 (2:27b)]

4. Bean and Millet in the *Mo-tzu*

Fifth to sixth centuries B.C.

When the Worthy governs the state, he leaves home early and returns late, and makes people farm and cultivate in order to store up beans and millet. Therefore beans and millet are plentiful, and people are sufficient in food. [*MT* 2:4a; cf. translation, Y. P. Mei, *The Ethical and Political Works of Motse,* p. 37]

5. Bean and Spiked Millet in the *Mencius*

Fourth century B.C.

As the Sage governs the world, may the bean and the millet be

6. This verse is quoted from the *Hsü-Han-shu* by *TPYL.*

7. This passage is written on the front of the strip.

The average area per man is 14.5 to 24.4 mu. Roughly, each person could take care of 20 mu of a variety of field crops. It reminds us of the size of the field each person was assigned to cultivate as suggested by Chao Ch'ung-kuo when he proposed establishing the *t'un-t'ien* system in 61 B.C. (*HS* 69:11a).

8. This passage is written on the back of the strip, and is quoted by Lao Kan in his "Han-chien-chung-ti Ho-hsi ching-chi sheng-huo," *Bulletin of the Institute of History and Philology* 11: 64-65. Lao says that the strips of the Chin dynasty (A.D. 270-310), were found by Aurel Stein in the ruins of Lou-lan, the ancient state located in the present Sinkiang Province (Chinese Turkistan).

as commonplace as water and fire. Beans and millet being [as easy to obtain] as fire and water, how could the people not be compassionate to each other? [*The Mencius*; cf. James Legge, trans., *The Chinese Classics* 2:463]

6. Bean and Millet in the *Hsün-tzu*

Third century B.C.

The artisan and the merchant do not need to till the field, yet they are sufficient in beans and millet. [*HT* 5:6a; cf. H. H. Dubs, *The Works of Hsüntze,* pp. 133-34]

7. Green Lentils for Soil Amelioration

Ca. Third to first centuries B.C.

1. The method of pit cultivation for melons: In the sixth month, after the rains, lentils are sowed. In the mid-eighth month, they are plowed under and killed. In the tenth month, once again turn the soil over. And then sow melons in the tenth month. [*Ch'i-min-yao-shu,* p. 130:14.11.1]

2. [For the cultivation of mallow], if the manure is not obtainable, lentils are sowed by broadcasting in the fifth or sixth months and plowed under and killed in the seventh and eighth months. It is as if the fields were fertilized with manure; in its excellence it is not different from manure. Furthermore, it saves labor. [Ibid., p. 150: sec. 17.5.10]

3. In the field designated [for onions], lentils must be sowed in the spring and plowed under and killed in the fifth month. In the seventh month, till the field several times.[9] [Ibid., p. 169: sec. 21.2.2]

8. Ratio of Farm Products

Second and first centuries B.C.

What sustains people's livelihood is clothing and food. What produces food is water and soil. Therefore, there are the principal [items]

9. These passages are quoted in the *Ch'i-min-yao-shu,* in the sixth century. The source is not given. Since there was a horticulturalist named Yin Che in the Western Han period whose book included chapters on the cultivation of melon, mallow, onion, mustard, etc., it is suggested that these sections are likely quoted by Chia Ssu-hsieh, author of the *Ch'i-min-yao-shu,* from the now lost book on horticulture. Ch'en Liang-tso, "Wo-kuo li-tai nung-t'ien shih-yung-chih lü-fei," *Ta-lu tsa-chih,* 46, no. 5:4. Cf. Ma Kuo-han, *Nung-chia i-shu chi-pen chiu-chung,* p. 23, and the opinion of Shen Ch'in-han quoted by the annotator Wang Hsien-ch'ien, *HS* 30:49a.

to enrich the people and there is a certain ratio [among items] to feed the people. In general, thirty mu is sufficient [to produce food] for the whole year. The average annual harvest of good and bad years is one *shih* [of grain] per mu. Hence, for each [thirty mu] there are thirty *shih* [of grain]. Melons and vegetables[10] make the equivalent value of ten *shih* [of grain]. [Fodders such as] husks and chaff, and [incomes from] the six domestic animals[11] make [another] equivalent of ten *shih* [of grain]. Then the people will have fifty *shih* [of grain or its equivalents]. The [revenues from] silk and hemp fabrics and textiles, as well as other fringe incomes, are not included. Therefore, the state has a surplus for storage and the people have a surplus for consumption.[12] [*Kuan-tzu,* SPPY ed., 17:11a, hereafter cited as *KT*]

9. Indigo as a Cash Crop

I [Chao Ch'i] went to Yen-ssu[13] to receive medical treatment. *Ca.* A.D 150-180
On the way, I passed through Ch'en-liu.[14] In this place, people all engaged in the cultivation of indigo and the industry of blue and purple dying. The indigo field stretched to the limit of sight, while no millet was planted. With lamentation that [people] forget the fundamental and only have the secondary in mind,[15] I composed a verse on indigo-blue.[16] [*CHHW* 62:5]

10. Soils in the Yü-kung

Yü, in dividing [China] into nine regions, followed the mountain *Ca.* 280 B.C.

10. Here, melons actually include all fruits, melons, and gourds. The "other vegetables" include all the herbaceous plants, such as onions, garlic, mustard, etc.

11. The six animals actually include poultry; they are chicken, dog, pig, horse, ox, and sheep (or goat).

12. This passage is quoted from the chapter "Chin-ts'ang," the date of which is suggested to be that of the reign of Emperor Wu (140-86 B.C.); Yu Yü, *Ch'ien-Han ch'ien-ch'i-ti chung-nung hsüeh-shuo,* quoted in *Chung-kuo nung-hsüeh-shih,* p. 105.

13. Yen-ssu is a county in the present province of Honan.

14. Ch'en-liu is also a county in the present province of Honan.

15. The "fundamental" means agriculture; the "secondary" means trade, manufacturing, and crafts.

16. This is a short preface to a poem written by Chao Ch'i (fl. ca. A.D. 150-180), a Han scholar whose major contribution is the standard early commentary on the *Mencius.*

courses and drained the rivers. According to the production of each region, he determined its [share of] contributions.[17]

Yü divided the land. Following the mountain courses, he marked places with wood poles and [thus] he distinguished high mountains and great rivers [as boundaries of regions].[18]

In Chi-chou,[19] . . . its soil is white loam. Its contributions were ranked upper-upper [in combination with some proportion of the middle-upper] grades. Its fields were of the middle-middle class.[20]

Between the Chi River and the Yellow River was the [region of] Yen-chou.[21] . . . Its soil is dark and rich. The grass thereof grew luxuriantly and the trees tall. Its fields were of lower-middle class. Its contributions were of the lower-lower grade.[22] It was not required to do as other regions did until the fields had been cultivated for thirteen years. Its tribute articles were lacquer and silk. Carried in the baskets were ornamental fabrics.[23]

17. The date of *Yü-kung* is suggested by Ku Chieh-kang, who argues on evidence derived from the text. He figures that this particular chapter was compiled sixty years before the unification of China under Ch'in; see Ku, "Yü Kung" in *Chung-kuo ku-tai ti-li ming-chu hsüan-tu*, ed. Hou Jen-chih et al., 1:3-5. Ch'ü Wan-li suggests much earlier dates for the Yü-kung compilation, in the Ch'un-ch'iu period; see Ch'ü, "Lun Yü-kung chu-ch'eng-ti shih-tai," *Bulletin of the Institute of History and Philology*, vol. 35 (1964). I prefer the date suggested by Ku Chieh-kang. The first passage translated here is one part of the prefaces to the *Shu-ching.*

18. The translation here differs from that of Legge, *Chinese Classics* 3:92. Our interpretation is based upon Ku, "Yü Kung," p. 7, n. 3, who prefers the theory first suggested by Cheng Ch'iao.

19. Chi-chou was regarded as the region where the capital was. It is the area embracing the present provinces of Shansi, north and western Hopei, and northern Honan.

20. Throughout the "Yü-kung" text, nine grades are assigned to both the fields and the contribution classifications: the upper, middle, and lower divisions are each subdivided into three classes. Hence our translation. What puzzles the reader is the obvious discrepancy between the field level and the contribution level. It has been suggested that classification of fields has nothing to do with field fertility and that it indicates the height of the concerned region. Other scholars have insisted that the classification of fields was a measure of fertility; see comments by Sun Hsing-yen, *Shang-shu chin-ku-wen chu-shu*, SPPY ed., 30A:2b. A compromise is that a dense population could raise the grade of contribution even from an area of relatively less productive land; see Ku, "Yü Kung," p. 9, n. 3.

21. South and east of the present province of Hopei.

22. Corruption of the original text makes the reading difficult. The characters for "lower-lower" seem to have been collapsed into one character. Hence the translation made by Legge, *Chinese Classics* 3:99. Our translation is based on the correction made by Chin Jen-shan, see Tseng Yung-ch'ien, *Shang-shu cheng-tu*, p. 54.

23. Tseng Yung-ch'ien suggested that whatever is carried in the basket is

Between the sea and T'ai Mountain was Ch'ing-chou.[24] . . . Its soil is white and rich; near the seaside was an extensive area of salty land. Its fields were of the lower-upper class. Its contributions were of upper-middle grade. Its tribute articles were salt, fine grass cloth, and various kinds of marine products, and, from the T'ai valley, silk and hemp, lead, pine, and strange rocks. . . . Carried in the baskets was silk from the mountain mulberry. . . .

Between the sea, T'ai Mountain, and the Huai River was Hsü-chou.[25] . . . Its soil is red, claylike, and rich. Plants and trees advance and encircle. Its fields were of the middle-upper class. Its contributions were of the middle-middle grade. Its tribute articles were colorful clays,[26] pheasant feathers from the Yü valley, the solitary *t'ung* tree from I-yang, the musical stones from the Ssu River. . . . Carried in baskets were black, checkered, and pure white silk fabrics. . . .

Between the Huai River and the sea was Yang-chou.[27] . . . The various kinds of bamboo spread [into groves]. The grass grew long and thin; the trees tall. The soil is muddy and miry. Its fields were of the lowest class. Its contributions were of the upper-lower grade, in combination with some proportion of the grade one step higher. Its tribute articles were three metals,[28] precious stones, bamboo, elephant ivory, hide, feathers, hair, and timber. . . . Carried in baskets were shell knits.[29] Wrapped for presentation were oranges and pomelos. . . .

Between Ching Mountain and the southern Heng Mountain was Ching-chou.[30] . . . Its soil is muddy and miry. Its fields were of the middle-lower class. Its contributions were of the lower-upper grade. Its tribute articles were feathers, hair, ivory, three kinds of

---

the weaving materials produced by women; ibid., p. 54.

24. Eastern part of the present province of Shantung, possibly including one portion of the Liaotung Peninsula.

25. The northern part of the present province of Kiangsu, one part of Shantung, and one portion of Anhwei.

26. At a certain place in Shantung there are clays of five colors. Such clays were used as symbols of the five directions—the four cardinal directions and the center, which formed a ceremonial altar where investitures of feudal vassals were held in imperial days.

27. Most of the present provinces of Kiangsu and Chekiang, possibly even extending to Fukien.

28. Gold, silver, and copper.

29. In the past, such a shell knit was often regarded as a decorative fabric. Now, with more ethnographic knowledge, we realize that many people in the Pacific basin do produce a crochet incorporating shell beads.

30. The whole middle reach of the Yangtse River and areas south of the Yangtse.

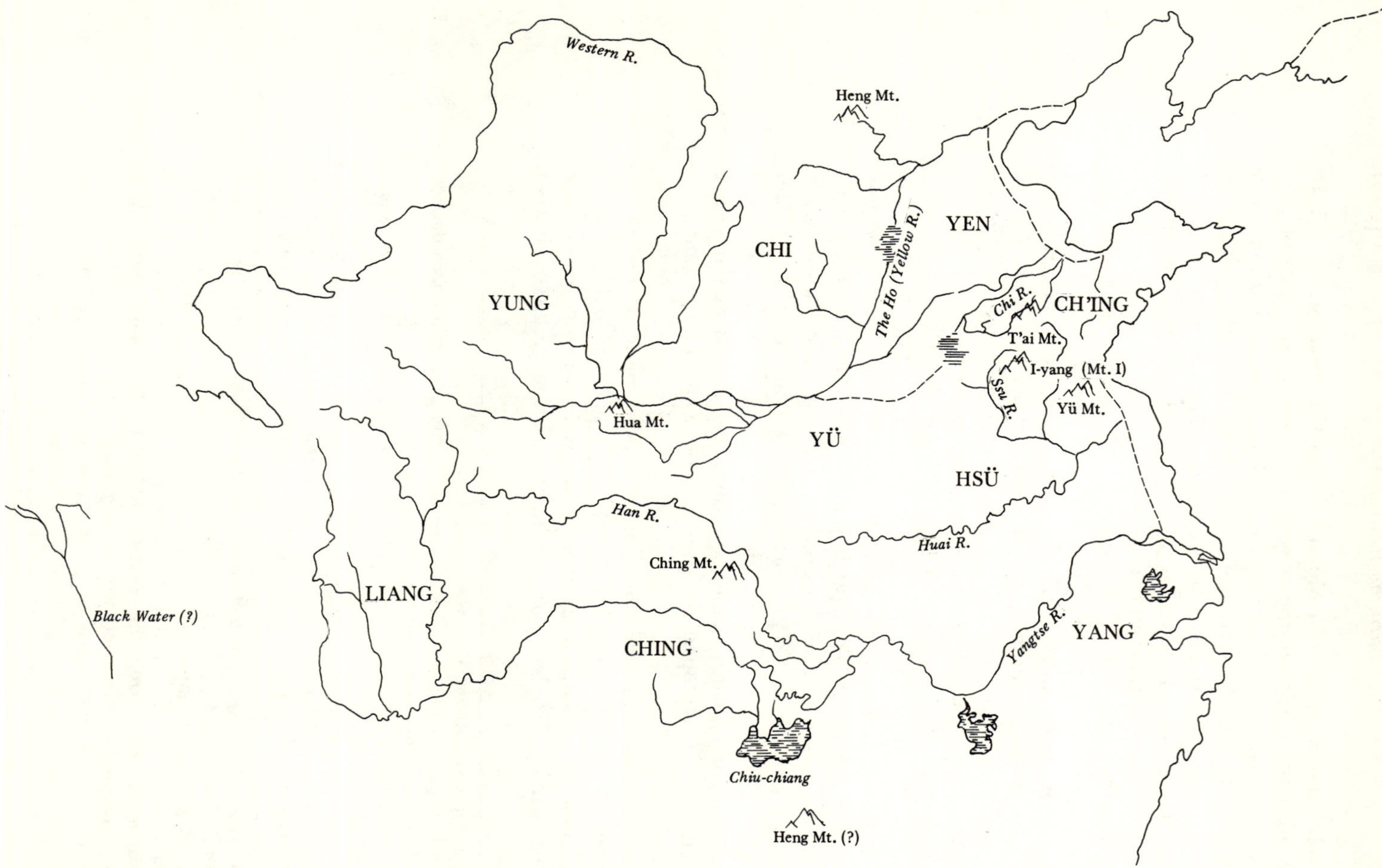

NINE REGIONS OF THE *YU-KUNG*
(Adapted from Ku Chüeh-kang, "Yü-kung," *Chung-kuo ku-tai ti-li ming-chu hsüan-tu*, p. 54)

metals, the timber of hardwoods and of cedar and cypress, grindstones, whetstones, arrowhead stones, and cinnabar. The bamboo and wood [good for arrow shafts] were presented especially by three states. Rush [for ceremonial usage] was wrapped and contained in cases. Carried in baskets were dark and purple silk fabrics and strings of pearls. From Chiu-chiang large tortoises were presented.[31] . . .

Between Ching Mountain and the Yellow River was Yü-chou.[32] . . . Its soil is loam; the lower parts are rich and dark with lumps. Its fields were of the upper-middle class. Its contributions were of the middle-upper grade, in combination with some proportion of the highest. Its tribute articles were lacquer, hemp, fine and coarse hemp clothes. Carried in baskets were fine silk fabrics and fine floss silk. Polishing stones for musical stones were presented.[33] . . .

Between the southern side of Hua Mountain and the Black Water was Liang-chou.[34] . . . Its soil is bluish and black. Its fields were of the upper-lower class. Its contributions were of the middle-lower grade, in combination with some proportion of one grade higher or lower. Its tribute articles were gold,[35] iron, silver, steel, arrowhead stones, musical stones, bears, great bears, foxes, and wildcats, and articles woven with animal hair. . . .

Between the Black Water and the Western River was Yung-chou.[36] . . . Its soil is yellow loam. Its fields were of the upper-upper class. Its contributions were of the lower-upper grade. Its tribute articles were gems and precious stones. [Sun Hsing-yen, *Shang-shu chin-ku-wen chu-shu*, SPPY ed., 3A:1; 3B:12; 30A:2; cf. James Legge, *The Chinese Classics* 3: 92-127; Bernhard Karlgren, "The Book of Documents," pp. 12-15]

11. Soil Amelioration and Seed Treatment

The weedman[37] is in charge of the amelioration of soil. Matching the appropriate crop to the soil, he decides the crop to be sown. (Cheng **A.D. second century**

31. The tortoise shell was used for oracle divination.

32. The middle reach of the Yellow River, the heartland of the loess plain.

33. A kind of coarse stone used to polish the surface of "music stones," which were played by hitting a series of stones of different sizes and thicknesses to render different musical tones.

34. The southern part of the present provinces of Shensi and most of Szechwan, and all southwestern China.

35. Although the character is written with the jade radical, it is gold of good quality; Ku, "Yü Kung," p. 26, n. 1.

36. Northern part of Shensi, Kansu, and Ninghsia.

37. *Ts'ao-jen*, literally translated, is the "weed man," a low rank agent whose duty is here described.

Hsüan's comment: To ameliorate the soil means to improve the soil, as the methods prescribed by Fan Sheng-chih. To match the appropriate crop to the soil means to observe the characteristics and colors of the soil and to select the crop thereof. The yellow and the white soils would fit millet, for instance.) To treat seeds: use oxen bones for the red, hard soil, sheep bones for the pink soil, tailed deer bone for the loam soil, deer bone for the dried swamp, badger bone for the salty and alkaline soils, fox bone for the fluffy soil, pig bone for the clayey soil, hemp for the hardened soil, and dog bone for the light soil. (Cheng Hsüan's comment: To treat seeds means to boil bone for decoction. Cheng Chung said: to use oxen means to steep the seed in the oxen-bone decoction.)[38] [*Chou-li chu-shu* 16: 4b, hereafter cited as *CLCS*]

12. Land Reclamation by Debarking the Trees and Cutting the Grass

A.D. Second century

The woodman[39] is in charge of felling bushes, woods, and forests. On the summer solstice, the trees at the *yang* side shall be debarked and burned. On the winter solstice, the trees at the *yin* side shall be debarked and treated with water.[40] (Cheng Hsüan's comments: Debarking means stripping off the tree bark just above ground level. Trees growing on the south side of a hill are called trees at the *yang* side; trees growing on the north side of a hill are called trees at the *yin*

38. The text here bears a notion that sounds rather strange and even mythical. The wild animals mentioned were not always easily available. If the original idea was to fertilize the soil with such bone decoctions, fertilization would require a large supply of the wild animals. The suggestion made by Cheng Chung, a scholar senior to Cheng Hsüan, certainly helped to clarify the situation. It also suggests that seed treatment in bone decoction possibly was a common practice in Han times. The reason for using all these animals, however, should be attributed to the application of a complicated cosmology in the *Chou-li.* Cf. Shih Sheng-han, *Fan Sheng-chih-shu,* p. 60.

Since our concern is the concept held by the Han people, the date given here is the date of the comments instead of that of the *Chou-li* text. Cheng Hsüan lived from 127 to 200 A.D.

39. *Tso-shih* is the title of a low-ranking agent listed in the *Chou-li.*

40. The emphasis on debarking on these two solstice days perhaps was ritualistic. Nevertheless, the practice could be carried on through both seasons. It is not clear how the debarked tree was treated with water. One can speculate that water frozen around the debarked tree could help to speed the death of the tree. It is also possible that using fire on the summer solstice and water on the winter solstice bears magico-religious significance in the *yin-yang* and the Five Elements schools of cosmology, which flourished during the Han period.

side. Burning and treating with water both mean to end their growth.) If transformation is desired, in the spring and autumn apply fire and water in order to change it. (Cheng Hsüan's comment: "Transformation" means vitality. It implies that it is time to grow grain crops. Application of fire and water means to treat with water what has been burned and to burn what has been treated with water in order to ameliorate the soil.)[41] For whatever relates to tree cutting, he is in charge of the laws and the regulations. (Cheng Hsüan's comment: There are specific seasons for felling trees.)

The Mower[42] is in charge of killing grass. In spring the grass is hoed at germination; on the summer solstice, slashed at ground level; in autumn, cut at its seed bearing; on the winter solstice, turned with a digger. (Cheng Hsüan's comment: "Hoed at germination" means to cut the sprout with a hoe. "Slashed" means to slash at ground level with a curved sickle. "Cut at seed bearing" means to cut with scythes. . . . Being cut at the seed-bearing time, the seed will not mature. "Turned with digger" means using a digger to turn up the frozen soil to uproot it.) If transformation is desired, apply water and fire to change it. (Cheng Hsüan's comment: It means to burn the grass, which has been cut down, and to flood it with water so that the soil is ameliorated.[43] The "Yüeh-ling" [chapter of the *Li-chi* says] "In the last month of summer, to burn, to cut down, and to flood with water are advantageous in killing the grass; they are like using boiling water." Here they are set forth for a single season.) He is in charge of all the laws and the regulations pertaining to killing grass.[44] [*CLCS* 37:2b-3b]

13. Weeds Being Raised as Green Manure

In the last month of summer[45] . . . the soil is full of moisture; [the weather is] humid and hot. Torrential rains come often. Burn, cut, and flood with water the grass in order to kill the weeds. If treated Fifth century B.C.

41. Please note the different interpretations Cheng Hsüan provided here and in the next paragraph, texts of both of which are essentially the same. One can only attribute them to the differences in practice that Cheng Hsüan observed. In the woody area, the debarked trees probably were either first burned and then flooded with water or first decomposed with water (and ice) and then burned to ashes.

42. *T'i-shih* is also a low-ranking agent listed in the *Chou-li.*

43. Cf. note 41. Here the grass is either burned or flooded, a much more simplified method than the treatment of trees.

44. The quotation of the "Yüeh-ling" is shortened from the original text; see IV, 13. The date given here is based on the comment made by Cheng Hsüan (A.D. 127-200), who used the agricultural techniques of his own time to explain the otherwise ambiguous text.

45. I.e., the sixth month.

with boiling water, it can fertilize the field and ameliorate the soil.[46] [*Li-chi Cheng-chu*, SPPY ed., 5:15]

14. The Shao Reservoir Restored

A.D. 83 The following year [Wang Ching][47] was transferred to the post of grand administrator of Lu-chiang commandery. Formerly the inhabitants did not know how to till the land with oxen. Consequently, even though the potentialities of the land were superabundant, food was usually insufficient. In the commandery were the Shao Reservoir[48] and paddy fields that had been established by the chancellor of Ch'u, Sun-shu Ao.[49] Thereupon, Ching exhorted and led the officials and the people to reconstruct the abandoned reservoir and taught the use of the ox plow for tilling. Because of this, the cultivated land was more than doubled, and within the territory there was an abundant sufficiency. He had carved a stone on which an oath was engraved in order to make the people familiar with the regular prohibitions.[50]

He also instructed the people to raise silkworms and to weave, setting up for them regulations, which were all made public in the districts and cantons. The people of Lu-chiang passed on his writings. [*HHS* 76:6b]

46. Cf. IV, 12. An identical passage appears in *LSCC*. Although the date of compilation of the latter is around 240 B.C., the star data included in the "Yüeh-ling" are recognized as those of the seventh century B.C. or the span of the eighth to the fifth centuries B.C. The most conservative date is the fifth century B.C.; Joseph Needham, *Science and Civilization in China* 3:195. Cf. Kao P'ing-tzu, *Hsüeh-li san-lun*, p. 160.

47. Wang Ching, a man known for his knowledge of astronomy and mathematics and for his ingenuity, was recommended to Emperor Ming as a person capable of constructing canals. He was instrumental in designing the Chün-i Canal and in A.D. 69-70 reconstructed the Pien Canal.

48. The Shao Reservoir was located in a region south of and adjacent to modern Shou-hsien in central Anhui. It was still in existence at the time of Li Hsien, who says that its diameter was a hundred *li* and that it irrigated ten thousand *ch'ing* of land. See IV, 15.

49. Sun-shu Ao was the most outstanding statesman in the history of the state of Ch'u. During the reign of King Chuang (613-591 B.C.) he was chancellor of state.

50. One section of this reservoir was excavated in 1959. The excavators discovered a lock and a dam, beyond which was a spillway, a pool, and a low dam embanking the pool. The lock and spillway were constructed of log pilings together with layers of hay-stalks and layers of well-sifted earth. The design permitted excessive water to flow over the lock and then be stored up again in the spillway pool, from where the water could be redistributed. See Yin Ti-fei, "Anhui-sheng Shou-hsien An-feng-t'ang fa-hsien Han-tai cha-pa kung-ch'eng i-chih," *Wen-wu*, 1960, no. 1, pp. 61-62.

15. THE SHAO RESERVOIR

The Fei River flows north from Ti-ch'iu, passing west of the former prefectural seat of Ch'eng-te. . . . It flows further north to the east of the Shao Reservoir through the Ching gate of the reservoir to connect with the reservoir along the Shao Reservoir canal. Therefore the text says, "The Fei River enters the Shao Reservoir."[51] The Fei River continues north, merging with the Yen-chien River on the right. . . . The reservoir receives the Chien River to the south of the Five Gates Pavilion. . . . [Northeast-bound] it accumulates in a lake, which is called the Shao Reservoir. The circumference of the reservoir is more than 120 *li*. It is located 80 *li* south of the prefectural seat Shou-ch'un. It is said that the reservoir was constructed by the Ch'u chancellor, Sun-shu Ao. . . . The reservoir has five water gates to receive and to pour out river waters. To the northwest is the Hsiang-men Reservoir. Water from the reservoir flows northbound, passing the shrine for Sun-shu Ao, and thence it is called the Shao Reservoir canal. Further northbound, it is divided into two courses, one of which flows eastward into the Li-chiang River. . . . The Fei River again receives, on the left, the Shao Reservoir canal, which has two courses separated at Li-chiang. . . . The [Fei River] flows westbound, merging with the Yang-t'ou Creek on the left, which receives water from the Shao Reservoir . . . and northbound pours into the Fei.[52] [*Shui-ching-chu*, SPPY ed., 32:5b-8a, hereafter cited as *SCC*] Han

16. CANALS CONSTRUCTED BEFORE THE HAN PERIOD

Hsi-men Pao[53] mobilized the people and dug twelve canals *Ca.* Fourth century B.C.

51. This sentence obviously is the original text.

52. *Shui-ching-chu*, the *Classic on Water*, together with its annotation is the earliest geographical treatise on river systems in China. The authorship is attributed to Sang Ch'in of the Han dynasty. The annotator Li Tao-yüan of the fifth century, however, added to the text tremendous amounts of detailed information, especially those of historical significance, such as the sites of ancient towns, the locations of old battlefields, the remains of ancient constructions, etc. The annotation and the original text, through a period of transition, have been hopelessly blended together in the version we have today. Scholars have tried to restore the original. The best versions so far are the ones reorganized by Tai Chen, Chao I-ch'ing, and Wang Kuo-wei. The most recent attempt was carried out by the late Dr. Hu Shih.

53. Hsi-men Pao was appointed magistrate of Yeh prefecture (modern Lin-chang County in northern Honan) by Marquis Wen of Wei (424-387 B.C.) around 406 B.C.

The story is also recorded in *SC* 29:5-6. In *HS* 29:5a, which tells the tale with fuller historical detail, the name is given as Shih Ch'i. In IV, 17, it is stated that Shih Ch'i replaced Hsi-men Pao as chief of Yeh prefecture and built the canals to bring water from the [Chang] River[54] to irrigate the people's fields. All the fields were irrigated. At that time the people regarded the construction of canals as burdensome; they did not like to do it. Pao said, "The people can [only] enjoy what has been achieved; one cannot plan with them for the beginning. Though at the moment the elders and the young men look upon me as an evil, nevertheless I intend to make the descendants of these elders a hundred years from now think of my words."

To the present[55] day everyone receives benefits from the water, whereby the people have sufficiency and property. [*SC* 126:31]

17. IRRIGATE SALINE SOIL TO GROW RICE

334-319 B.C. King Hsiang of Wei had a banquet with his court. After much drink, the king bestowed his blessing upon the court, and said that everyone should have his own wish fulfilled. Shih Ch'i stood up and answered, "In the court, some are worthy, while some are not. It is all right to have those who are worthy to fulfill their wishes, yet it is not all right to have those who are not worthy to fulfill their wishes." The king said, "Let all of the court be like Hsi-men Pao[56] in serving as ministers over men." Shih Ch'i answered, "In the state of Wei, the farm is [normally] regulated to be of 100 mu in size. It is only in Yeh that [the size is regulated] as 200 mu, because of the poor quality of the soil. The Chang River is adjacent to Yeh. If Hsi-men Pao was not aware of the advantages of it, he was unintelligent. If he was aware of it, yet did not advise [the ruler], he was unfaithful. Neither unintelligence nor unfaithfulness are to be taken as a model." The king of Wei had no way to respond. The next day [the king] summoned Shih Ch'i and asked, "Is it still possible to irrigate the Yeh fields with the Chang River?" Shih Ch'i answered, "Yes." The king said, "Why don't you do it for me?" Shih Ch'i said, "I am afraid that Your Majesty will not be able to see it through to completion." The king said, "If you indeed are able to do it for me, I will fully abide by what you say." Shih Ch'i promised with reverence and said to the king, "If I do it the people are going to resent me very much. I will be

---

in the reign of King Hsiang (334-319 B.C.) of Wei. Wang Hsien-ch'ien was of the opinion that the construction of the canals for irrigation in this region was begun by Hsi-men Pao and continued by Shih Ch'i. *HS* 29:5a.

54. Both *SC* 29:5-6 and *HS* 29:5a record that the river from which the water was taken was the Chang.

55. Up to the time when Ch'u Shao-sun (fl. 48-4 B.C.) added this story to the *SC*.

56. A magistrate of Yeh during the reign of Marquis Wen of Wei (424-387 B.C).

punished either with capital punishment or with the penalty of being branded. Though I may suffer from death or being branded, may Your Majesty commission another person to finish the work?" The king said, "Yes." Shih Ch'i was appointed the magistrate of Yeh. Shih Ch'i went to Yeh and carried out the project. The people of Yeh resented it very much. They wanted to have Shih Ch'i branded. Shih Ch'i hid in order to stay away [from the purge]. The king then appointed another person to finish the work. When the water was already flowing [into the irrigation system] and the people had been greatly benefited, they chanted to each other, "There is a sage magistrate at Yeh whose name is Shih Ch'i. Water from the Chang River is tapped to irrigate the neighborhood of Yeh. Now on the soil, saline ever since antiquity, rice and *liang* millet are produced."[57] [*LSCC* 16: 11b-12b; also see *HS* 29:5a]

18. The Cheng Kuo Canal

Then Han,[58] learning that Ch'in was inclined to undertake projects, and wishing to exhaust it to prevent it from attacking the eastern states, sent a hydraulic engineer, Cheng Kuo, to inveigle Ch'in into tapping the Ching River and constructing a canal from Mount Chung[59] westward to Hu-k'ou.[60] It was to parallel the Pei Shan[61] and to flow eastbound to the Lo River,[62] a distance of over three hundred *li*.[63] It was to be used to irrigate the fields. 246 B.C.

57. In this passage, Hsi-men Pao is stripped of all the credit of having constructed the irrigation system; cf. IV, 16. The work, however, is generally acknowledged as a project initiated by Hsi-men Pao and improved by Shih Ch'i. Cf. Wang Hsien-ch'ien's annotation in which Wang cites ancient sources by which Hsi-men Pao and Shih Ch'i are both credited; *HS* 29:5a.

58. The state of Han, eastern neighbor of Ch'in.

59. Mount Chung was east of and on the Ching River in modern Shensi.

60. Hu-k'ou was in the modern Ching-yang County, Shensi.

61. Pei Shan, the Northern Mountains, seems to have been a collective term for the entire chain of mountains west of the Lo River and northwest of the Ching River (*HS* 22:19b).

62. The Lo River, which is to be distinguished from another Lo River in present Honan Province, had its source in the northern part of Shensi Province, ran southward and joined the Wei River near the point where the latter joined the Yellow River.

63. The Cheng Kuo Canal takes off from the Ching River east of Chiu-tsung Mountain, proceeds eastward along the base of the Pei Shan until it enters the western border of modern Fu-p'ing County, where it turns south for some distance, then northeast or north.

According to a modern scale the shortest distance between the canal's place of origin and the point at which it enters the Lo River is approximately two hun-

When the work was half done, Ch'in became aware of the conspiracy and wanted to put Cheng Kuo to death. Cheng Kuo said, "At first I did indeed intrigue; nevertheless, once the canal is completed, it will certainly benefit Ch'in."[64] Ch'in, considering that this was so, finally had him finish the canal.

When the canal was finished,[65] it was used for conducting silt-laden water to irrigate over forty thousand *ch'ing* of saline land. The produce in the area everywhere rose to one *chung*[66] per mu. Consequently, Kuan-chung (present Shensi) became a fertile country without bad years. Ch'in thus grew rich and powerful, and finally conquered all other feudal states. Because of this it was named the Cheng Kuo Canal. [*SC* 29:6-8; similar passage in *HS* 29:5b-6a; cf. Burton Watson, *The Records of the Grand Historian of China* 2: 71-72]

19. TU-CHIANG DAM

306-251 B.C. [Li] Ping[67] then blocked the river with a dam. He dredged channels for the Pei River and the Chien River. As separate branches they flowed in a pair past the commandery capital; ships thereby moved in them.[68] . . .

Furthermore, three commanderies were irrigated, and rice paddies were opened. Hence the Shu region had a thousand *li* of fertile land, which was known as Sea on the Land.[69] During drought, water was

---

dred *li,* or three hundred *li* of the Ch'in period. Since the canal follows a curved course, the length "over three hundred *li*" given in our text, although indefinite, is obviously quite correct.

64. According to *HS* 29:3a, Cheng Kuo ended his defense by saying, "As to Han, I have prolonged its existence for a few years, but, as for Ch'in, I have created a work that will endure for ten thousand generations."

65. There are no records that give definite information about the completion of the canal. According to *SC* 15:113, the construction began in the first year of the reign of King Cheng of Ch'in (246 B.C.). *HS* 28B:18b also dates it at the beginning of his reign. Ten years later (237 B.C.) it was still under construction (*SC* 6:11-13).

66. The *chung* was a measure of capacity in the Ch'in and Han periods. Meng K'ang and Yen note that a *chung* was equal to six *hu* (bushels) and four *tou* (pecks) (*HS* 24B:1b and 29:3a). Since ten *tou* made one *hu,* the yield was equal to sixty-four *tou* per mu, which was considered a large yield.

67. Li Ping was governor of Shu, the western part of the present province of Szechwan, in the reign of King Chao of Ch'in (306-251 B.C.).

68. This statement agrees with a passage in the *SC,* in which the water-control project carried out by Li Ping is said to have been for the purpose of improving this waterway transportation. *SC* 29:5-6.

69. Traditionally, this expression is interpreted to mean that resources converge

tapped to moisten [the soil]. [If it had excessive] precipitation, the water gate was blocked. Therefore, there was the saying, "Being wet or being dry it is up to the wish of man. People know no starvation." There are no famine years; [the region of Shu] is known as the Heavenly Treasury.[70] [Ch'ang Ch'ü, *Hua-yang-kuo-chih,* SPPY ed., 3: 4a]

20. Grain Transportation and Irrigation

At that time, Cheng Tang-shih, the grand minister of agriculture, said, "Formerly grain from the east of the pass was shipped along the Wei River. Grain shipment takes approximately six months. Besides, along more than nine hundred *li* of waterway, frequently there are difficult places.[71] We should dig a canal that will draw water from the Wei River, one that will begin near Ch'ang-an and run below the southern hills. The distance to the Yellow River is about three hundred *li.* The route is so easy that the grain shipment will only take three months.[72] The fields along the canal, more than ten thousand *ch'ing,* then also can be irrigated. This will reduce grain shipment, save corvée labor, and increasingly fertilize the fields in Kuan-chung, from which grain is harvested." The emperor considered the plan correct and ordered the hydraulic engineer, Hsü Po from Ch'i, to install marks [along the projected route]. Several ten thousand laborers were mobilized to dig the grain-shipping canal. It took three years to complete. The grain shipment was then greatly facilitated. Later on, grain shipped gradually increased. Meanwhile, the people along the canal were quite able to use it for irrigation. [*HS* 29:6b-7b; *SC* 29:10; cf. Watson, *Grand Historian* 2:73] 130 B.C.

---

in great abundance as water pouring into the sea, e.g., *HS* 65:7b. I interpret it to mean an extensive flat plain in a basin.

70. For an eyewitness account in English on the operation of the Tu-chiang Dam, see Joshua Vale, "Irrigation of the Ch'eng Tu Plain," *Journal of the North China Branch of the Royal Asiatic Society* 33 (1899-1900): 22-36.

Hua-yang-kuo-chih is the oldest local history in China. Ch'ang Ch'ü, a Szechwan native, composed the history of his local region, from antiquity to the fourth century

71. The text gives the Wei River as the water source. A comparable passage in the *SC,* however, gives *ts'ao* ("shipping canal") instead of Wei, name of the river.

72. Wang Hsien-ch'ien cites the opinion of Liu Feng-shih, who doubts the validity of this event because the distance from the juncture of the Wei and the Yellow rivers is only three hundred *li,* not nine hundred *li.* But Shen Ch'in-han, drawing upon evidence from the T'ang, suggests an alternate route the length of which is compatible with this *HS* statement (*HS* 29:7a).

21. Canals between the Fen and Yellow Rivers

126-124 B.C. Later the administrator of Ho-tung commandery, P'an Hsi,[73] said, "That which is transported by water from Shan-tung to the west is annually more than one million piculs. In going through the difficult passage of Ti-chu[74] the destruction and losses are very heavy, and expenses are also great. If a canal is dug and the Fen River is tapped to irrigate the region below P'i-shih and Fen-yin,[75] and if the Yellow River is tapped to irrigate the region below Fen-yin and P'u-pan,[76] I estimate that 5,000 *ch'ing* of land could be reclaimed along the Yellow River. These 5,000 *ch'ing* have hitherto all been abandoned lands on the riverbank where the people collected fodder and grazed their cattle. If now it is irrigated and cultivated, I estimate that more than two million *shih* of grain[77] can be obtained. This grain, if sent up via the Wei River, would be no different from that produced in Kuan-chung. And we could do without further transportation east of Ti-chu."

The emperor considered the plan correct. Several tens of thousands of laborers were mobilized to make the canals and the fields.[78] [*SC* 29:10-11; cf. *HS* 29:7b; cf. also Watson, *Grand Historian* 2:73-74]

22. Well-Canals

Ca. 119 B.C. Thereafter, Chuang Hsiung-p'i said, "The people of Lin-chin[79] want to tap the Lo River to irrigate, from Ch'ung-ch'üan[80] eastward, over ten thousand *ch'ing* of soil that has been alkaline for some time. If indeed water can be obtained, one mu can be made to yield ten *shih* [of grain]."

Thereupon, more than ten thousand laborers were mobilized to dig the canal that conducted the water of the Lo River from Ch'eng[81]

73. P'an Hsi, fl. 126-124 B.C.

74. Ti-chu, a rock formation in the middle of the Yellow River, was halfway between modern P'ing-lu County, Shansi, and Shan County, Honan.

75. P'i-shih was located where the Fen joins the Yellow River; Fen-yin was located below the confluence of the Yellow and the Fen.

76. P'u-pan was in the southwestern corner of Shansi.

77. According to P'an Hsi's estimate, the produce from one mu of the projected fields would be not less than four *shih*.

78. This document is continued in V, 14.

79. Lin-chin is roughly identical with modern Ta-li County, eastern Shensi.

80. Ch'ung-ch'üan was located in eastern Shensi.

81. Ch'eng was located south of the modern capital of Ch'eng-ch'eng County, eastern Shensi.

down to the foot of Shang-yen [Mountain].[82] But as the banks [of the canal] tended to collapse, wells were dug, the deepest being over forty *chang*.[83] The wells were constructed one after another and were connected at the bottom so as to make the water flow along. The water ran down and thus by-passed Shang-yen [Mountain], penetrating more than ten *li* into the mountain range to the east. The origin of well-canals started from this.

When the canal was being dug, dragon's bones were found. Therefore, it was named the Dragon-Head Canal.

Work was carried on for more than ten years[84] and the canal carried water fairly well, but benefits derived from the project were not abundant. [*SC* 29:12-13; *HS* 29:8b-9a; cf. Watson, *Grand Historian* 2:75]

23. The Metropolitan Region

At this time, [Tung-fang] Shuo,[85] who was at the side [of the throne], approached [the emperor] and admonished him, saying, 138 B.C.
"The Nan Shan are the bulwark of the empire. To the south are the Yangtse River and the Huai; to the north are the Yellow River and the Wei.[86] As for the land east of Ch'ien and Lung[87] and west of Shang and Lo,[88] its soil is fertile and rich. When the Han arose, it abandoned the land of the three Ho [commanderies], settled west of the Pa and Ch'an rivers,[89] and made its capital south of the Ching

82. According to Ying Shao, *shang-yen* was the name of a mountain. Shang-yen is also identified as Shang-yüan and Hsü-yüan Mountain near modern Ch'ao-i; see the commentary to *HS* 29:8b.

83. One Han *chang* consists of ten *ch'ih*.

84. It is not known exactly when work on the Dragon-Head Canal started. However, its construction followed that of the Pao-yeh road during 120-116 B.C. (*HS* 19B:18a).

85. Tung-fang Shuo was famous for his humorous writings. In the early days of Emperor Wu's reign, he was a regular attendant gentleman. In 138 B.C., after the events in this passage, he was promoted to grand palace grandee, which office, however, he held for a short time only. He remained without official position until he was made palace gentleman in 122 B.C. (*HS* 65).

86. These four rivers were the largest in the Han Empire.

87. Ch'ien is the Ch'ien River, a watercourse identical with the modern Ch'ien River in western Shensi; Lung is Lung-ti, the Lung Mountains in the border regions of modern Shensi and Kansu.

88. According to Fu Ch'ien, Shang was Shang prefecture (located in modern Shang-nan County, southeastern Shensi), and Lo was Shang-lo prefecture (located in modern Shang County, northwest of Shang-nan County).

89. The Pa River had its source in the mountains southeast of modern Lan-t'ien County in southeastern Shensi and flowed north through eastern Ch'ang-an

and the Wei rivers. This is what is called the region of the continental ocean of the empire, and is the terrain whereby Ch'in subjugated the western barbarians and unified the area east of the mountains.

"Its mountains yield jade, precious stones, gold, silver, copper, iron, the silky spicebush, camphor trees, sandalwood, wild mulberry trees, and other unusual products that cannot be completely enumerated. These are what the artisans use to obtain their supplies and what the people depend upon to satisfy their wants.

"There is also an abundance of *keng* (glutinous rice) and *tao* (non-glutinous rice), pears, chestnuts, mulberries, hemp, and big and small[90] species of bamboo. The soil is suitable for [the cultivation of] ginger and taro,[91] and the rivers abound in edible frogs[92] and fish. The poor are thereby able to provide sufficiently for their families without worries over hunger and cold. Therefore the region of Feng and Hao[93] is called 'earth fat.' Its value is one *chin* of gold per mu.[94] Now you plan to make the area into a park, thus cutting off the advantages from the reservoirs and streams and taking away the people's fertile land; upwardly this will decrease the resources of the state and downwardly it will deprive people of enterprises in agriculture and sericulture. This is to give up what has been achieved, to head for disaster, and to reduce [the production of] the five grains: this is the first reason why it should not be done. . . ." [*HS* 65:7b-8a]

24. The Six Auxiliary Canals

111 B.C. One hundred and thirty-six years elapsed from the beginning of the Cheng Kuo Canal to the sixth year of Yüan-ting (111 B.C.), and Ni K'uan, as governor of the Left Capital District, presented a memorial, asking that the six auxiliary canals be dug to extend irrigation to the high land alongside the Cheng Kuo Canal.[95]

County to join the Wei River. The Ch'an River had its source in the mountains south of Lan-t'ien County. It joined the Pa River north of Lan-t'ien County.

90. *Chien* is a species of small bamboo. Its leaves are small, but because the stem is very hard it is particularly suited to the making of arrows.

91. The *yü* is a plant whose leaves resemble those of the *ou-ho*, or Indian lotus, but are longer. Its root is pure white and edible. Today *yü* is commonly identified as taro.

92. Yen identifies *wa* with *hsia-ma* (toad), but smaller, with longer legs.

93. Feng was the capital of the Chou people before they established their dynasty. It was built by King Wen and located east of the modern city of Hu in central Shensi. Hao was the capital of the Western Chou. It was built by King Wu and was located southwest of the modern city of Sian.

94. According to *HS* 24B:22a, one catty of gold in A.D. 9 equaled ten thousand cash.

95. The six auxiliary canals were six small canals that were dug south of the upper course of the Cheng Kuo Canal to support the latter in irrigating the fields.

The emperor said, "Agriculture is the foundation of the world. Irrigation from springs and streams is the means by which the five grains are raised. In the territories under the command of the left and right clerks of the capital region, streams and springs of the famous mountains are very numerous. The common people do not understand the advantages from these. Therefore canals are built and water is stored up for them in embanked reservoirs; these are the means for taking precautions against drought. Now the tax on rice fields in the capital districts is heavy and differs from that in the [other] commanderies. Let measures for reducing it be discussed. Let the officials and the people stress farming so as to exploit fully the bounties of the land. Let the labor service be equalized and the water be made to flow along so as not to let them miss the proper season."[96] [*HS* 29:11b-12a]

25. The Po Canal

Sixteen years after this,[97] in the second year of T'ai-shih [95 B.C.], 95 B.C.
the palace grandee of the kingdom of Chao,[98] Elder[99] Po, again presented a memorial concerning the digging of a canal to tap the Ching River. It began at Ku-k'ou and ended at Li-yang, where it discharged its waters into the Wei River.[100] It was 200 *li* long and irrigated more that 4,500 *ch'ing* of land. Hence it was named the Po Canal. The people received its benefits and sang of it:

Where do we till the fields?
At Ch'ih-yang and Ku-k'ou.
The Cheng Kuo came first,
The Po Canal came after.
Raising spades[101] we make clouds,
Digging canals[102] we make rain.

96. *HS* 58:11b states that after requesting the construction of the six auxiliary canals, Ni K'uan promulgated a number of measures for water distribution that placed more fields under irrigation. According to Yen, these measures regulated the use of the water and assured benefits to everyone (*HS* 29:12a).

97. After Ni K'uan presented the memorial concerning the six auxiliary canals; see IV, 24.

98. According to Pan Ku, in Han times the courts of the kingdoms were organized like their counterparts at the imperial court (*HS* 19A:8b-9a). It was the duty of a palace grandee of the Han court to advise the emperor.

99. According to Yen Shih-ku, the word *kung* here was a term of respect for an older man, and is not to be understood as "duke."

100. The canal must have run north of and parallel to the Ching River, joining the Wei River in the Li-yang region.

101. According to Yen Shih-ku, the *ch'a* was a spade used for digging canals.

102. The waters in the canal were actually a substitute for the much needed rain.

One picul of the Ching water
Has several pecks of silt.[103]
It irrigates; it fertilizes;
It makes grow our *ho* and *shu* millets.
We clothe and feed the imperial capital,
A hundred million persons.

This describes the bounties of the two canals. [*HS* 29:12a-b]

26. Water Resources to Be Emphasized

3 B.C. [Hsi-fu] Kung[104] also said, "Ch'in opened up the Cheng Kuo Canal and thereby enriched the state and strengthened its military force. [The same area is] now the capital region, the soil of which is highly fertile. It is advisable to survey the terrain and the water resources thereof so that the benefits of irrigation are extended."

The emperor had Kung carry the staff of authority[105] to take charge of the Office of the Supervisor of Waters of the three metropolitan districts. Kung set up markers with the intention of cutting through the city of Ch'ang-an for having the water transportation [route] directed to a point beneath the Great Granary to reduce transportation costs. As the proposal was not feasible, it was dropped. [*HS* 45:15b]

27. Irrigation in the Yellow River and Wei River Valleys

A.D. 9-23 At the time of Wang Mang experts on the Yellow River were summoned by the hundreds. . . . A clerk to the grand minister of war, Chang Jung of Ch'ang-an,[106] said, "The nature of water is to go down. If it runs quickly, it scours and clears the riverbed, gradually making it deeper. The water of the Yellow River is very muddy. It is

103. Ju Shun states that the silt left when the water receded could be used as fertilizer.

104. Hsi-fu Kung was a literary man in the court of Emperor Ai; he died in prison after losing imperial favor.

105. The possession of the staff of authority *(chieh)* gave an official the right to perform his official duties with imperial sanction (*CLCS* 15:5a-8a).

106. In Huan T'an's *Hsin-lun*, p. 20a, which contains statements similar to those of the above text, the name is given as Chang Chung-i who is said to have held the position of grand minister of war. Yen Shih-ku, quoting the *Hsin-lun*, states that Chang Jung's style was Chung-kung, and that he was skilled in irrigation works.

said that one *shih* of water contains six *tou* of silt. Now from the western commanderies to the imperial capital and farther to the east, the people all conduct water from the Yellow River, the Wei River, and mountain streams to irrigate their fields. Spring and summer are the times of dryness and water scarcity, which cause the river current to be slow, silt accumulates and the rivers gradually become shallow. When rain falls in abundance and water rushes in violently, the rivers overflow and break their embankments. Yet the government frequently builds dikes to fill the breaches; hence the riverbeds have been gradually raised higher than the [surrounding] flat land. This is as if walls were built to contain the water. It is advisable to follow the nature of each [river] and to do away with irrigation so that all the rivers will flow freely, and the waterways will flow smoothly. No calamities by overflowing or breaking their dikes hence would be caused." [*HS* 29:20b-21a; cf. translation by Timoteus Pokora, trans., *Hsin-lun,* p. 128, n. 12.]

[The secretary of] the grand minister of war, Chang Jung, whose style was Chung-kung, and who was skillful in matters pertaining to irrigation, submitted his opinion: "Water in the Yellow River is so muddy that there are six *tou* of silt in one *shih* of water.[107] Yet the people strive to irrigate with water from the Yellow River, which causes the river not to flow rapidly. Every year in the third month when the spring flood arrives the river will overflow because of the impediments to its flow. People should be forbidden from leading water from the river."[108] [*CHHW* 15:1a-b] *Ca.* A.D. First century

28. Stele Inscription to the Deity "Lord Hsiao K'ang"

The worst part of the Chung-t'iao Mountains[109] is a high cliff . . . between which hundreds of streams converge and fill up the whole valley. When there is heavy rain . . . the flood flows to Hsiao K'ang where a reservoir takes the water northward to the Wei River[110] so that the water is stored up without causing damage. In previous generations people enjoyed such a benefit. . . . The local official there- A.D. 181

107. One *tou* is one-tenth of one *shih.* Therefore, this phrase means that 60 percent of the water is silt.

108. This is a quotation preserved in the Sung encyclopedic work, *TPYL,* 61, as a part of the work of Huan T'an. See also translation in Pokora, *Hsin-lun,* p. 112.

109. Chung-t'iao is located in the southern part of Shensi Province.

110. Wei is a tributary of the Yellow River, which flows through southern Shensi.

fore erected a shrine to repay the debts to the deity. Since the collapse of [Wang Mang's] Hsin dynasty, the sacrifice has been neglected and the reservoir blocked. The dikes were broken and fields were washed out. Markets in the towns were flooded. The deity was angered and the people complained. . . . The local officials lost their offices because of this. . . . In the fourth year of the Kuang-ho reign period [A.D. 181] the prefect [name missing] . . . again repaired the reservoir from which the water makes its way and restored the old worship by enlarging the shrine. . . . The stele was made on the *hsin-wei* day of the sixth month of the fourth year of the Kuang-ho reign period. [Following are the names of 402 local functionaries who were natives to the area. These were the persons who contributed to the erection of the stele.[111] [*Li-shih*, SPTK ed., 2:14a-21a, hereafter cited as *LS*]

29. Standing Water Prevents Sowing

26 B.C. [Tu Ch'in told the Grand General Wang Feng that][112] if [the embankment] cannot be accomplished this winter, the spring flood, which inundates when the peach blossoms, is certainly going to overflow. There will be the calamity of mud covering the land. The consequence is that no seed will be sown in several commanderies. People then will scatter and bandits will arise. . . . [*HS* 29:15a-b]

30. Irrigation Advocated

7 B.C. [Chia Jang memorialized:][113] " . . . Digging canals has three benefits; not digging them has three detriments. When the people are constantly exhausted by preventing floods, half of them lose their livelihood. When the water overflows the land and when the accumulated moisture evaporates, the people are made ill by the

111. It was a custom in Han times for all contributors to a stele to have their names engraved on the stele. Sometimes the money contributed is also listed.

112. On this occasion, Tu Ch'in advised Wang Feng, who was the actual regent, to take caution for a prompt embankment of the Yellow River, a dike of which had broken again after a major project had been completed only two years earlier.

113. Chia Jang, a specialist on flood control, presented this memorial in answer to Emperor Ai's decree soliciting opinions on river conservancy on the Yellow River and at the suggestion of the Imperial Chancellor K'ung Kuang and of the Minister of Works Ho Wu, who held these positions in 7 B.C. (*HS* 19B:48a-b). In an effort to prevent further destruction by the Yellow River he formulated three policies, the best in his opinion being the construction of dikes on a large scale; the medium, digging new irrigation canals; and the least advantageous, repairing the old dikes.

humid atmosphere. All the trees rapidly rot away, and the soil, turning alkaline, does not produce grain. When the river breaks the dikes and overflows, destruction ensues, and [the victims] become food for fish and turtles. These are the three detriments.

"If there are canals for irrigation, then the salt is washed down to the marshy ground and the spreading of the silt increases fertility. Where formerly millet and wheat were raised, even nonglutinous and glutinous rice can be produced; the productivity is increased fivefold in the high-lying land and tenfold in low-lying land. Furthermore, there is the advantage of transportation by water. These are the three benefits.

"At present, the number of functionaries and conscript laborers for embankments along the Yellow River in each commandery is several thousand, and the costs of cutting and buying wood and stone are several tens of millions yearly [an amount] that is sufficient to dig canals and construct water gates. Furthermore, when the people benefit from irrigation, they will urge one another to make canals, and they will not be weary even if the work is strenuous. At the same time the people's fields will be cared for and the dikes on the Yellow River will be completed. This will indeed enrich the state, make the people secure, create profit, and do away with calamities, and it will endure for several hundred years. Therefore, I consider this a medium policy. Repairing the old dikes by means of increasing the height and thickness would cost limitlessly, and we would frequently encounter calamities. This would be the worst policy." [*HS* 29:20a-b]

31. Irrigation Canals

From this time on,[114] those who were concerned with such matters vied with each other in advocating irrigation. In Shuo-fang, Hsi-ho, Ho-hsi, and Chiu-ch'üan commanderies, the Yellow River and the mountain streams were tapped to irrigate fields.[115] In Kuan-chung, the auxiliary canals[116] and the Ling-chih Canal[117] got their water *Ca.* 109 B.C.

114. After 109 B.C., when the breach in the Yellow River near Hu-tzu was closed.

115. There were various canals in these regions.

116. The auxiliary canals apparently refers to the six auxiliary canals discussed in IV, 24. The *HS* (29:11b) records the Ch'eng Kuo and the Wei canals instead of the auxiliary canals. Cf. IV, 24.

117. Pan Ku locates the Ling-chih Canal in Chou-chih prefecture (*HS* 28A: 32b), which is a county of the same name in Shensi.

from various rivers. In Ju-nan and Chiu-chiang[118] water was drawn from the Huai River. In Tung-hai it was taken from Chü-ting [Lake].[119] At the foot of Mount T'ai water was drawn from the Wen River.[120] In all these cases canals were dug to irrigate fields, which in each instance amounted to more than ten thousand *ch'ing*. Other smaller canals and embankments that were led through mountains cannot be described in full. [*SC* 29:17-18; cf. Watson, *Grand Historian* 2: 77-78]

32. Flood Control Project along the Yellow River

A.D. 69 In the twelfth year of the Yung-p'ing period of the Emperor Ming [A.D. 69] . . . several hundred thousand laborers were mobilized to build dikes. By imperial edict [Wang] Ching[121] and Wang Wu, an internuncio of the court architect, were instructed to open channels and construct dikes and dams along the Yellow River from Yung-yang to Ch'ien-ch'eng[122] where the river entered the sea. The length was a thousand *li*. [Wang] Ching, according to the terrain, chiseled through mountains, opened creeks, reinforced the spots where floods were strong, and cleared out obstructions. He constructed water gates every ten *li* so the waters could circulate.[123] There was no longer a problem of leakage. [*SCC* 5:7b-8a]

33. Irrigation in Nan-yang

Before 33 B.C. [Shao] Hsin-ch'en[124] was promoted to be grand administrator of

118. Reservoirs had been constructed in both these commanderies. Li Tao-yüan, quoting the *Shih-san-chou-chih*, states that in A.D. 96 many reservoirs were built in Fu-p'o prefecture in Ju-nan to irrigate the rice fields. In fact, the name means "rich in reservoirs" (*SC* 30:8b-9a). For the most famous one, the Shao Reservoir, see IV, 15.

119. Tung-hai was a commandery in modern eastern Shantung.

120. The Wen River flowed through the southern part of the T'ai-shan area, joining the Chi River in modern Tung-p'ing County in western Shantung. According to Li Tao-yüan, the upper and middle courses of the Wen River had formerly been used for irrigation (*SCC* 24:16b-19b).

121. Wang Ching was from Lo-lang, in present-day North Korea.

122. Yung-yang was a commandery in present-day Hunan Province; Ch'ien-ch'eng a seaport in present-day Shantung Province.

123. This sentence apparently means that water was returned to the river after it had been used for irrigating crops. The precise manner of operation of this system is not clear.

124. Shao Hsin-ch'en was one of the most distinguished administrators in Han.

Nan-yang. . . . Hsin-ch'en was a person of energy and plans; he took an interest in creating benefits for the people and regarded it as his urgent task to enrich them. Crossing in and out of the fields, stopping and resting even at remote villages and cantons, and having very little time for quiet living, he personally encouraged farming.

As he traveled about, he inspected the waters and springs in the commandery. He dug canals and ditches and built water gates and dikes[125] in several tens of places in all to expand the irrigated land, which increased year by year to as much as 30,000 *ch'ing*.[126] The people obtained benefits from this and had a surplus of stores.

Hsin-ch'en formulated regulations for the people concerning the equitable distribution of water. They were inscribed on stones and set up at the boundaries of the fields to prevent disputes over the distribution [of water].

He forbade extravagance in weddings and funerals, emphasizing frugality and simplicity in these things. He abruptly dismissed those subordinant commandery and prefectural staff whose children or brothers liked to loaf and play and did not take up cultivation of the land as their occupation. He punished the worst [offenders] for their violations of the law to make clear the good and the bad. His influence spread widely. There was no one in the commandery who did not exert himself in the cultivation of the land. The people flocked to him; the population doubled. Banditry and lawsuits declined and [finally] ceased. The officials and the people so loved Hsin-ch'en that they called him Father Shao.

Because the inspector of Ching division reported that Hsin-ch'en created benefits for the people and that the commandery thereupon became very prosperous, [Hsin-ch'en] was granted forty catties of gold and promoted to the post of governor of Ho-nan commandery. His administration was always ranked first. His position in the official hierarchy was raised several times and he was granted gold frequently. [*HS* 89:13a-b]

125. Ch'i Shao-nan presents additional data concerning Shao Hsin-ch'en's activities in promoting irrigation in Nan-yang. He reports that he neglected no waters that could be put to work. His most important achievements were the construction of the great Ch'ien-lu Reservoir and the Liu-men Reservoir, both of which were located within the territory of modern Teng County and Hsin-yeh County in the extreme southwestern part of present Honan.

126. The largest single irrigated area of Ch'in and Western Han, that abutting the Cheng Kuo Canal, covered only 40,000 *ch'ing* of land.

34. THE HUNG-HSI RESERVOIR

A.D. 39-43 Formerly there was in Ju-nan the Hung-hsi Reservoir. In the time of Emperor Ch'eng, the imperial chancellor, Chai Fang-ching presented a memorial and had it destroyed.[127] During the Chien-wu period, Grand Administrator Teng Ch'en,[128] wanted to repair and restore this work. On hearing that [Hsü] Yang[129] knew about water-courses, he summoned him to discuss the matter.

Yang said, "Formerly Emperor Ch'eng accepted Fang-chin's proposal. He then had a dream that he had ascended to heaven. The Heavenly Emperor was angry and said, 'Why did you destroy my Lake of the Bathing Dragon?'[130] After [its destruction] the people lost the benefits from it, and many of them became hungry and distressed. At that time there was a ditty that said:

The one who destroyed our reservoir
Was Chai Tzu-wei.[131]
We have only big beans to eat;
We have only taro roots to cook.
If conditions are to be reversed
The reservoir should be restored.[132]

"In the past the Great Yü dug the Yangtse and dredged the Yellow River to benefit the world. Now Your Excellency restores the abandoned works to enrich the country and make the people secure. The words of the ditty will come true. I sincerely desire to exert my energy until death."

127. According to *HS* 84:20b, Chai Fang-chin suggested that this old reservoir be destroyed because the constant inundation from it had created disaster in the local area. The commissioner dispatched by the office of the chancellery, after investigating the situation, proposed to drain away the water and take advantage of the fertility of the base soil of the reservoir. Chai's tenure in the chancellery was 15-8 B.C. Cf. *HS* 84:20a-b.

128. Teng Ch'en (d. A.D. 49), Emperor Kuang-wu's brother-in-law, was one of his earliest associates. From the year A.D. 25 on he held various offices.

129. Hsü Yang was a native of Ju-nan. From his youth he liked to study magic and mathematics. He had been the chief commandant of Chiu-ch'üan, but when Wang Mang ascended the throne he changed his name and engaged in sorcery and medicine and went into hiding. Since he was familiar with water works, he was appointed head of the bureau for the supervisor of waters by Teng Ch'en to repair and restore the destroyed Hung-hsi Reservoir. Hsü's biography is in *HHS* 82a: 4b-5b, most of which is translated in this document.

130. Here the dragon is associated by extension with artificial rain irrigation.

131. Tzu-wei was the style name of Chai Fang-chin.

132. For a slightly different wording of the ditty and the commentaries, see *HS* 84:20b.

Ch'en was greatly pleased. He thereupon appointed Yang the head of the bureau for the supervision of waters to take care of this matter. Following the elevation of the terrain and conformation of the land, Yang built an embanked reservoir of over four hundred *li* [in circumference]. After several years it was finished. The people enjoyed its advantages; there were good harvests in the years that followed. [*HHS* 82A:5a]

Ch'en restored the Hung-hsi Reservoir [that irrigated] several thousand *ch'ing* of land. The territory of Ju[-nan] thereby became prosperous. Its abundance of fish and rice flowed in large quantities to other commanderies. [*HHS* 15:8b]

35. Water Gates to Control Irrigation Reservoirs

[Pao Yü] . . . later was appointed grand administrator of Ju-nan. *Ca.* A.D. 62-74
This commandery had many reservoirs, which burst open year after year. . . . The annual [repair] costs often amounted to over thirty million [cash]. Yü then built square-beam stone walls on the reservoirs that kept water in abundant amount at all times. Thus the acreage of the irrigated fields was doubled, and the people became wealthy. [*HHS* 29:8a]

36. An Old Canal Repaired

[Ho Ch'ang][133] also repaired the old canal in T'ung-yang prefecture.[134] *Ca.* A.D. 90
The people relied upon its benefits. The cultivated land was increased by more than 30,000 *ch'ing*. Both the officials and the people engraved a stone tablet eulogizing Ch'ang's merits and virtues. [*HHS* 43:21a]

37. Irrigation in I-chou Province

[Wang Mang] appointed Wen Ch'i of Kuang-han commandery A.D. 19-25
to the position of grand administrator [of I-chou commandery].[135] He built embanked reservoirs, set up an irrigation system, and brought over 2,000 *ch'ing* of land under cultivation. He led and trained military

133. At this time Ho Ch'ang, fl. A.D. 90, was the grand administrator of Ju-nan commandery.

134. T'ung-yang was located in modern Hsin-ts'ai County in the extreme southeast of Honan.

135. Wen Ch'i was appointed after the unsuccessful military expedition headed by Lien Tan against the rebellious barbarian tribes of this commandery. This expedition took place in A.D. 19 (*HS* 99C:4b).

forces, constructed fortifications, subdued and gathered together various barbarian tribes, and achieved to a great degree a harmony [between them and the government]. [*HHS* 86:14a]

38. Irrigation Canals in the Southeastern Part of the Empire

A.D. 41-43 Wherever [Ma] Yüan[136] passed he always created commanderies and prefectures, built city walls, and dug canals for irrigation to benefit the people. [*HHS* 24:9b]

39. Meritorious Water Works

A.D. 90 [Lu P'ei] was transferred to the position of grand administrator of Tung commandery. In both commanderies[137] P'ei constructed watercourses for irrigation for the people. The people became very wealthy. [*HHS* 25:10a]

40. Reservoirs Restored in Hsia-P'ei

A.D. 86 [Chang Yü][138] was transferred to the position of chancellor of Hsia-p'ei Kingdom. In the northern part of the Hsü prefecture was the P'u-yang Reservoir.[139] Alongside it there was much good land, but it had gone to ruin and had not been repaired. Yü constructed water gates for it and conducted water for irrigation. Thus several hundred *ch'ing* of well-cultivated land were developed. He encouraged and led the officials and the people, lent them seeds and provisions, and personally urged them to work and exhorted them. Consequently, there was a large harvest of grain. The poor people from neighboring

136. In A.D. 41, Ma Yüan was sent to subdue the rebels in Chiao-chih, which was located in today's North Vietnam, returning in A.D. 43, after having successfully carried out his mission.

137. Lu P'ei had previously held the position of chancellor of the kingdom of Chao. In Han times kingdoms were considered equal to commanderies, hence the phrase "both commanderies."

138. Chang Yü, an able administrator, was made grand commandant in A.D. 94, grand tutor in charge of the affairs of the master of documents in 106, and in 107 was again made grand commandant. He held the title of Marquis of An-hsiang with a fief of 1,200 households (*HHS* 44:2a-6a).

139. Hsü prefecture was located southeast of modern Ssu County in northeastern Kiangsu. According to a statement in the *Tung-kuan-chi*, a collective work composed by a group of Eastern Han scholars, the P'u-yang Reservoir was twenty *li* wide and a hundred *li* long. To the east were some ten thousand *ch'ing* of land.

commanderies came to him to the number of more than a thousand households. Their houses and huts were clustered together. A market came into being where they lived. Two years later there were over a thousand odd *ch'ing* under cultivation.[140] The needs of the people were sufficiently provided. [*HHS* 44:2b-3a]

41. Irrigation in Kuang-ling

Ma Leng[141] was transferred to the position of grand administrator of Kuang-ling commandery.[142] At that time grain was expensive and the people were starving. He made a request and had the Office for Salt [in the commandery] abolished to benefit the people. He distributed relief to the poor and weak and lowered the taxes. He restored reservoirs and irrigated over twenty thousand *ch'ing* of land.[143] The officials and the people engraved in stone a eulogy for him. [*HHS* 24:24a] A.D. 87

42. Old Canals Repaired

. . . it was decreed that . . . the branch canals that Hsi-men Pao[144] had constructed by diverting water from the Chang River should be repaired to irrigate the people's land. . . . It was decreed that [the authorities of] the three metropolitan districts, [as well as those of the commanderies of] Ho-nei, Ho-tung, Shang-tang, Chao Kingdom, and T'ai-yüan[145] were each to repair the old canals and open up watercourses to irrigate the government and private fields. [*HHS* 5:10b] A.D. 115

140. The *Tung-kuan-chi* has a similar statement. It further says that this thousand-odd *ch'ing* of land produced more than a million *hu* of grain.

141. Ma Leng, a grandnephew of Ma Yüan, was transferred to the position of grand administrator of Han-yang in A.D. 90 and a few years later to that of Tan-yang, K'uai-chi, and Ho-nei in succession.

142. Kuang-ling is the area between the Huai and the Yangtse rivers on the seaboard.

143. According to the *Tung-kuan-chi,* through his promotion of irrigation, Ma Leng increased the annual land tax by more than one hundred thousand *hu.*

144. On Hsi-men Pao, see IV, 16 and 17.

145. The commandery seat of T'ai-yüan was Chin-yang, which is equivalent to modern T'ai-yüan County in central Shansi. According to the *Shui-ching-chu,* this canal must have been an old one that branched off from the Chin River, flowed through the city of Chin-yang, and joined the Fen River, southeast of the city (cf. *SCC* 6:39b–40a).

43. Canals Dug

A.D. 175 The prefect of the Office for Imperial Stationery[146] was sent to Yen-chien[147] to have canals dug to benefit the people. [*HHS* 8:6a]

44. Canals Repaired

Ca. A.D. 36 In the Ho-hsi[148] region for a long time there had been very little rainfall. He [Jen Yen] set up for the people waterworks officials to repair the canals. All profited from this. [*HHS* 76:5a]

45. Cascade Reservoir

A.D. Third century Emperor Wu[149] of Wei again constructed a dam on the Chang River so that the watercourse would again run eastbound. This was called "the Dam of the Heavenly Well." Within the distance of twenty *li* twelve cascades were constructed with distance of 300 paces between each of the cascades. Water was therefore made to irrigate and to return then to the original stream. One water source was divided into twelve branches on each of which there was a sluice gate.[150] [*SCC* 10:6b]

46. Drainage Canals

A.D. Third century [Ssu-ma Fu[151] memorialized]: "Your subject was commissioned by an edict to promote works of irrigation in the commanderies of Honei.[152] Arriving at the locality, I made an inspection trip of the Ch'in River. . . . I followed a winding route over a watercourse of 900 *li* between the [area] west of the T'ai-hang Mountains and the [area] east of Wang-wu Mountains through rocks and cliffs. During heavy rains, water runs into valleys and small rocks are tossed about,

146. The *shou-kung-ling* was an official whose duty it was to manage the supply of paper, brushes, ink, and sealing clay (*HHC* 6:4b).

147. Possible at the salt lake Yen-ch'ih in An-i of Shensi.

148. This region consisted of commanderies west of the Yellow River, in today's Kansu Province.

149. Ts'ao Ts'ao; the text mistakenly reads King Wu.

150. The project was carried out in the southern part of present-day Hopei Province.

151. Ssu-ma Fu was at that time the general of the palace gentlemen in charge of agriculture in the commandery, Yeh-wang, in the southern part of present-day Shansi Province.

152. Honei is a general term embracing all the commanderies in the present-day southern Shansi Province.

[causing] the wooden sluice gates to become broken and ruined. The rice fields are thus flooded and there is crop failure. I investigated and noticed that there were several tens of thousands of stone blocks [at a site] about five *li* away from the dam. I believe that the stone blocks can be piled up to construct sluice gates. If there is drought the augmented dam will send water [into the canal system]. If there is heavy rain and the water reservoir is full, allow the dikes to be closed and the water blocked. The empty canal can drain the water [from the fields] as a river. . . ."

An edict granted permission. Hence on both banks stone blocks were piled to make gates replacing the wooden sluice gates.[153] [*SCC* 9:9b-10a]

47. Irrigation Network

A.D. second century

Yen-chu is a water reservoir for storing flowing water; *fang* is a side dike of the reservoir; *sui* is a small trench at the head of a field for receiving the water; *lieh* is a furrow; and *kuei* is the large trench at the end of the furrow for draining [the water] away. . . . The *sui* is opened so that the water may enter the furrows. Hence, [the water] spreads over [the field], raising and carrying off the grass that was cut in the preceding year. The field is cleaned up and rice is planted.[154] [*CLCS* 16:5a]

48. Mechanical Devices for Irrigation

A.D. 168-69

Furthermore, [Pi Lan][155] made the "turning wheel"[156] (or water wheel) and the "thirsty crow" (or the siphon) at the west of the bridge to sprinkle the roads to the northern and southern suburbs. Thereby the cost of having the commoners sprinkle the street was reduced. [*HHS* 78:20b]

153. The significance of this particular passage is not only in showing the use of stones to replace the wooden sluice gates that were not strong enough to resist pebbles, but also in showing that there was a technique for using canals as drainage for the excessive water not needed in the fields.

154. This passage is by the commentator Cheng Hsüan. Cf. V, 11.

155. Pi Lan was ordered by the then powerful eunuch, Chang Jang, to create various kinds of mechanical devices for the comforts of the capital city.

156. The same device is mentioned again in a third-century record as a labor-saving machine, easily operated by children irrigating a garden in the city. *San-kuo-chih pu-chu*, I-wen ed., 29:9a-b.

49. Scale of a Provincial Irrigation System

A.D. Third century

The stele inscription for Liu Ching[157] . . . says: "In the second year of the Chia-p'ing reign period (250) a dam was constructed in the river . . . on which was inscribed: 'The Kao-liang River is one of the tributaries of the Lu River from Ping Chou.[158] A long bank, steep and firm, directly intercepted the middle of the watercourse. Baskets of stones were piled up to construct the main bank, which was one *chang* high, thirty *chang* long from east to west, and seventy paces wide from south to north. On the north bank [of the river] a sluice gate was erected; the gate was four *chang* wide. There was erected a bank ten *chang* long. When there was flooding from the mountain rivers, it overflowed the dam to the east. When the usual flow was maintained at a constant level then it entered [the irrigation network] by the northern sluice gate. Each year it irrigated two thousand *ch'ing*. . . . '

"Thirty-six years after the construction of the dam[159] in the fifth year (of the Yüan-k'ung reign period, 295) in the summer in the sixth month [of the year] the flood came suddenly and ruined three-fourths [of the dam]. . . . Two thousand soldiers constructed the long bank, paved the canals, restored the main dam, and repaired the sluice gate. The gate had a breadth of four *chang*. The standing water was five *ch'ih* deep. . . . The construction was restored to its original scale. The work required the labor of some forty thousand working days.[160] . . . " [*SCC* 14:7a-8b]

50. Stele Inscription of a Water Dam in Mien-Chu

A.D. 176

In the fifth year of the Hsi-p'ing reign period (176) in the fifth month of which the date *hsin-yang* was the first day. . . . [This passage has been so corrupted that it is impossible to translate word by word. The first paragraph seems to indicate that some people in Mien-chu prefecture contributed their personal money for the erection of something, presumably the dam. However, it is not impossible that the money was contributed to erect the stele.]

157. Liu Ching was then a governor-general in Hopei Province, which included the present-day province of Hopei and part of Shansi.

158. The Lu River was a small river flowing from present-day Shansi to northern Hopei.

159. At this time the military governor-general was Liu Hung, the younger son of Liu Ching.

160. Here the scale of such a water-control project and the size of the area benefited by the work is indicated.

[The second paragraph begins with an almost complete sentence indicating the arrival dates of two officials to assume local office. One was the grand administrator of Kuang-han commandery, named Shen, who arrived in the third year of Hsi-p'ing in the tenth month. Another was the prefect of Mien-chu prefecture, named Fan, who arrived in the fourth year in the third month. Following are fragments of sentences that state that originally there was no water system for irrigation and flood control, and that these two officials instructed two or three native-born functionaries to take charge of the construction of water-control systems. The officials held the titles of chief of the Office of Water Control and deputy of the Office of Water Control. The construction work seems to have included the opening of a watercourse through rocks and the erection of a dam across the river.]

[The final paragraph is the least well-preserved section of the stele. However, the fragmentary remains convey the idea the people praised a good crop after the completion of this particular project.]

[The end as usual should be a list of the names of those who contributed to the erection of the stele. In its present form, however, only two names appear and neither bears any connection with the names appearing in the stele text. Between the two possibilities mentioned previously, it is more likely that people contributed money for the erection of the dam than that they contributed money for the stele.][161] [*LS* 15:7b-9a]

51. Stele Inscription on a Public Well

[This stele was found in Sui-ch'eng prefecture in Hopei Province. A.D. 180
The whole inscription originally consisted of twenty-three lines with thirty-seven characters each. The text has been so corrupted that the only legible part consists of the date, which is the third year of Kuang-ho (A.D. 180), and a section that states that the stele was erected upon the completion of a public well. At the end of the stele were figures of cash such as 40,000 and 1,000,000. There also remain phrases to indicate that "some were richer and some were poorer." On the back of the stele a long list of names of no less than ninety-one people is preserved. Each name is followed by a figure ranging from 3,000

161. From this badly corrupted text it seems that the construction of the water control system although possibly initiated by local officials, was actually the real responsibility of native-born local functionaries who held office created for such purposes. If the deduction about the nature of the account is correct, then it is highly probable that the financing of such a work was also handled by local people.

to 20,000 cash. Therefore, it seems that the well itself cost several hundred thousand cash that were contributed by people according to their own means.][162] [*LS* 15:15b-18b]

162. The *Li-shih* quoted the *Shui-ching-chu,* saying that this particular well is located southeast of the county seat. Water from it forms a stream, which eventually flows into the Cha River. The *LS* therefore explained the unusually high cost of the well by its peculiarity of being a well for irrigation purposes rather than an ordinary well used merely for drinking. The whole work was taken up by ordinary people who held no position or office nor rank higher than honorary titles of the ninth grade, a very common rank for ordinary citizens in the Han dynasty. Therefore it seems to have been a collective undertaking initiated and operated by the local population themselves.

FARMING METHODS AND TECHNIQUES

1. Arable Land in the State of Ch'in

An area of 100 *li* square[1] consists of 10 percent mountains and woods, 10 percent glades and morasses, 10 percent valleys, dales, and streams, 10 percent settlements and roads, 20 percent poor fields, and 40 percent fertile fields—from which 50,000 farmers are fed.[2] Mid-third century B.C.

Now the state of Ch'in comprises 5,000 *li* square; yet the arable land is no more than 20 percent and the fields are less than 1,000,000 mu. The natural resources of the glades and morasses, valleys, dales, and of its famous mountains and rivers are not all exploited. The population is not in proportion to the territory.[3] [*Shang-chün-shu*, SPPY ed., 4:1a-b; cf. J. J. L. Duyvendak, *The Book of Lord Shang*, p. 266]

2. Farm Size in Early Han

The land of 10,000 households is an area of fifty *li* square of cultivated fields, which should be sufficient. . . . Wherever the field has been completely cultivated, yet the people have no saving, is a small county with little arable acreage. Wherever the field is half cultivated, yet the people have food surplus and the grain storage is plenty, is a county of large land and extensive arable acreage.[4] [*KT* 5:3a] Second century B.C.

3. Five Crops in Four Years

[In the area] to the east of Ch'ang-shan and between the Yellow *Ca.* Mid-second century B.C.

1. The total size was 9,000,000 small mu (Yu Yü, "Yu Hsi-Chou tao Ch'ien-Han-ti keng-tso chih-tu yen-ke," *Nung-shih yen-chiu chi-k'an* 2 [1960] : 6).

2. The calculation goes, according to ibid.: Total 9,000,000 mu; 60 percent for arable land, i.e., poor land plus fertile land, 5,400,000 mu; divided by 50,000 farmers, 108 mu for each.

3. This is one passage of the chapter of "Encouragement of Immigration" in *The Book of Lord Shang*. It probably refers to the situation in the reign of King Chao-hsiang of Ch'in (306-251 B.C.). Cf. ibid.

4. This passage is from the chapter "Pa-kuan." Ch'en Heng-li suggests that this chapter was composed in the early part of the reign of Emperor Wu (140-87 B.C.); cf. Ch'en, "Kuan-tzu Mu-min, Ch'eng-ma, Pa-kuan teng-pien yen-chiu," *Hsin-chien-she*, 1956, no. 5.

River and the Ju River,[5] plants grow early and wither later. [It is the place where] the five grains are raised and harvested. In four years there are five harvests.[6] The annual yield is two *shih* for each mu. One hundred mu of land produces two hundred *shih* of grain.[7] [*KT* 15:15a]

4. Fan Sheng-chih-shu

1. *Basic Principles of Farming*

Late first century B.C.

A.1. The basic principles of farming are: be timely, pulverize the soil, attend to its fertility and moisture, hoe early, and harvest early.

A.2. By taking what is appropriate to the time and choosing what is fitting to the soil, a harvest of 10 *shih* per mu can be obtained even from very poor land.

A.3. In springtime, after thawing, the breath of earth[8] begins to come through, so that the soil is completely pulverized and loosened. At the summer solstice, the breath of heaven begins to grow hot, and the *yin* breath begins to thrive, the soil is again loosened. Ninety days after the summer solstice, duration of the day equals that of the night, and the breaths of heaven and earth are in harmony. By cultivating in these proper seasons, one operation is worth five. Such conditions are called *kao-tse*;[9] they are all achievements of appropriate timing.

A.3.a. In the spring, when the breath of earth comes through, hard heavy lands and black soils may be plowed. Then harrow to level the clods, and let grasses sprout. After the sprouting of grasses, plow again. And plow again after a drizzle. Pulverize the soil, not allowing

5. In the southern part of the present province of Hopei and the eastern part of the present province of Honan.

6. The T'ang commentator Fang Hsüan-ling interpreted this sentence to mean five grains are planted in four seasons. The agricultural historian Yu Yü argues, however, that winter is not a season to plant and that the term *chung,* to plant, should be interpreted as "year." According to his definition, any system that produces more than one harvest annually qualifies as a multicrop system. Yu Yü, "Yu Hsi-Chou tao Ch'ien-Han-ti keng-tso chih-tu yen-ke," p. 13.

7. This passage is from the "Chih-kuo" chapter of the *Kuan-tzu.* It is dated to the reigns of Emperors Wen (189-157 B.C.) and Ching (156-141 B.C.); cf. Wan Kuo-ting, *Chung-kuo nung-hsüeh-shih,* p. 105.

8. Breath of the earth or *yin* breath means the complex conditions of low temperature and high humidity of the soil. Breath of the heaven, on the other hand, indicates warm and dry conditions prevailing under sunshine.

9. *Kao-tse,* literally means "greasy moisture," but it connotes rich benefits or sweet blessings.

any clods to remain, and wait for the proper time [to sow]. This is what is called "to make the heavy soils light."

A.3.b. In the springtime, watch for the breath of earth to begin to come through. Sharpen a wooden peg 1 *ch'ih* and 2 *ts'un* long, bury 1 *ch'ih* of it, and let the remaining 2 *ts'un* appear above the ground level. After *Li-ch'un,*[10] the clods of earth disintegrate, and will rise up and cover the peg; then old roots can be pulled out. This is the proper time [to plow].

Twenty days later, the mellow breath [of earth] is gone, and the soil hardens. One plowing in time is worth four, while four plowings after the mellow breath is gone will not equal one.

A.4. Light soils are to be plowed when apricot trees come in blossom.[11] Plow again when the blossom fades, and roll the soil every time after plowing. After the grasses sprout, plow and roll it again when it rains and the soil is moist. With soils that are too light, use cattle to tread them down. The soils will then harden. This is what is called "to make the light soils heavier."

A.5. In the spring, when the breath of earth has not yet come through, the soil will be lumpy [when plowed] and will not retain moisture; throughout the year good crops will be unlikely. Unless there is fertilizing, the situation cannot be relieved.

A.5.a. Be careful not to plow too early; wait till grasses sprout. When the time arrives when planting can be done and there is rain, then the seeds and the soil will be in good contact. Only seedlings will grow while grass and weeds will rot [beneath the soil], and a good field results. This is what "one plowing worth five" means.

If a field is plowed too early, the clods are hard, and seedlings and weeds come out from the same crevices. No hoeing can be done, and a bad field results.

A.6. If one plows in autumn when it does not rain, the breath of earth is cut off, and the soil will be hard and arid. This is called *la-t'ien* (a field like dried meat). If one plows in the height of winter, the *yin* breath is broached and the soil will be dry and parched. This is called *fu-t'ien* (a field like jerked beef).[12] Both dried fields and jerked fields have been damaged.

A.7. If a field gave a poor crop in two consecutive years, fallow it for one year.

10. *Li-ch'un,* beginning of spring, the first of the twenty-four subseasons of a year. It normally falls on February 4 or 5.

11. About the first ten days in April.

12. The special condition of loess earth in North China necessitates a continuous concern that moisture not be lost from the soil.

A.8. Wheatfields should always be plowed in the fifth month. Plow again in the sixth month. Don't plow in the seventh month, but diligently harrow it level, and wait for the time to sow. One plowing in the fifth month is worth three; one in the sixth month, two; but five in the seventh month is not worth one.

A.8.a. In the winter, whenever the snow stops falling, roll the snow [into the ground]. Do not allow the snow that covers the ground to be blown away by the wind. After later snowfalls, roll the snow again. Thus at *Li-ch'un* (beginning of spring) the moisture will have been retained and insects will be frozen to death. The coming year ought to have good crops.

2. *Choosing [the Days and Crops] for Planting*

B.1. Regarding lesser beans, avoid *mao* (the fourth in the set of every twelve) days; rice and hemp, *ch'en* (fifth in twelve) days; millet, *ping* (the third in the set of every ten days); glutinous millet, *ch'ou* (second in twelve); spiked millet, *yin* and *wei* (third and eighth in twelve); wheat, *shu* (eleventh in twelve); barley, *tzu* (first of twelve); large beans, *shen* and *mao* (ninth and fourth). All the nine cereals have their days to be avoided. If one plants them without circumventing the days to be avoided, there will be much damage and failure. This is not empty talk. It is natural. If one burns the chaff of the glutinous millet then one harms the [growing] gourds.[13]

B.2. Regarding wheat, avoid *shu* days; barley, *tzu* days. *Ch'u* days are not suitable for planting.[14]

B.3. Wishing to know what is the most suitable crop for the coming year, fill cloth bags each with millet and other things you might plant. Put equal measure in each and bury them in shady places. Fifty days after the winter solstice, dig them out and measure them. That which has increased the most will be the most suitable for the coming year.

3. *Treating the Seeds*

C.1. Take horse bones and chop them finely. Decoct each *shih* of them in three *shih* of water, bringing to a boil three times. Strain off the sediment and steep five stalks of aconite in the liquid.

13. Notice the tone of the Han theory that forces in the universe are being interfered with; this cosmology heavily influenced the thinking and behavior of the Han people.

14. *Shu* and *tzu* are both days of the duodecimal (earthly stems, *ti-chih*) series; *ch'u* is a day of the twelve spirits of *chien* and *ch'u* cycle (see III, n. 8).

After three or four days, remove the aconite and mix in equal amounts of silkworm and sheep excrement; stir well so as to form a paste like thick gruel.

Twenty days before sowing, immerse the seeds [in the paste]; it will have the appearance of boiled wheat. Always immerse them when the weather is dry and dry them immediately. Thinly spread the treated seeds; repeatedly rake them to cause them to dry quickly.

On the next day, immerse them again, but if the day is overcast or if it is raining, then do not immerse them. Stop after six or seven immersions; then dry them in the sun. Store them carefully, not allowing them to become damp again. When the time comes when they can be planted, if you use the extra liquid to immerse them and then plant them, then the crop will not be infested with locusts and other insects.

C.1.a. If there are no horse bones, you can also use melted snow. Snow is the essence of cereals; it causes crops to be drought-resistant. Always take advantage of the winter to store snow; fill containers and bury them in the ground.

Soak excrement of polyvoltine silkworms in melted snow. Wait five or six days until the excrement becomes softened, and crumble it with your hands. Mix grain [seed with this paste]; it will have the appearance of boiled wheat. When planted it will be drought-resistant. Therefore it is said that "snow is the essence of cereals."

C.2. If seeds are treated like this, the harvest regularly will be doubled. In regulating moisture,[15] either get [more] or reduce it, emulating Shen-nung.[16] Also add to the seeds bone decoctions and liquefied manure.

C.2.a. Chop finely 1 *tou* of horse bones, or ox, sheep, pig, elk, or deer bones, cook it with 3 *tou* of melted snow, bringing it to a boil three times. Drain off the liquid and use it to soak aconite. The rate is 5 stalks of aconite to each *tou* of liquid. After soaking for five days, remove the aconite. Pound equal parts of elk, deer, and sheep manure, put them into the liquid, and stir thoroughly until they are blended. Wait for a clear and warm day, then [treat the seeds][17] by immersing and drying them, according to the method of Hou-chi. Only when all have been immersed in the liquid and dried does one stop.

C.2.b. If one has no bones, boil silkworm cocoons, [prepare the] liquid, mix [the manure] and immerse [the seeds].

15. The meaning of this paragraph is highly uncertain.

16. "Shen-nung" perhaps refers not to the god of agriculture but to a text ascribed to him.

17. This phrase has been interpolated by Shih Sheng-han.

C.2.c. Having carried out these steps, sow the seeds in shallow pits. In case of drought, water the seeds. Yield will be 100 *shih* or more per mu, that is, tenfold that claimed by Hou-chi.

C.3. When it is time to plant wheat, if the weather has been dry and the fields are not damp, then slightly soak the wheat seed in sour rice drink[18] and silkworm manure. Soak the seeds by midnight and sow[19] quickly with the dawn, so that they will enter [the ground] with white dew. Sour rice drink renders the wheat drought-resistant; silkworm excrement renders the wheat cold-tolerant.

4. *Planting*

D.1. There is no fixed date for sowing spiked millet; the timing depends upon conditions of the ground.

D.1.a. In the third month, when the elm pod appears, spiked millet may be sown in heavy soils on high land whenever it rains.

D.1.b. On poor land that cannot be manured, if seeds of spiked millet are mixed with excrement of polyvoltine silkworms and planted, then the millet will not be infested with insect pests.

D.1.c. For early spiked millet, always watch at midnight eighty to ninety days after the summer solstice. If there should be frost or white dew, then at dawn let two persons facing each other drag a long rope horizontally through the crop to scrape away the frost or dew. Stop only after sunrise. If this is done the crops will not be injured [by frost].

D.2. Glutinous millet, *shu,* means [a plant of] "*shu,*" the heat of summer. To plant it, one must wait for the heat of summer. Twenty days before summer solstice when there is rain is the proper time. In heavy soils one can sow 3 *sheng* per mu of glutinous millet.

D.2.a. When the heart of glutinous millet has not yet grown, if rain pours into the heart, the heart will be injured and there will be no fruit.

D.2.b. When the heart of glutinous millet has first grown, one is fearful of dew. Let two persons facing each other drag a long rope through [the crop] to remove the dew. Stop only after sunrise.

D.2.c. For glutinous millet the thickness of the soil layer to cover

18. "Sour rice drink" *(ts'u-chiang)* is identified by Shih Sheng-han (p. 13, n.j.) as a cold fermented rice drink.

19. The verb "to sow" here is *t'ou,* which literally means "to throw"; the reference is apparently to broadcasting as a means of planting.

the seeds sown and the hoeing and care are all like the methods used for spiked millet. But the planting ought to be thinner.

D.3. For every field, for which there are six alternates, wheat is the first to be planted. Timely sowing of wheat insures successful harvest. Winter wheat may be sown seventy days after the summer solstice. If one sows early, the ears will be solid and the stalks firm; if one sows late, the ears will be small and the fruit will be reduced.

D.3.a. If the wheat is yellow when it sprouts, it is being injured by overcrowding. If it is crowded, hoe and thin it.

D.3.b. In autumn, hoe, and drag thorny branches through the wheat so as to bank up its roots. The proverb goes: "If you wish to be rich, cover with yellow gold." "Cover with yellow gold" means to hoe the wheat and to bank up the wheat roots by dragging branches [through the plots] in the autumn.

D.3.c. In winter, whenever the snow stops falling, use [some] thing to roll over the wheat; do not allow the snow that covers the ground to be blown away by the wind. If one repeats this with later snowfalls then the wheat will be drought-tolerant and will have much fruit.

D.3.d. When the spring thaw comes, drag thorn branches through the wheat plots so as to detach dried leaves. Wait until the wheat is growing, then hoe it again. During the period when elm pods grow, after every downpour of rain, watch the ground. When the surface turns pale, hoe again. If these things are done, the yield will certainly be doubled.

D.4. In the spring, when it thaws, plow and pulverize the soil and sow spring wheat. After the wheat sprouts and the roots are flourishing, hoe out the weeds as with winter wheat.

D.5. To plant rice: in the spring when it thaws, plow to turn over the ground. The field for planting rice should not be too large; if it is large, the depth of the water will not be suitable [throughout the field].[20]

20. The following note, with little change, is from Shih (p. 17, n. k.). A rice paddy may be visualized as a shallow artificial pool with a flat bottom surrounded by mud ridges (i.e., the paddy ridges of D.5.c.). The bottom must be uniformly level so that there is little difference in depth of water standing upon the basin floor and hence in the level of immersion for individual rice plants. The depth of the standing water, usually 10-25 cm., can be regulated by gaps made in the ridges. These gaps can be opened or stopped with mud at will. If the inlet and outlet gaps of a single field are opposite to each other on a straight line—especially if they are made at the narrower part, then, because the path of flow is comparatively short, the total amount of water in motion and hence the change in temperature

D.5.a. In the third month, sow ordinary rice; in the fourth month, sow glutinous rice.

D.5.b. One hundred and ten days after winter solstice, rice may be sown. When the rice land is good, use 4 *sheng* of seed per mu.

D.5.c. When first planted, rice needs warmth. For warmth, break the paddy ridges so that the route of the water will directly cross [the field]. After summer solstice, there is intense heat; cause the route of the water to mix [the fresh and standing waters].

D.6. Soya beans to assure a good year are easily raised. It is fitting to follow the means by which the ancients prepared for a year of famine.

Carefully calculate the number of people in the family and plant soya beans at the rate of 5 mu per capita. This is "the basic" for farming.

D.6.a. In the third month, when the elm pods grow and when there is rain, one can plant soya beans on highland fields. When the soil has been pulverized and there are no clods, 5 *sheng* [should be planted] per mu. If the soil is not pulverized, then increase the amount planted.

Soya beans may still be planted as late as twenty days after the summer solstice.

D.6.b. Soya bean seedlings come out as though with a helmet on the top, so there is no need to plow deeply. The soil over the seeds cannot be thick—only enough to cover the bean. If it is thick then it bends the necks [of the bean stalks], which cannot penetrate upwards and they bend under the soil and die.

D.6.c. Soya beans must [be planted] evenly and sparsely.

D.6.d. Flowers of soya beans dislike seeing sunshine; if they see the sun they rot and wither.

D.7. Lesser beans will not assure a good crop; they are difficult to grow.

D.7.a. It is fitting, when mulberries are darkening and when there is a downpour of rain, to plant [lesser beans], 5 *sheng* per mu.

---

for the whole body of water in this field will also be less. In late spring or early summer, when both the intensity and duration of insolation is less, water in rice fields tends to be cool. At such times, less flow would result in having warmer water in individual fields. When the weather becomes warm in later summer, water in the field, exposed to longer and stronger sunshine, may be too warm for plant growth; then greater amount of flow and hence greater change in volume would mean better temperature and aeration.

D.7.b. When beans sprout and send out leaves, hoe them. When they have produced five or six leaves hoe again.

D.7.c. Neither soya nor lesser beans can be completely defoliated.[21] In antiquity the reason that they were not completely defoliated is that when the beans sprouted and sent out leaves, the beans were nourished. If they are completely defoliated then the nourishment is harmed. If it is harmed then [the crop] does not reach full development. But people completely defoliate; therefore, the harvest is diminished. Hence the saying: "Beans cannot be completely defoliated."

D.7.d. In raising [beans?], on good land one can harvest 10 *shih* per mu; on poor land, one can still harvest 5 *shih* per mu.

D.8. In planting hemp, ameliorate the ground thoroughly beforehand.

D.8.a. In the last ten days of the second month or the first ten days of the third month, plant it shortly before a rainfall. When hemp sprouts and sends out leaves, hoe [to thin it] to a space of 2 *ch'ih* per plant.[22]

D.8.b. When the plants grow to 1 *ch'ih* high, fertilize them with silkworm manure, 3 *sheng* per plant. If there is no silkworm manure, use well-ripened fertilizer from a cesspool to fertilize them; it is also good. Use 1 *sheng* per plant.

D.8.c. When the weather is dry, water with flowing water, 5 *sheng* per plant. If there is no flowing water, expose well water to the sun to kill its cold breath and use it to water them. When rain and soil moisture are timely and proper, do not water. Watering should not be too frequent.

D.8.d. If one raises hemp like this then on good land one will harvest 50 and even up to 100 *shih*; on poor land one will still harvest 30 *shih*.

D.9. In planting male hemp, after the spring thaw plow and dress the soil. After the sprouting of spring grasses, manure the field and plow again, harrowing to level it.

D.9.a. If you plant hemp too early [the male hemp plants] will be hard and rigid, with thick skins and many knots. If you plant late then the skin will not be hard. It is better to err in being early than to err in being late.

D.10. The method for planting gourds is, in the third month,

21. Bean leaves were used as greens; see G.5.

22. The text reads 9 *ch'ih* apart; Shih corrects the figure, on the basis of the *Ch'i-min-yao-shu*, to 2 *ch'ih*.

plow 10 mu of good land and dig shallow pits 1 *ch'ih* in diameter, 1 *ch'ih* deep, and 6 *ch'ih* apart. Using a pestle, pound the bottom of the pit to make it retain moisture. In each pit plant 4 seeds. Mix 1 *tou* of silkworm manure with compost. Water them with 2 *sheng* of water. In places that dry up, water again.

D.10.a. After 3 fruits have set [on one vine], use a horsewhip to knock off the top of that vine; do not allow the vine to spread—the more fruits per vine, the smaller they will be in size. Place some straw underneath every fruit to prevent direct contact with the earth; otherwise there will be scars on the skin.

D.10.b. When they are considered good enough to make dippers, use your hands to rub the fruit from the stem end to the top and remove the fur; they will grow only in thickness but no more in length. In the eighth month, after a mild frost, gather them up.

D.10.c. Dig a pit 1 *chang* deep; line the bottom and sides with straw, 1 *ch'ih* thick. Place the fruits, tips downwards, in the pit. Cover each layer with earth 3 *ch'ih* thick.

In twenty days, remove them; the color will be yellow and they will be good. Split them to make dippers. The white flesh within can be used to feed pigs and make them fat; the pulp can be used to make candles to produce light.

D.10.d. Each vine yields three fruits, so each pit yields twelve fruits, 1 mu produces 2,880 fruits, and 10 mu produces 57,600 dippers. Each dipper is worth 10 cash; the total value is 576,000 cash. You will have used 200 *shih* of silkworm manure, which together with cattle and manpower spent in cultivation will cost 26,000 cash. The remainder is 550,000 cash, exclusive of the profits from the fattened pigs and bright candles.

D.11. The method for planting taro is: one should select rich and loose soil situated near water and soften and fertilize it. In the second month, with downpours of heavy rain, one can plant taro at the rate of 1 rootstock every 2 *ch'ih*.

D.11.a. After sprouting, the roots will want to go in deeply, so hoe around the plant to loosen the soil. Water if it is too dry. If there are weeds, hoe them; do not become weary of doing it many times. If you take care of taro like this, yields will always double.

D.12. Water darnel *(Panicum)* is highly tolerant of both drought and flood; there is no time when the crop will not ripen. Also, it is especially prolific and easy to grow. A yield of 20 to 30 *hu* per mu can be obtained from a good field badly crowded with weeds. It should be planted to prepare for bad years.

D.12.a. There is a kernel in every darnel grain. When it is ripe, pound it to obtain the kernel. When steamed and eaten, it is not inferior to millet. It also can be fermented to make beer.

D.13. The method for planting mulberry trees is, in the fifth month, collect [ripened] mulberries, put them in water, and then use your hands to scrub them. Flush with water to cleanse; take the seeds and dry them in the shade.

D.13.a. Prepare 10 mu of rich fields—the best are fallowed fields that have not been plowed for a long time—plow and prepare well. In each mu plant together a mixture of 3 *sheng* each of millet and mulberry seeds.

D.13.b. The millet and the mulberry will grow together. Hoe them. Thin the mulberry and arrange them suitably.

D.13.c. When the millet is ripe, reap it. The mulberry plants will have grown to the same height as the millet. Accordingly, with a sharp sickle reap both close to the ground. Sun them to dry them. Later, with a good breeze, stir them and set fire to them. Always start the fire against the wind. The mulberry, when spring comes, will grow; 1 mu will feed 3 frames of silkworms.

5. *Harvesting*

E.1. In harvesting one cannot but be fast; always regard urgency as fundamental. When the awns are extended and leaves are yellow, reap promptly without delay.

E.1.a. The method to harvest millet is when over half [of the crop] is ripe, cut it down.

E.2. The method to harvest beans is when the pods are dark and the stalks are gray to reap immediately, for the beans will drop to the ground and you will lose them. Therefore, it is said, "beans ripen on the threshing ground." Harvesting beans on the threshing ground means with green pods above and darkened pods below.

E.3. The method to harvest female hemp plants is, after a frost, when the seeds are mature, cut them quickly. For larger plants, use a saw to saw them down.

E.3.a. The method to harvest male hemp plants is when the flowers change color so they are like dust, uproot the plants.

E.3.b. Twenty days after summer solstice, soak the male hemp plants in water [to soften] ; the hemp will be as pliable as silk.

6. *Storing Seeds*

F.1. In choosing wheat seed, watch for [wheat] to become ripe when it can be harvested; select [plants with] large and solid ears, mow, tie [into bundles] and stand them in a high and dry place on the threshing ground. Let sunshine dry it out to the utmost. Do not allow any silverfish [on the grain] ; if there are any, shake the grain to take care of them. Use dry moxa to mix [with the seed] to store it; for 1 *shih* of wheat,[23] use 1 handful of moxa. Store in earthen or bamboo containers.

If these seeds are planted at a favorable time, the yield will always be doubled.

F.1.a. In choosing millet seed, select those that are tall and large, cut below the first node down, tie them into bundles, and hang in a high and dry place. When they sprout they never fail.

F.2. If seeds are injured by moisture and high humidity, then they will produce insects.

F.3. Lead a horse to the grain pile and let the horse eat several mouthfuls; then have the horse walk through it. As seed, such grain will have no boring or other insects.[24]

7. *Cultivation in Shallow Pits*

G.1. Emperor T'ang was faced with a drought. [The Prime Minister] I Yin created the fields of shallow pits *(ou-t'ien)*; he taught the people to fertilize seeds and to carry water for irrigating the crops.

G.1.a. Fields of shallow pits take the power (*ch'i,* "breath") of fertilizer as its fine quality; it is not necessary to have good land. Mountains, mounds, high places in the vicinity of towns, slopes, hillocks, and the tops of the city walls can all be used for fields of shallow pits.

G.1.b. In fields of shallow pits, do not plow the adjacent land; thus one may exhaust the productivity of the land.

G.1.c. In planting in shallow pits one does not first prepare the land; one uses fallow land as it is.

G.1.d. Take 1 mu as a standard: let a piece of land of 1 mu be 18 *chang* long and 4 *chang* and 8 *ch'ih* wide. Horizontally divide the 18 *chang* into 15 parcels; between the parcels will be paths for people to walk on.

23. The *shih* is a measure of weight of approximately 21.336 kg.

24. This is obviously one more part of the text that reflects Han cosmology, in which all the parts of the universe are subject to influence from each other.

The paths are 1 *ch'ih* and 5 *ts'un* wide; the parcels, 1 *chang* and 5 *ts'un* wide and 4 *chang* and 8 *ch'ih* long.

Dig trenches straight across the width of the parcels. Trenches are 1 *ch'ih* wide and also 1 *ch'ih* deep. Pile the dirt [from the trenches] between the trenches, which will also be 1 *ch'ih* apart. (Try to make each dirt pile 1 foot across; if that much dirt cannot be accommodated, enlarge the place for the dirt pile to 2 *ch'ih*.)

G.1.e. To plant spiked and glutinous millet, sow them in 2 rows along the flanks of a trench. Individual plants are 5 *ts'un* apart; rows, 5 *ts'un* apart and 2.5 *ts'un* away from the side of the trench. One trench will hold 44 individual plants, and the whole mu will have 15,750 plants altogether.[25]

G.1.f. When planting millet, cover with one *ts'un* of earth; the amount cannot be either more or less than one *ts'un*.

G.1.g. When weeds appear within the plots, uproot them. Weeds that grow between the plots may be eliminated with a sharp spade or a hoe. After the crop has grown too high to weed, sharp billhooks may be used to cut the weeds at ground level.

G.1.h. With cultivation in shallow pits, one must always water the crop in days of drought. One can always harvest one hundred *hu* per mu.

G.1.i. Experience shows that the yield from good land may attain 19 *shih* per mu, from medium land 13 *shih* per mu, and from inferior land 10 *shih* per mu.

G.2. For superior land,[26] the *ou* or shallow pits should be 6 *ts'un* square, 6 *ts'un* deep, and 9 *ts'un* apart. One can accommodate 3,700 pits in one mu.[27] In one day one can make 1,000 pits.

Twenty millet seeds are sown in one pit; use 1 *sheng* of good

25. Based on the figures given in the text the total number of plants should be 15,240. Each trench requires 1 *ch'ih* of land for the trench itself plus another *ch'ih* for the mound. Thus each parcel, 48 *ch'ih* long, will accommodate 24 trenches. Forty-four plants per trench times 24 trenches per parcel times 15 parcels per mu equals 15,240 plants. Errors of this kind in the text, perhaps due either to careless mathematics or sloppy copying, are not uncommon; cf. Shih Sheng-han, *Ch'i-min-yao-shu* 1:43 ff.

26. "Superior land" *(shang-nung-fu)* could be translated literally as "best farmer." Shih identifies *fu* as a reference to a plot of land. Superior land requires no fallowing *(Fan Sheng-chih-shu,* p. 31, n.q.).

27. Note that the system described here is based upon square pits, not trenches, as in G.1.d and G.1.e. If we assume a mu of the same dimensions as in G.1.e then a mu would have 3,840 pits, not 3,700 as given in the text. Also, note that this pit system allows for no footpaths except the narrow 9 *ts'un* spacing between pits.

manure per pit and mix well with the soil. For 1 mu 2 *sheng* of seeds suffice. The harvest in autumn is 3 *sheng* per pit and 100 *hu* per mu. When an able-bodied man and an adult woman work 10 mu of land, the yield of the 10 mu will be 1,000 *shih*. Annual food consumption per capita is 36 *shih,* so the yield of 10 mu in one year suffices for 26 years.[28]

G.2.a. For medium land (fallowed every other year), the pits are 7 *ts'un* square, 6 *ts'un* deep, and 2 *ch'ih* apart. One mu can accommodate 1,027 pits. Using 1 *sheng* of seed you will harvest 51 *shih* of millet. In one day one can make 300 pits.

G.2.b. For inferior land (fallowed for two consecutive years), the pits are 9 *ts'un* square, 6 *ts'un* deep, and 3 *ch'ih* apart. One mu can accommodate 567 pits. Using 6 *sheng* of seeds, one will harvest 28 *shih.*[29] In one day one can make 200 pits.

G.3. Whenever one plants wheat in shallow pits, allow the rows to be 2 *ts'un* apart. One trench will accommodate 52 plants, and 1 mu 45,550 plants. The layer of soil over the wheat [seed] should be 2 *ts'un* thick.

G.3.a. In cultivating wheat in shallow pits, the pits are of the same size as those for medium land. After harvesting the spiked millet, plant wheat seeds in the pits. Use 2 *sheng* of seeds per mu. Cover the seeds with soil 2 *ts'un* thick; pack the soil down with your feet to bring the seed into close contact with the soil.

After the wheat germinates and takes root, hoe the autumn weeds between the pits. Drag thorn branches through the rows to bank up the wheat roots. If the autumn is dry,[30] water [the pits] when the mulberry leaves fall. With timely autumn rainfall and proper soil moisture, do not water.

28. The yield of 100 *hu* per mu is incredibly high. The suggestion has been made that the area of 1 mu is the total of all the individual pits, which add up to 1 mu in area. In each mu, the area made into pits constitutes 15.42 percent of the whole mu. The real yield from 1 mu of 3700 pits would be 100 x 15.42 percent = 15.42 *hu*. The figures fall reasonably within the range of 10 to 19 *shih* per mu given in G.1.i. Cf. Li Meng-yang and Liu Yu-chu, "Mou shou-pai-shih-chih-mi," *Wen-shih-che,* 1964, no. 2, pp. 54-55. This argument, however, is defective, for the figure of 100 *hu* per mu is apparently computed and rounded from the yield per pit figure. Three *sheng* per pit times 3,700 pits totals 11,100 *sheng,* which converts to 111 *hu*—here rounded to 100 *hu*. Thus the basic error in the text seems to be in the 3 *sheng* per pit.

29. The yield figures for these three grades of land vary sharply. For the superior land, 1 *sheng* of seed produces 5,550 *sheng* of harvested grain; for the medium land, the ratio is 5,100:1; and for the inferior land 466.6:1. Even the latter figure seems unacceptably high.

30. In the Yellow River region, autumn is usually the season of heavy rainfall.

With thawing in the spring, drag thorn branches through the rows to detach dry dead leaves. Hoe the weeds that have grown between the pits.

An adult man and an adult woman together can work 10 mu. Harvest in the fifth month. From 1 mu of pits one will obtain 1 *shih* or more; from 10 mu one will obtain 1,000 *shih* or more.

G.4. To plant soya beans in shallow pits make [the trenches] 1 *ch'ih* and 2 *ts'un* apart. One trench will hold 9 plants, and 1 mu, 6,480 plants.

G.4.a. The method for planting soya beans in shallow pits is to make the holes 6 *ts'un* square, 6 *ts'un* deep, and 2 *ch'ih* apart. In 1 mu one will have 1,680 pits.

When the holes are completed mix 1 *sheng* of good manure with earth from the pit, and put it in the hole. Just before planting, water each hole with 3 *sheng* of water. Put 3 beans in the hole. Cover with earth, but do not cover thickly. Press down with the palms of the hand to insure close contact of the seeds with the earth. For each mu use 2 *sheng* of seeds and 16 *shih* and 8 *tou* of manure.

When the bean plant has produced five or six leaves, hoe. Water if drought occurs—3 *sheng* of water per hole.

A full-grown man can work 5 mu. At the autumn harvest, one mu will produce 16 *shih.*

G.5. To plant melons in shallow pits, make 24 holes in 1 mu. Each hole is to be 3 *ch'ih* in diameter and 5 *ts'un* deep. In each hole use 1 *shih* of manure; mix the manure well with an equal amount of earth. Bury an earthen jar of 3-*tou* capacity in the center of the pit, leaving the mouth of the jar level with the ground. Fill the jar with water. Plant 4 melon seeds around the jar. Cover the jar with a tile. When the amount of water falls, immediately add more; always keep the jar full.

Always plant the melon seeds 90 or 100 days after winter solstice, on a *wu* or *ch'en* day.

Also plant 10 scallions around the jar outside the melons. In the fifth month, when the melons begin to ripen, the scallions can be pulled and sold. The scallions and the melons will not interfere with each other.

Furthermore, lesser beans, 4 to 5 *sheng* per mu, may be sown among the melons; the bean leaves can be sold as greens.

This method fits well with level lands. When melons are harvested, one will receive 10,000 cash [worth of melons] per mu.

G.6. The method for planting gourds in pits is to collect seeds from big gourds. Seeds from a gourd 1 *tou* in size will produce a harvest of gourds 1 *shih* in size. Gourds 1 *shih* in size will produce a harvest of gourds 10 *shih* in size.

G.6.a. First, excavate a pit 3 *ch'ih* in diameter and in depth. Mix equal amounts of silkworm dung and earth. Put the mixture in the pit and tamp with your foot to make it firm. Water and, only when all the water has soaked in, place 10 gourd seeds. Again using the foregoing manure mixture cover them.

After the vines have grown to 2 *ch'ih* or more, bring all 10 of the stalks together in one place, wrap 5 *ts'un* of the stalks with cloth, and also seal them together with mud. After a few days the stalks at the place where they were wrapped will fuse into one. Retain the vine that is the strongest and pluck off all the others. When the extended creeper sets fruit, those branches that do not have fruit should also be plucked off. Do not allow the vine to spread.

G.6.b. The method for retaining fruit is to eliminate the first 2 or 3 fruit; they will not be good. Take the fourth, fifth, and sixth. It is quite sufficient to retain 3 fruits for each pit.

G.6.c. When the weather is dry you must water them. Encircle the bank of the pit with a small ditch 4 to 5 *ts'un* deep; the water will be held in it so that the plant will be moistened from a distance. Never apply water directly into the pit.

G.7. To plant taro make the pits 3 *ch'ih* square and 3 *ch'ih* deep. Place beanstalks in the pit and pack down with your feet so that they are 1 *ch'ih* and 5 *ts'un* deep. Mix the moist soil from the pit with manure, and put into the pit on top of the beanstalks so that it will be 1 *ch'ih* and 2 *ts'un* deep. Water and pack down with the feet to preserve the moisture.

Take taro tubers and place them in the 4 corners and the center [of the pit]. Pack down with the feet.

When dry frequently water them. The beanstalks will rot as the taro grows; the fruit will all grow to be 3 *ch'ih* long. From each pit one will harvest 3 *shih*.

G.8. To plant Perilla in shallow pits, make the pits 3 *ch'ih* apart; for sesame, 1 *ch'ih* apart.[31] [*Fan Sheng-chih-shu*]

31. The translation is largely that by Shih Sheng-han in his *On Fan Sheng-chih-shu*. Much revision has been made here, however. The text is restored from the quotations in the *Ch'i-min-yao-shu*, since the original work attributed to Fan Sheng-chih has long been lost. For the life of Fan Sheng-chih and the transmission of text, see Shih, *On Fan Sheng-chih-shu*, pp. 40-43.

5. Multicrop System (1)

The current custom[32] is to say [the crop] after the wheat is [the crop] after reaping, which means the wheat has been reaped and because it has been reaped one plants spiked millet and beans.[33] [*CLCS* 34: 5a] A.D. Second century

6. Multicrop System (2)

Now it is said that the wheat next to spiked millet is the wheat after the reaping, which means after the spiked millet is reaped, the wheat is sown.[34] [*CLCS* 16:5a] A.D. Second century

7. Chao Kuo's *Tai-t'ien*

At the end of his reign, Emperor Wu regretted the military campaigns he had undertaken. He then bestowed upon the imperial chancellor the title of "marquis enriching the people" and issued an edict, saying: "The most urgent task of today is to strengthen agriculture." He made Chao Kuo the chief commandant for grain.[35] 89 B.C.

Kuo knew how to make *tai-t'ien,* or "alternating land," in which one mu had three furrows whose position was exchanged yearly.[36]

32. Both this passage and the next are notes made by the Han annotator Cheng Chung (died 183) on the *Chou-li.* Cheng explains the text by reference to practices of his own time.

33. This passage comes as a note to the office of *t'i-shih*, the field cleaner.

34. This passage comes as a note to the phrase on cleaning the paddy field by flooding to suffocate the weeds.

35. This office was under the grand minister of agriculture to be in charge of agricultural production in the army (*HS* 19A:15a).

36. The mu in our text must be distinguished from mu as a measurement of land; here it is used as a technique for arranging the fields by means of the ridge *(lung).* Before the time of Ch'in and Han a standard mu was a strip of one pace *(pu),* equal to six *ch'ih* ("feet") wide and a hundred paces long. The *ch'üan* (furrows) extended the length of the mu. Since between the *ch'üan* there must be a ridge, called *lung* in the present text, three furrows, each being one *ch'ih* wide, occupy just half the width of one mu, the other being occupied by three ridges. Every other year the furrows and ridges were alternated. On the definition of *ch'üan* and *lung,* see Yang Chung-i, "Tsai-lun i-mou-san-ch'üan-sui-tai-ch'u," *Shih-huo* 2, no. 4 (1935):183 ff. A new interpretation to include six furrows in the strip is presented by Itō Norio. See his "Daidenhō no ichi kōsatsu" in *Shigaku zasshi,* vol. 69, no. 11 (1960), and the rejection of his theory by Nishijima Sadao, in *Chūgoku keizaishi kenkyū*, pp. 79 ff. and 145 ff. Cf. Hsia Wei-ying, *Lü-shih ch'un-ch'iu Shang-nung teng-ssu p'ien chiao-shih,* pp. 42 ff.

Hence it was named alternating land. This was an old method, Hou-chi being the first to arrange fields with furrows.[37] Two *ssu*[38] were paired,[39] making a trench one *ch'ih* wide and one *ch'ih* deep, which was called a "furrow." It extended the length of a mu. With three

37. In *LSCC*, "It is desired that the mu (*lung*, 'ridge') be broad and flat and the *ch'üan* narrow and deep" (*LSCC* 26:9a). This is probably the earliest reference to *ch'üan* (furrow) together with mu and as a constituent part of the method of such an arrangement of the fields. It is interesting to note that the method of cultivation described in the *LSCC* is practically the same as the "alternating lands," except for the width of the *ch'üan*. We may, therefore, assume that as early as the third century B.C. in certain areas of China the *ch'üan* method of cultivation was practiced. If the *ch'üan* in the *Kuo-yü* stood for the same thing as in the *LSCC*, the method might have been employed in the early days of the fourth century B.C. But however early the introduction of this method may have been, Pan Ku's statement that Hou-chi was the first person who adopted it is hardly conceivable, as Hou-chi is a legendary figure. It is possible that Chao Kuo, instead of making an innovation, in fact perfected the *tai-t'ien* farming, which had been in use in the advanced agrarian areas, and introduced it into other areas under government sponsorship. Nevertheless, Pan Ku, as an historian who recorded the details of the method, interpolated his own interpretation by means of quoting the ancient classics, the *SC*, and attributed the credit to Hou-chi, a cultural hero who became a common target of agricultural inventions. Cf. Amano Motonosuke, "Daiden to Oden" *Matsuyama shōdei ronshū*, vol. 1; and Ōshima Riichi, "Tonden to daiden" *Tōyōshi kenkyū*, vol. 14, nos. 1-2 (1955). I do not agree with Ōshima, however, on his arbitrary divisions of the so-called Hou-chi school and the Shen-nung school of agrarian theoreticians; the interesting speculation lacks substantial evidence.

38. The earliest documentary record of the word *ssu* in its present form is in the *SC* (*MSCI* 8:1/5b; 19:4/4a, 5b). In his commentary Lu Te-ming (7th c.) merely refers to it vaguely as a "farming implement" (*MSCI* 19:4/5b and *Ching-tien shih-wen* 7:402). In the writings of the Chan-kuo period this character, although frequently used alone, is generally mentioned in combination with *lei*, as in the phrase *lei-ssu*. From the Han period on, this combined form gradually came to designate the *li*, "plow." In the text here, the *ssu* still refers to a kind of prototype of plow evolved from a simple digging stick with a handle to push. Hsü Chung-shu, "Lei-ssu-k'ao," *Bulletin of Institute of History and Philology*, vol. 2, pt. 1, 1930. Sun Ch'ang-hsü, *Lei-ssu-ti ch'i-yüan chi ch'i fa-chan*, and Sekino Takeshi, "Shin rai-shi kō," *Tōyōbunka kenkyūjō kiyō*, vol. 19 (1959).

39. The word *ou* literally means "pair." According to the old tradition as recorded in the "Chou-sung" of *Shih-ching* (*MSCI* 19:2/11b and 4/3b) and the *Lun-yü* (*Lun-yü chu-shu* 18:2b) and as preserved in the *Chou-li (CLCS* 15:14b; 42:1a), the fields were tilled by two men working shoulder to shoulder, each with a *ssu* or *lei*. There has been much debate on the ways these two persons worked together. Other than the parallel position, there is the possibility of one in front while the other follows. There is also the possibility that they faced each other. I believe the actual pattern of cooperation between these two persons should allow more than two possibilities. For the summary of the debate, see my article, "Liang-Chou nung-tso yü chi-shu," *Bulletin of the Institute of History and Philology*, vol. 42, pt. 4 (1971).

furrows to a mu, a *fu*[40] contained three hundred furrows. The seeds were sown in these furrows. After the shoots sprouted more than three leaves, the ridges were lightly weeded. While this was being done, the soil was moved down to cover the roots of the shoots. Thus an ode says of it:

> Some weed *(yün)*, some gather earth around the roots *(tzu)*;
> The *shu* and *chi* millet shoot up luxuriantly.[41]

Yün means "to remove weeds"; *tzu*, "to cover the roots." That is, as the shoots grew sturdier, at each weeding earth was always piled around the roots. When full summer came, the ridges were all leveled, and the roots were sunk deeply, making it possible for them to withstand wind and drought. Hence they flourished luxuriantly. The implements for tilling, weeding, and sowing the fields all had advantages and skillful contrivances. Generally twelve *fu*[42] consisted of one *ching* plus one *wu* of land. This, in terms of the present mu, is equal to five *ch'ing*.[43] A plow[44] with two oxen[45] and three men[46] was used. One

40. One *fu* was 100 mu.

41. Cf. James Legge, trans., *The Chinese Classics* 4:377.

42. According to the *HS*, in the land system laid out by the "sages," six feet *(ch'ih)* equaled one pace *(pu)*; 100 square paces, one mu; 100 mu, one *fu*; three *fu*, one *wu*; three *wu*, one *ching*. Swann, *Food and Money*, pp. 116-17.

43. In early times one mu was composed of one hundred paces, while in Han times it consisted of two hundred and forty paces. Twelve *fu*, i.e., twelve hundred mu, old style, would thus be equal to five *ch'ing* or five hundred mu of the Han period after the reign of Emperor Wu.

44. In the text the term "plow" is affixed with the modifier *ou*, "pair." Swann therefore translated the term as "double-shared plow," p. 189. Amano Motonosuke agrees with the interpretation by citing the twin-share plow in today's Manchuria and India for illustration. See *Chūgoku nōgyōshi kenkyū*, pp. 740-41. The problem is that thus far no double-shared plow has been depicted in Han mural reliefs of farming scenes. On the other hand, it is common for such reliefs to depict two oxen pulling one plow—for instance, on Han reliefs in Hsü-chou, in Ping-lu, and in Mi-chih. It is possible therefore the *ou* ("pair") here implies the teaming up of two oxen instead of double plowshares. Cf. Shensi-sheng po-wu-kuan, et al., "Mi-chih Tung-Han hua-hsiang-shih-mu fa-chüeh chien-pao," *Wen-wu*, 1972, no. 3, p. 71.

45. It is probable this statement has led some historians to claim that Chao Kuo was the first man to use the ox to draw a plow. Obviously, such an inference cannot be scientifically upheld. The *HS* text makes no such assertion, and historical data reveal that oxen were used to pull plows as early as the Chan-kuo period.

46. The labor division of these three persons also constitutes a much debated point. Swann suggested two persons at the plow handles and one person leading the oxen (see Swann, *Food and Money*, p. 187, n. 259). T'ien Ch'ang-wu imagined that one was at the plow while two persons each led an ox. Amano seems to adopt this hypothesis (see Amano, *Chūgoku nōgyōshi kenkyū*, pp. 740-41). Han tomb

year's crop regularly exceeded by more than one *hu* per mu that of the nontrenched fields.[47] The best managed fields even doubled this.[48]

Chao Kuo sent men to teach cultivation in the territories under the jurisdiction of the grand minister of ceremonies[49] and in the three metropolitan districts. The grand minister of agriculture provided artisan slaves to participate in the manufacture of agricultural implements. The two-thousand-picul officials sent prefects, chiefs, the thrice venerables, diligent farmers,[50] and village elders who were experienced in cultivating land to receive the agricultural implements and to learn the methods of tilling, sowing, and bringing up sprouts.

Some of the people suffered from lack of oxen and had no way

---

reliefs discovered at Hsü-chou and at Chia-yü-kuan show the twin oxen scene. At the handle of the plow was one farmer, behind him there was another person sowing the seed. Suppose there was a third person leading the oxen. The team then consisted of three persons. Kiangsu sheng wen-wu-kuan-li wei-yuan-hui, ed. *Chiang-su Hsü-chou Han-hua-hsiang-shih,* plate 63 and p. 12; Chia-yü-kuan wen-wu hsiao-tsu, "Chia-yu-kuan Han-hua-hsiang-chuan-mu," *Wen-wu,* 1972, no. 12, plate 8, no. 1.

47. Yen Shih-ku explains *man-t'ien* as those fields that were not arranged in furrows. It actually is a kind of overseeding farming by which the seed was scattered at random in the field and the farmer merely removes some plants from the overcrowded clusters.

48. According to Yen Shih-ku, *shan-che-pei-chih* means that the yield of those fields that were best arranged into furrows exceed by more than two bushels per mu that of the plain fields. The unit of land for this comparison is one mu, and the increased crop was due to the system of "alternating lands," not, as Ch'en Hsiao-chiang believes, to the rotation of crops. Cf. the bibliography Ch'en, "Kuan-yü i-mou-san-ch'üan wen-t'i-ti shang-chüeh p'ing-ta Yang-chün," *Shih-huo,* vol. 2, no. 1 (1935).

49. The grand minister of ceremonies was a high official who besides conducting the ceremonies of the imperial ancestral temples also had jurisdiction over the temples and the imperial burial grounds that had been declared prefectures (*HS* 19A:6b-7a). The annotator Su Lin comments that the people who lived in the prefectures of the imperial mausoleums were also given the new type of agricultural instruction.

50. The grand administrators of the commanderies, whose position in the official hierarchy was that of "two thousand piculs." The "thrice venerable" *(san-lao)* was a local functionary either in a district or in a prefecture who promoted the peoples' education and morals. The "diligent farmer" or the "diligent cultivator of the soil" *(li-t'ien)* was an honorary title given because of a man's proficiency in agricultural work. He encouraged the people to cultivate their land. See chap. 1, p. 25, and I, 22, 23.

to avail themselves of the precipitation.[51] The former prefect of P'ing-tu,[52] Kuang, advised Kuo to use men to pull plows. Kuo presented Kuang to the emperor for appointment as his assistant.[53] He taught the people to help each other in pulling plows. Generally when there were many people at work, thirty mu of land could be tilled per day; when there were fewer, thirteen mu. For this reason much land was brought under cultivation.

Kuo experimented by having the guards at the detached palaces till the side lots of the palaces.[54] A check of their harvest showed that they all obtained over a *hu* per mu more than adjoining fields. He ordered that the soldiers teach their relatives to cultivate the government lands of the three metropolitan districts.[55] He also taught [the troops of] the border commanderies and of Chü-yen city.[56] Later he also taught the people of the border cities, of Ho-tung and Hungnung commanderies, and the three metropolitan districts, and of the territory under the jurisdiction of the grand minister of ceremonies; all found the *tai-t'ien* system advantageous. They expended less labor and obtained more grain. [*HS* 24A:17a-19a; cf. Nancy Lee Swann, *Food and Money in Ancient China*, pp. 184-91]

51. According to Yen Shih-ku, the character *ch'ü* means "to hasten to, to attain," and *tse* is the moisture caused by rain. The quicker the fields were tilled after a rain the better, and the peasant who lacked oxen was therefore at a decided disadvantage. As Fan Sheng-chih declared, one of the rules of thumb for North China, dry-land farming is that it is essential to preserve the moisture of spring rains before it evaporates (V, 4: A.1).

52. P'ing-tu is the modern An-ting County, northern Shensi.

53. This person was formerly the prefect of P'ing-tu; hence the modifier *ku*. Under the Ch'in and Han dynasties, the *ch'eng* acted as assistant to the top-ranking man in an office either central, regional, or local. In the present context it must designate the assistant to the chief commandant for grain since this was the position held by Chao Kuo.

54. According to Yen Shih-ku, *kung-yüan-ti* are the unused side lots lying between the inner and outer walls of the detached palace.

55. The original text here could be read in different ways. Our reading is based on the annotation of Li Ch'i, who suggested reading *ling* as "to be ordered," and *ming* as "to teach." Another annotator, Wei Chao, however, suggested reading *ming* as a modifier to *chia*, which would mean "the households with titles and ranks." Cf. Nishijima, *Chūgoku keizaishi kenkyū*, pp. 116-30.

56. The prefecture of Chü-yen of Chang-yeh commandery was located northwest of modern Chang-yeh County in western Ninghsia. Discovered here in the early twentieth century was a large number of Han wood strips, remnants of Han garrison units' official records, on which there is a great deal of information regarding the *tai-t'ien* system. See Michael Loewe, *Records of Han Administration*.

8. A Song on Plowing

188 B.C.

Let us plow deep and sow the seed thick.[57]
Let the little plants be set out to allow growing space.[58]
What does not grow from the seed
We hoe out and weed away.[59]

[*SC* 52:4; see translation in Burton Watson, *Records of the Grand Historian of China* 1:407]

9. Use of a Seeder

Ca. A.D. 166

Emperor Wu appointed Chao Kuo the chief commandant for grain; he taught the people farming. According to his method, three plows are pulled together by one ox. One person leads the ox, handles the plows, and hauls the seeder—all taken care of by himself. In one day he plants 100 mu. Down to today the three metropolitan districts still rely upon the benefits [of his method]. Now the cultivating plow in the commandery of Liao-tung[60] has a beam four *ch'ih* long that hinders it in turning about. Two oxen are used; two persons lead the oxen, and one handles the plow. One sows the seed, and two haul the seeder. In all, they use two oxen and six persons; in one day they can only plant 25 mu.[61] Such is the disparity.[62] [*CHHW* 46:11a-b]

57. The phrase *chi-chung*, according to the annotator Yen Shih-ku, means "thickly sow the seed." Nishijima pointed out that the word *chi* also implies "to widely scatter" and "to generously cover." Meanwhile, in another version of the same song, *chi-chung* is given as *kuang-chung*, or "broadly sow the seed." Nishijima Sadao, *Chūgoku keizaishi kenkyū*, p. 96.

58. It means that after growing the young plants had to be set apart from each other to allow ample space for growth.

59. This song was sung by a Han prince, Liu Chang, a grandson of Liu Pang, the founder of the dynasty. It bears political implications as the Empress Lü, the widow of Liu Pang, planned to cultivate the power of her own Lü clan members at the cost of the Liu's. The empress, acknowledging that Chang's father was familiar with farming, expressed doubt about the prince's knowledge. When Liu Chang assured her of his knowledge, she told him to sing his song. In spite of being a political satire, there is no reason that Liu Chang should fabricate the principles of farming.

60. Ts'ui Shih was appointed the grand administrator of Liao-tung commandery (the southern part of the present Liaoning Province), in or about A.D. 166. For the date of this event, see Shih Sheng-han, *Ssu-min yueh-ling chiao-chu*, p. 87.

61. The ratio of efficiency of the two systems is: $\frac{1 \times 1}{100} : \frac{2 \times 6}{25} = 1:48$.

62. This passage is one part of the now-lost book, *Cheng-lun* [Essays on politics] written by Ts'ui Shih. The passage is preserved in a quotation in the *Ch'i-min-yao-shu* and the *TPYL*. For Ts'ui Shih's life and career, see III, 1.

10. Water-weeding in the Rice Paddy

In the second year of the Yuan-ting reign period (115 B.C.) . . . in the fall, the ninth month, it is decreed, " . . . Now floods have shifted to the south of the Yangtse River and the severity of winter is approaching. We are afraid [that the people are] suffering from starvation and the cold so much [that they may] not survive. In the land south of the Yangtse River, people till with fire and hoe with water.[63] Now we have sent grain from Szechwan to Chiang-ling. . . .[64] [*HS* 6:18a; cf. Dubs, *HFHD* 2:72] 115 B.C.

11. Preparing Paddy Land

The rice-man is in charge of cultivation of the lowlands. The reservoir *(ch'ü)* is for storing the water. The dike *(fang)* is for preventing the water [from inundation]. The canal *(kou)* is for the water to flow in. The small trench *(sui)* is for distribution of water. The ditch *(lieh)* is for tapping water into the field. The large trench *(kuei)* is for draining the water from the field. [After the water is drained, the farmer] is then able to walk across the field to clear up the grass that had been cut down in order to prepare the paddy [for a new crop].[65] Whenever one cultivates in the swamp, water is used in the summer to A.D. Late second century

63. The annotator Ying Shao (A.D. late second century) said: "They burn the grass, let in the water, and then sow the rice seed. Grass grows together with the rice. [As the rice] reaches seven or eight *ts'un* in height, cut down all [the grass]. Once again let in water to cover the grass. The grass dies and only the rice grows. This is what is called 'tilling with fire' and 'hoeing with water.'"

The T'ang commentator Chang Shou-chieh repeated virtually the same interpretation in *SC* 129:27, but he did not mention the grass being cut so that only water seemed to be in use to flood the grass. Chang's mistake is his assumption that the rice would outgrow the weeds, which is unjustified. Ying's interpretation is supported by the practice mentioned in *Ch'i-min-yao-shu*, pp. 111-12 (see our V, 13, 11.4.1 and 11:6:2). Other interpretations either paraphrase Ying's theory and the *Ch'i-min-yao-shu* paragraph or simply misread the text; cf. Nishijima, *Chūgoku keizaishi kenkyū*, pp. 187-90. Gale translates the phrase as "clearing by fire for farming and water weeding," with no elaboration other than the note, "The field was flooded, destroying the weeds but not harming the rice plants. This was done after the grass and brush had been burned." E. M. Gale, trans., *Discourses on Salt and Iron*, p. 19, and n. 2.

64. Chiang-ling is in the mid-reach of the Yangtse River. From Chiang-ling, the grain could be shipped to the lower part of the Yangtse as well as to the valley of the Tung-t'ing Lake in the present province of Hunan.

65. The annotator Cheng Hsüan (A.D. 127-200) said: "One hence walks across [the field] to clear up the grass that had been cut down in the preceding year and prepares the paddy for growing rice."

flood the weeds and exterminate them. The swampland that had grown grasses is then ready to grow spiked grain.[66] [*CLCS* 16:5a-b]

12. No More Water-weeding in the South

A.D. 278 [Tu Yü again memorialized]: Those who would see the water reservoirs restored all argued for the expediency of the fire-tilling and water-hoeing.[67] It is not that they are wrong. Nevertheless, this is a method to be applied in new land unclaimed and segregated from the residential areas. In the past the South was newly opened and the population was sparse. Due to such conditions, one could take the advantage of burning [the grass in] the field.[68] [*CS* 26:6b; cf. Yang, *Studies in Chinese Institutional History*, pp. 175-76]

13. Two Systems of Rice Farming

A.D. Sixth century 11.2.1. For rice cultivation, there are no specific conditions. Annually changing [the field for fallow] is good. In selecting the land, it is desirable to be close to the upper reaches of a stream. (Original note: since there is no distinction of the quality of land, the rice will grow well as long as there is clean water).

11.2.2. Sowing in the third month is the superior time; the first ten days of the fourth month is a mediocre time; the middle ten days is an inferior time.

11.2.3. First, let water into the field; ten days later, pull a roller over it ten times. (Original note: regarding the number of times, the more the better.)

11.3.1. After the field is well prepared, wash the seeds. (Original note: if the floating ones are not rejected, darnel will grow in the fall.)

66. Cheng Hsüan said: "To grow grain on swampy land, one must take advantage of heavy rain, which prevails in the sixth month in the summer, so that water is utilized to damage and extinguish the grass that grows later. In fall, when the water is dried up, let the grass be cut. The next year, the field can be planted." The spiked grains, according to another annotator, Cheng Chung, are rice and wheat.

This is a passage of the *Chou-li*, under the heading of "Tao-jen" (Rice-man). Cf. IV, 47.

67. Lien-sheng Yang translates this phrase, *huo-keng-shui-nou* as "to till the land with fire and hoe it with water," which indeed is an accurate rendering. The translation given in the text here, however, is an effort to keep the original word order.

68. Tu Yü in A.D. 278 memorialized the abolishment of numerous new water reservoirs in the mid-Yangtse valley where great floods inundated large areas. Cf. his biography in *Chin-shu* 34:8-12.

Soak [the seeds in water] ; after five days, drain them and put them in woven straw containers where the seeds are kept warm and moist to help germination. After three more days, the seeds are budding. When the buds reach two *fen* in length, let them be broadcast into the field at a rate of three *sheng* per each mu.

11.3.2. For three days, make someone drive away birds.

11.4.1. As the rice seedling grows to a height of seven or eight *ts'un,* the grass re-emerges. Use a sickle to cut down the grass under the water level so that all the grass will suffocate. As the rice seedling is growing, one must weed again. After weeding, let the water be drained; exposing the roots to the sun will cause them to become strong. Assess the amount of water and irrigate accordingly.

11.4.2. When the rice is almost ripe, once again drain the water [from the field] .

11.5.1. At the time of *Shuang-chiang,*[69] the rice is reaped. (Original note: if mowed too early, the grain will be green and unsolid; if too late, the grain falls and there will be loss.)

11.6.1. On the highlands in the north, there are no shores and marshes; one prepares paddies by following the crooks and bends. In the second month, as the ice thaws and the land is dry, burn [the grass off] the ground and till it.[70] As before, tap water into the field. After ten days, the earthen clods are dissolved. Using a wooden mallet, pulverize [the clods] and smooth [the surface of the field] .

11.6.2. Put the seed [in woven straw containers] according to the aforementioned method. After the seedling grows to the height of seven or eight *ts'un,* uproot [the seedling] and transplant. (Original note: since the fields are not fallowed annually, weeds and darnel both grow; they are not even extinguished by cutting them down. It is necessary therefore to transplant the seedling to have thorough weeding.) Irrigation and harvest are according to the aforementioned method.[71] [Shih Sheng-han, *Ch'i-min-yao-shu chin-shih,* pp. 110-11]

69. *Shuang-chiang,* one of the twenty-four subseasons in the Chinese calendar. It falls on or around October 23 or 24.

70. Shih Sheng-han *(Ch'i-min-yao-shu chin-shih,* p. 112) suggests the advantages of burning the grass off the ground are fourfold: (a) to elevate the ground temperature; (b) to eliminate some bacteria that are harmful to plants; (c) to speed up nitrification; and (d) to ameliorate organic conditions.

71. This section is one part of the chapter on rice in the *Ch'i-min-yao-shu,* a work compiled by Chia Ssu-hsieh in the sixth century. Our text here is taken from the edition annotated by Shih, *Ch'i-min-yao-shu chin-shih.*

14. YÜEH PEOPLE CULTIVATING IN THE NORTH

126-124 B.C. Several tens of thousands of laborers were then mobilized to make the canals [to connect the Fen and the Yellow rivers] and the fields. After several years, the Yellow River changed its course, and the canal provided no benefits. Those who tilled were unable to recover the cost of their seed. After a while, the irrigated fields in the commandery of Ho-tung were abandoned. The fields were given to the Yüeh people, who were ordered to make small payments to the Privy Treasury.[72] [*SC* 29:10-11; *HS* 29:7b; cf. Watson, *Grand Historian* 2:74]

15. WOODEN AGRICULTURAL IMPLEMENTS (1)

A.D. second century [Following are some of the names of wooden agricultural implements under the *lei* 耒 radical given by Hsü Shen in his *Shuo-wen-chieh-tzu*]:

lei [耒] : A crooked wooden [implement] pushed by hand.[73]

ou [耦] : A *lei* five *ts'un* in breadth is called a *fa*.[74]

kuei [耝] : . . . [An implement] used for cutting wheat. It is used in the commandery of Ho-nei.[75] [*SWCT* 4b:8a-b]

16. WOODEN AGRICULTURAL IMPLEMENTS (2)

A.D. second century [Following are some of the names of wooden implements under the *mu* radical given by Hsü Shen in *Shuo-wen-chieh-tzu*]:

jou [槈] : A kind of weeding implement.

chu [𣡌] : A kind of hoe that depicts the wood penetrating the earth.

72. The annotator Ju Shun pointed out that the Yüeh people were migrants from the south. Yüeh was an ethnic group scattered along the south and southeast coastal regions of China before the third century B.C. They were dispersed in the early years of the Han period. Cf. IV, 21.

73. According the the Ch'ing linguist Tuan Yu-ts'ai the *Shuo-wen* text was 耒耕曲木, i.e., the word 手 does not belong here. Hence, the translation should read: "*lei*: a crooked wooden [implement] for tilling."

74. According to Tuan Yü-ts'ai the *Shuo-wen* text 耦耒廣五寸為伐二伐為耦 should read: "*ou*: a *lei* five *ts'un* in breadth is a *fa*; two *fa* are an *ou*."

75. Ho-nei commandery was in southern Shansi Province.

hua [茉] : A hoe with two prongs.

ssu [梠] : A hoe . . . some say it is an implement to carry earth. . . .

hun [楎] : A plow with six forks. . . . Some say it is the curved shaft on the plow.

yu [櫌] : Tool used in smoothing [the surface] of the field.

pa [杷] : Tool used in harvesting wheat.

fo [柫] : A cluster of wooden pieces to flail grain.

ch'u [杵] : A grain-husking pestle. [*SWCT* 6a:5b-6a]

17. Metal Implements

[Following are some of the names of metal implements given by Hsü Shen in *Shuo-wen-chieh-tzu*] : A.D. second century

ch'ien [錢] : [literally: money]. A weeding tool (*tiao* 銚). An ancient agricultural implement.

chüeh [鑊] : A large hoe.

ch'ien [鈐] : A large plow. Some say it is the metal tip on the [wooden] plow.

po [鏺] : A two-edged blade with a wooden handle used for cutting grass.

ch'u [鉏] : A hoe used for uprooting grass while standing.

lien [鐮] : A sickle.

chao [鉊] : A big scythe.

chih [銍] : A short sickle for harvesting rice.

[*SWCT* 14a:2b-3a]

18. Agricultural Implements

A junction where four roads meet is called *ch'ü* (衢). In Ch'i and Lu,[76] a rake with four teeth is called *ch'ü* (鑺); when one rakes the soil with the *ch'ü*, there are four lines.[77] [*Shih ming*, SPTK ed., 9a] *Ca.* post-A.D. 100

76. Present province of Shantung.

77. In the lexicon *Shih-ming*, compiled by Liu Hsi, words of similar pronunciation are always brought in to explain the etymological connections of the word meaning. These explanations, however, are usually superficial or arbitrary; they have been omitted in the second half of this document.

ssu (耜 , spade) . . . resembles the cutting of things by the incisors.

li (犂 , plow) . . . if it is sharp then it turns over the soil and cuts the roots of the grass. . . .

chü (鋤 , hoe) . . . eliminates weeds and helps the growth of the shoots. The people of Ch'i call the handle *chiang* (橿); *chiang*-like means upright and straight. Its head is [called] "the crane" because it resembles the head of a crane.

chia (枷 , flail). . . . One adds sticks to the head of a handle to beat the ears of grain and get out the grain. It is sometimes called *lo-chia* (羅枷) when three sticks are fastened to it; it is also called *ya* (丫) when two sticks are fastened to it as a fork, which revolves at the head of the handle.

ch'a (鍤 , spade) . . . is for digging and turning the soil over. . . .

nou (耨 , weeding hoe) . . . is for the weeding of grain fields. . . .

kou ([illegible] , spade for digging trenches). . . . Having cleared the grass from the mounds [between the furrows] one again pushes down the soil in order to bank up the roots of the sprouts. One makes the area below the mounds a trench to receive the water flow. [*Shih-ming,* 49b]

19. The State Monopoly of Implement Manufacturing

First century B.C.

[The grand secretary said] : "Now the government casts agricultural implements [for the people] so that people pursue the fundamental profession and not engage in the secondary professions. Therefore, there will not be the problem of starvation and of being cold [due to lack of adequate clothing]. What is the harm [caused] by the monopoly of salt and iron that it should be abolished?" The worthies said, "Agriculture is the great enterprise of the world, and iron implements are the great appliances of the people. When the tools are convenient [to use], the expended labor will be little and gains will be numerous. The farmer enjoys his work and results are encouraged. When the tools are not sufficient, the fields will be wasted and the grains will not prosper. Labor is expended but the results are reduced by a half. Therefore, the convenience of having good tools [when compared with the] inconvenience of not having good tools will double the efficiency or [multiply it] by ten times. Of the iron implements cast and manufactured by the government, most of them generally are large implements in order to meet the quotas. They are not proper for the people's use. [The implements] used by the people are so dull and worn out that the grass cannot be cut. Therefore, the farmer works strenuously, but his gains are small. The people suffer from it." [*YTL* 6:15a-b]

20. The Invention of the Water Bellows

[Tu Shih, grand administrator of Nan-yang], was very good at planning and frugal with the people's labor service. He created the water bellows[78] for casting agricultural implements.[79] The energy expended was reduced and the results increased. The people found it advantageous.[80] [*HHS* 31:3a] A.D. 31

21. Regional Varieties of Implements

Now in Ch'in, Ch'u, Yen, and Ch'i[81] the quality of the soil differs. There is a variety in the method [of cultivation] of heavy and light soils. The use of large or small, the suitability of straight or curved [plows], is different according to districts and customs. Each follows what is convenient. But as the government establishes monopolies [of the manufacturing of implements] the iron implements lose their suitability, and the farming population loses what is convenient to them. When the tools are not suited to their use, the farmer grows fatigued in the fields, and grass and weeds are not kept down. When the grass and weeds are not kept down, then the people are wearied and in want. [*YTL* 1:12a-b; cf. Gale, *Discourses on Salt and Iron*, p. 33] First century B.C.

22. Wooden Implements for Poor Farmers

The prices of salt and iron are so high that the people are inconvenienced. The poor people sometimes have to till with wood, weed with bare hands, and eat insipid food. When the Office of Iron cannot dispose of implements it has to sell, it sometimes makes exactions on the people. [*YTL* 6:16a] First century B.C.

78. The *shui-p'ai*, "water-powered bellows" or water mills.

79. The Eastern Han dynasty seems to have continued the Western Han system of government management in iron production. Government control was still maintained, and the people were still forced to render service for the casting of iron implements produced by the Offices for Iron.

80. For more information on Tu Shih, see I, 69.

81. These represent the present-day provinces of Shensi and Hupei, Honan, Hopei, and Shantung, respectively.

23. Pounding Machine

First century A.D. When Fu Hsi invented the pounder and the mortar, the multitude of the people benefited. In later ages more skill was applied, using the weight of the body in order to add power by treading the pounder, which made it ten times more convenient than hand pounding. Again machines are used with the animal power of the ass, mule, ox, and horse. Even water power is used to pound and then it is a hundred times more convenient.[82] [*CHHW* 15:3b]

82. This is attributed to Huan T'an and is preserved in the Sung encyclopedic work, *TPYL* 762 and 829.

VI

THE NONAGRICULTURAL ALTERNATIVES

1. Agricultural and Nonagricultural Activities

[Huang Pa][1] made all the post stations and district headquarters raise chickens and pigs to aid the old widowers and widows and the destitute. Then he set forth regulations and instructions and appointed the elders, teaching and leading the chiefs of the Fives[2] to circulate among the people for the purpose of encouraging virtuous behavior, suppressing villainy, promoting farming and sericulture, reducing expenses, multiplying wealth, planting trees, raising cattle, and putting an end to the feeding of grain to horses; they were extremely detailed and complete. At the beginning they seemed vexing and trivial. However, Pa concentrated all his energy [on the task] and was able to put them into practice. [*HS* 89:5a-b] 66-58 B.C.

2. Textile Plants

Ts'u Ch'ung of Nan-yang, who replaced [Wei] Li[3] as [grand administrator of] Kuei-yang, was also good at managing the administration. He taught the people to plant mulberry trees, wild mulberry trees, hemp, and flax, and the like. He exhorted the people and had them breed silkworms and plait sandals. The people benefited from this. [*HHS* 76:2b-3a] A.D. 51

3. Chinese Agrarian Culture

Emperor Huan, considering that [Ch'en] Kuei's family for generations had been familiar with the customs of the frontier region, made him the "general who crosses the Liao River."[4] On departing for his *Ca.* A.D. 154

1. Huang Pa was the grand administrator of Ying-ch'uan between 66 and 58 B.C.

2. The *wu* or the Five was an organizational unit within the village during Ch'in and Han times. Each *wu* had a chief who supervised the conduct of the members of the five households in his charge, reporting on their actions, both good and bad.

3. Ts'u Ch'ung replaced Wei Li in A.D. 51.

4. Ch'en Kuei (fl. 140-150) held the position of general who crosses the Liao River in 154.

office, Kuei presented a memorial saying, " . . . The western division[5] is situated on the border and is uncivilized. The land is barren. [Its people] make the saddled horse their home and hunting their occupation. Men get but little profit from farming while women lack the wealth of the loom." [*HHS* 51:8a]

4. Auxiliary Productions

A.D. First century

Mulberry trees are planted around the huts [in the fields]. For vegetables there are garden plots; melons, gourds, and fruit are planted along the field edges. When chickens, dogs, and swine do not miss the proper times,[6] and women tend to sericulture and weaving, then those who are fifty years old and older can wear silk clothing and those who are seventy years old or older can dine on meat. . . . In the spring, people all go to [temporary huts in the] fields, while in the winter, all return to settlements. . . . During the winter, when the people have returned, women who dwell in the same neighborhood should gather together to spin in the evening. Thus, female labor of one month is equivalent to forty-five days.[7] The reason they must join together is to economize on the expense of lighting and heating, to equalize the level of skill, and to harmonize practices and customs.[8] [*HS* 24A:4a-5a; cf. Nancy Lee Swann, *Food and Money in Ancient China,* pp. 123-29]

5. Daily Production of Weaving

First century B.C.-A.D. First century

Now, there is a girl skillful in weaving who can double her own product each day. In five days, she produces five *ch'ih* of silk fabric. How much does she produce on each of the five days? The answer: 1 and 19/31 *ts'un* on the first day; 3 and 7/31 *ts'un* on the second day; 6 and 14/31 *ts'un* on the third day; 1 *ch'ih,* 2 and 28/31 *ts'un* on the fourth day; and 2 *ch'ih,* 5 and 25/31 *ts'un* on the fifth day.[9] [*Chiu-chang suan-shu,* SPTK ed., 3:3b-4a]

5. That is, Liang division.

6. The text does not specify what the "proper time" is. It seems to imply the appropriate time for breeding.

7. One evening's work is counted as a half-day's work. Swann estimated that there could be four months each year during which women worked on sewing and spinning (*Food and Money,* p. 129, n. 72).

8. This passage contains largely the same concepts presented in James Legge, trans., *The Chinese Classics* 2:149, but in greater detail. I do not agree with Swann, who assigns the date to 700 B.C. (*Food and Money,* p. 122). It is more likely that Pan Ku gives here a description of an idealized family life.

9. The fifth-day production rate, although good, still reflects the working

6. Weaving Record

Ca. A.D. 200

The peacock flies southeastbound.[10]
No clothes in the bitter cold.
I, your wife,
In my heart, grief and sympathy.
Night after night, I weave,
Unable to leave the loom.
In three days, a piece is finished,[11]
Yet I am still scolded for delaying.

["Ku-yen-ke" (Ancient Romance)
Quoted in *TPYL* 826]

7. Sheep and Wine for Special Occasions

Mid-third century B.C.

Lu Wan, a native of Feng,[12] was from the same hamlet as Kao-tsu. Wan's father and the father of Kao-tsu were close friends. Their sons, Kao-tsu and Wan, were born on the same day and people in the hamlet brought sheep and wine to congratulate the two households. When Kao-tsu and Wan grew up and attended school, they also became close friends. People in the hamlet, with appreciation of the close friendship of the two families, their sons being born on the same day, and also being close friends as they grew up, again congratulated them with sheep and wine. [*HS* 34:21b]

8. Consumption of Meat

First century B.C.

In antiquity, the common people ate coarse millet and wild vegetables. There was no wine nor meat unless it was during the formal occasions of community feasts or on religious celebrations. Therefore, feudal lords would not kill cattle or sheep without [adequate] reason. the great officers *(ta-fu-shih)* would not kill dogs or pigs without

---

capability of a new hand. If skilled labor can double the amount that a good new hand can put out, a daily rate of five *ch'ih* should be a reasonable estimate.

Chiu-chang suan-shu is a mathematics textbook; although there are some later interpolations, the basic content is generally regarded to be of the Western Han period. Joseph Needham, *Science and Civilization in China* 3:24.

10. This beginning line is the same as that of "The Peacock." See *Ku-shih-hsüan*, SPPY ed., 1:10-14.

11. "A piece," in the text, is a *p'i*, a unit of textile measurement, i.e., forty *ch'ih* in length. The paraphrased line in "The Peacock" gives five *p'i* for three days. I suspect the latter rendering is an exaggeration.

12. The native town of Han Kao-tsu, Liu Pang.

[adequate] reason. Now in local communities meat is hung out [for sale] and along country paths there is butchering and selling. Without adequate reason animals are killed and cooked. People gather together in the countryside, going with millet on their backs and returning with meat in hand. The price of the meat of each pig equals the crop of a middle-rate yield.[13] Fifteen *tou* of millet is equal to a half-month's provision for an adult male.[14] [*YTL* 6:4a-b]

9. Pig Raising as a Specialized Occupation

Ca. A.D. 32 After [Liang Hung][15] finished his studies [in the Imperial Academy], he then herded pigs in the Shang-lin Park. Once he carelessly left a fire that spread and set fire to a house. Hung then looked for the person who had suffered from the fire, inquired about his losses, and offered him all his pigs as compensation. The owner of the burned house still regarded the compensation as insufficient. Hung said, "I have no other property but I am willing to stay in your household and work for you." The owner consented to this. Thereupon he worked hard for him and was never idle either morning or evening. The elders of the neighboring families saw that Hung was an unusual man; they all came to reproach the owner and to praise Hung as an upright man. [The owner] thereupon began to respect him and to consider him extraordinary and gave him back all his pigs. Hung refused to accept them, but went away and returned to his native community. [*HHS* 83:8a]

10. Pig Raising on the Seashore

Before 160 B.C. Kung-sun Hung[16] was a native of Hsieh of Tzu-ch'uan. When he was young he served as a jail officer. Once he was found guilty and dismissed. As his family was poor he raised pigs on the seashore. [*HS* 58:1a-1b]

11. Pig Raising in a Swamp

Before A.D. 184 [Sun Ch'i's][17] family was poor. He served his mother in a most

13. The unit of the land concerned is not given here. Perhaps it refers to the the income from farming in an ordinary household.

14. This "fifteen *tou* of millet" probably refers to the amount of grain in exchange for the meat that was to be consumed for a meal or a day.

15. Liang Hung spent many years as a hermit and died as a laborer.

16. Kung-sun Hung was very much esteemed by the emperor and finally held the office of imperial chancellor from 125 B.C. until his death in 121 B.C. at the age of eighty (*HS* 19B:6a-b).

17. Sun Ch'i, a scholar of the *I-ching*, taught students in the field where he

filial manner. He herded pigs in a large swamp to provide for her. [*HHS* 79:7a]

12. Sheep Breeding in Ho-nan

Pu Shih was a native of Ho-nan commandery. He made farming and stockbreeding his occupation. [At the time of his parents' death] Shih had a younger brother of tender age. When his younger brother reached manhood, Shih went away and divided the household, taking only a hundred-odd sheep and leaving all the land, houses, and other properties to his younger brother. Before 121 B.C.

Shih went into the mountains to graze his stock. After about ten years his sheep amounted to more than a thousand. He bought land and houses. In the meantime his younger brother had completely ruined his patrimony. Several times Shih again shared his property with his younger brother.[18] [*HS* 58:8b]

13. Oxen and Sheep Abundant in China

[This year the empire was peaceful and tranquil] . . . cattle and sheep were all over the countryside. [*HHS* 2:13a] A.D. 69

14. Price of Sheep

There was for sale one head of sheep, adult and female; Tzu-ch'eng was buying it from Chün-chü; the price was 900 *ch'üan*.[19] There was for sale one head of sheep, adult and female; Tzu-ch'eng was buying it from Chün-chü; it [the sale] was not yet complete. There was for sale one head of sheep, adult and female; when Chün-chü left he ordered Ch'ü-hsiang[20] to avail himself of Wei Chung-t'ung who together Ca. 9-23 B.C.

---

was herding the swine.

18. These events occurred a number of years before Pu Shih offered one-half of his property to the government for use in the war against the Hsiung-nu, whose Prince Hun-yeh finally surrendered in 121 B.C.

19. Lao Kan notes that the use of the word *ch'üan* instead of the familiar Han term *ch'ien,* meaning "cash," dates this record to the period of Wang Mang's reign. The record is one of the wooden slips from the Han northwest frontier area.

20. The meaning of the text at this point is not clear. Ch'ü-hsiang could be a person's given name, as in the case of Tzu-ch'eng. But Ch'ü can be read "Ou" as a surname in which case the proper rendering would be Ou Hsiang. Furthermore, "to avail himself of" is a rendering of *yung*; *yung* could also be a part of the person's name as in Ou Hsiang-yung. The text is so terse that there is no way to determine which of these readings is correct.

with Tzu-ch'eng was to buy it; the price was 1,000 *ch'üan*.[21] [Lao Kan, "Han-chien-chung-ti Ho-hsi ching-chi sheng-huo," *Bulletin of the Institute of History and Philology* 11:68]

15. Chickens at a Farming Household

Before A.D. 76 [Kao Feng's][22] family occupation was farming, yet he devoted all his attention to study, never ceasing day or night.

One day his wife went to the fields. She had wheat drying in the courtyard and told Feng to protect it from the chickens. At this time a storm came up, but Feng, holding a stick [to scare away the chickens] continued to read the classics, not knowing that the rain water had washed away the wheat. [*HHS* 83:10a]

16. Shepherd Boy

208 B.C. Thereupon [Hsiang Liang][23] started to search for Hsin, the grandson of King Huai[24] of [the old state of] Ch'u, and found him among the common people tending sheep for someone. [Hsiang Liang] then set him up as King Kuai of Ch'u. [*SC* 7:12; *HS* 31:11a]

17. Swineherd Boy

First half, first century A.D. Ch'eng Kung,[25] whose style was Shao-tzu, was a native of Ku-mu prefecture of Lang-ya commandery. He lost his father when he was a child. At eight years of age he herded pigs for others. [*HHS* 27: 11a]

18. The Price of Wine

81 B.C. The office of wine monopoly is abolished. . . . Wine may be

21. This record is probably part of a court case or at least of a complaint. This is a full translation of all of the available material and yet the material seems to be incomplete. Of the four individuals in the text, only Wei is identified by surname; presumably the others were fully identified in an earlier and now missing part of the document. The simple and repetitive tone of the document suggests that Ch'ü-hsiang, the presumed author, was only semiliterate.

22. Kao Feng was a classical scholar.

23. Hsiang Liang was the uncle of Hsiang Yü, the rival of Emperor Kao during the latter's campaign for the throne (206-202 B.C.). The Hsiangs were natives of the state of Ch'u, their forefathers having been generals of the rulers of the state. Because of this, Hsiang Liang wanted a descendant of the ruling family to head the fight against the second emperor of Ch'in (*SC* 7:1a ff).

24. The King of Ch'u from 328 to 299 B.C.

25. Ch'eng Kung later became an official.

sold at the rate of four cash for each *sheng.*[26] [*HS* 7:5a-b; cf. Dubs, *HFHD* 2:161-62]

19. Cost of Wine

At one brewing, there should be used two *hu* of coarse, husked grain and one *hu* of malt to get six *hu* and six *tou* of liquor.[27] [*HS* 24B:25b; cf. Swann, *Food and Money*, pp. 345-46] A.D. 10

20. Kinds of Liquor

Fan: when it is made (i.e., when the brewing is finished), lees float [on the surface]. . . . It resembles the thick wine of I-ch'eng[28] today. *Li:* . . . when it is made, the liquor and lees are together; it resembles the sweet wine of today. *Ang:* when it is made is an onion-white color; it resembles the white wine of Tsan[29] of today. *T'i:* when it is made is red in color; it resembles the low-class wine of today. *Ch'eng:* when it is made, the lees settle [to the bottom]; it resembles the clear wine of today.[30] [*CLCS* 5:6a] A.D. second century

21. Cost for Operating Government Factories

[Kung Yü memorialized . . .Formerly, the three offices for textiles in Ch'i[31] [annually] presented no more than ten bamboo chests of their products. Now, each of the three offices for textiles in Ch'i has 48 B.C.

26. Lao Kan compares this figure and the wine price given on a Tun-huang tablet, which was ten cash for two *tou.* As one *tou* is ten times one *sheng,* while one *sheng* equals 200 cubic centimeters, the *HS* price seems too expensive. Lao Kan deduced that the character *sheng* must have been mistakenly written for *tou.* He also stated that the price of wine in the Han dynasty was customarily twice the price of grain. Thus, according to his estimate, the grain price in 81 B.C. should have been approximately 20 cash for each *shih* (or picul); Lao Kan, *Chü-yen Han-chien: K'ao-shih-chih-pu,* "K'ao-cheng," pp. 62-63.

27. It is clear that winemaking was a profitable enterprise since the natural increase in volume more than doubled, and the price of wine was double that of grain. Cf. VI, 18, n. 26.

28. I-ch'eng was a prefecture located in present-day Shantung Province.

29. Tsan is the name of a city in present-day Hupei Province.

30. This passage is Cheng Hsüan's annotation of the five categories of wine in the *Chou-li.* Cheng (A.D. 127-200) is a major annotator of the classics.

31. The text also can be read as "Office for the Garments of Three Seasons." Nevertheless, since the text also mentions "each," three offices, rather than one office for three seasons, is more logical. Cf. comments cited by Wang Hsien-ch'ien, *HS* 9:6a. Ch'i was in the present Shantung Province.

several thousand workers. The annual cost amounts to millions. [The offices for manufacturing] in Shu and Kuang-han[32] are in charge of gold and silver vessels. Annually for each [office], expenditures amount to five millions. For the three offices of manufacturing[33] the government spends fifty millions. Such is also the case with the East and West Weaving Chambers. . . . Your subject, Yü, once accompanied [on an imperial visit] to the Eastern Palace,[34] and had bestowed upon him vessels and plates that are all elegantly painted and decorated with gold and silver.[35] These are not what ought to be bestowed upon the subjects. [*HS* 72:10a-b]

22. Luxurious Items for Imperial House

A.D. 106 Formerly, the regular expenditures for the Office of Imperial Provisions and the Office of the Imperial Bakery amounted to almost two hundred million annually. The empress dowager[36] decreed to gradually reduce the conspicuous spending. Thereafter, several tens of millions were trimmed. . . . Gold and silver inlaid vessels, belts, and knives were no longer sent. Thirty-nine categories of painters were released [from government employment]. [*HHS* 10A:17b]

23. Self-defense Organized by Kinship Group

A.D. 23 The Keng-shih emperor,[37] just enthroned, wanted to appoint Fan Hung[38] as a general. Hung, kowtowing to decline the offer, said, "Being a student of literary learning, I am not familiar with military affairs." He was finally able to evade service and to return home.

32. Both Shu and Kuang-han are in the western part of the present province of Szechwan.

33. The text is ambiguous on the definition of the three offices for manufacturing. It is possible, as suggested in the comments quoted by Wang Hsien-ch'ien, that it refers to the chief and two deputies who were in charge of manufacturing by government agencies. Cf. *HS* 72:10b.

34. The Eastern Palace, normally the residence of the crown prince, was at that time a residence of the empress dowager.

35. Recent archaeological excavations of Han tombs have frequently turned up lacquer vessels with gold and silver inlay. Kung Yü must have referred to such items.

36. Née Teng, the widow of Emperor Ho. At this time, she was the actual regent as the new emperor on the throne, whom she had adopted from an imperial branch, was still an infant.

37. One of the contenders for the throne after the downfall of Wang Mang.

38. Fan Hung, the maternal uncle of Emperor Kuang-wu, was the son of Fan Chung; see III, 2.

Together with his kinsmen and relatives, he erected fortifications and a moat for their self-protection. The aged and the weak who came to them numbered over a thousand households. [*HHS* 32:1b-2a]

24. Self-defense Fortifications

At the end of the reign of Wang Mang,[39] bandits and outlaws arose. The kinsmen and neighbors strived to attach themselves [to Ti-wu Lun]. Lun then took advantage of the terrain to build up fortifications and walls. Whenever there were raids, [Lun] always encouraged his people to resist by pulling strong bows to the full [on the wall]. The Bronze Horse, the Red Eyebrows,[40] and their like, altogether several tens of groups, could not take the fort [of Lun]. [*HHS* 41:1a] A.D. First century

25. Hsü Ch'u and his Fortified Village

At the end of the Han period, Hsü Ch'u[41] gathered youths and his kinsmen together, several thousand households in all, to build fortified walls for self-defense against the raiders. At that time, more than ten thousand bandits of Ke-p'i of Ju-nan commandery attacked Ch'u's fort. Ch'u's men, inferior in number, were not the equals of [the bandits]. Fighting hard, the men became extremely exhausted. Their weapons and arrows were all used up. Ch'u then ordered the men and women in the fort to collect and arrange rocks the size of bowls and dippers and to put them at the four corners [of the fort]. Ch'u thereupon hurled the rocks, making them fly, and crushed whatever the rocks hit. The bandits hence did not dare to advance. [*SKC*, Wei-chih, 18:10b] A.D. Third century

26. A Fortified Community in the South

[T'ai-shih Tz'u said:][42] "The leaders of people at Po-yang[43] have separately established a kinship army; blockading troops guard their *Ca.* A.D. 200

39. A.D. 9-23.

40. The names of the rebel groups.

41. Hsü Ch'u, a man of tremendous physical strength, later served under Ts'ao Ts'ao as a general.

42. T'ai-shih Tz'u (A.D. 165-206) was a general under Sun Ts'e whose brother Sun Ch'üan later proclaimed himself Emperor of Wu, a state in the Yangtse valley during the Three Kingdoms period. This passage is allegedly a report of T'ai-shih Tz'u to Sun Ts'e after the former made a visit to Yü-chang.

43. Northern part of the present province of Kiangsi.

borders and refuse to accept the officials dispatched by Tzu-yü. . . .[44] Close to Hai-hun, there is a camp named Shang-liao where five or six thousand households have united under kinship ties to form military units. They merely remit to the commandery government certain land taxes and fabrics [as their share of the tax]. No single individual can be mobilized [by the commandery authority]. Tzu-yü does nothing more than look on." [*Chiang-piao-chuan*. Quoted by the commentator P'ei Sung-chih in *SKC*, Wu-chih, 4:8a]

27. A Colony in the North

Ca. A.D. 190

[T'ien] Ch'ou was able to return to the north.[45] He led all his kinsmen and other attached people, numbering several hundreds, . . . to enter the Hsü-wu mountain area where they settled down on a plain strategically located deep in the mountain area. [Ch'ou] tilled land to support his own parents. More than five thousand families joined him within a few years. Ch'ou told the elders among the people, "You gentlemen do not consider me unworthy; from distant places you have come to join me. Now the multitudes have formed towns, and yet they are not united. I am afraid that this is not the way to long-lasting security. May we elect a worthy elder to be our leader." All [the people] said, "Good." Unanimously they elected Ch'ou. . . . Ch'ou thereupon made laws setting forth binding agreements regarding homicide, injury, fighting, robbery, and litigation. The most severe of the laws called for capital punishment. Over twenty articles pertained to lesser crimes. He also instituted rituals regarding marriage. Schools and curricula were established. All these were publicized to the people. People felt them to be so beneficial that items left beside the road were not even picked up by others. [*SKC*, Wei-chih, 11:9b-10a]

28. Rent on Garrison Fields

Ca. First century B.C.

1. [Recorded at] right, in Unit 2 of the commanding officer,[46] two plots, 67[47] mu of cultivated land; rent is 26 *shih*.

44. The name of Hua Hsin, the grand administrator of Lu-ling.

45. T'ien Ch'ou (A.D. 168-214) was a native of Wu-chung, in the northern part of the present province of Hopei. This particular incident occurred after he was released by a local warlord who arrested him because T'ien Ch'ou was loyal to the provincial shepherd *(chou-mu)* whom the warlord had murdered.

46. Loewe gives no interpretation of this term which means the senior officer in command.

47. Loewe suspects that the figure may read 65 (Loewe, *Han Administration*,

2. . . . at 4 *tou*[48] per mu.

3. . . . at 4 *tou* per mu.

4. . . . land rent 16 *shih*.

5. [Recorded at] right, five plots, 65 mu of cultivated land; rent in large [measurements] is 21 *shih* and 8 *tou*.[49] [Michael Loewe, *Records of Han Administration* 2:296. Translation revised. Cf. ibid., p. 297]

29. *T'UN-T'IEN* IN THE YELLOW RIVER PLAIN

[It is decreed][50] " . . . when the Yellow Turban forces were defeated and the Hsü area[51] was settled, we then acquired the assets[52] of the [Yellow Turban] bandits. When the *t'un-t'ien* was being set up, the discussants of the time all suggested that rent in grain was to be paid according to the number of oxen [being leased]. . . . [Tsao] Chih explained that he thought that grain paid according to leased oxen would lead to no increase in grain [paid] in good harvest years, yet there would be exemptions in famine years caused by flood or drought. . . . At the time the former Military Libationer Hou Sheng said, 'the rent should be collected according to the oxen [being leased] for use on government land. [Tsao] Chih's proposal would be advantageous to the government, but disadvantageous to the tenants.' . . . [Tsao] Chih still was confident in his proposal, which he presented to me, and insisted that the land be divided [among the tillers]. I hence approved and appointed him the commandant of the farming to set up the [*t'un-t'ien*] fields."[53] [*Wei-wu ku-shih*; quoted by the commentator P'ei Sung-chih in *SKC*, Wei-chih, 16:1b-2a] A.D. 196

p. 295). Lao Kan simply renders the figure as 65 in "Han-chien-chung-ti Ho-hsi ching-chi sheng-huo," 11:67.

48. A *tou* is one-tenth of a *shih*.

49. The rate of rent in this tablet is 0.335 *shih*, much lower than the 0.4 suggested in tablets 2 and 3.

50. This is part of a decree issued posthumously by Ts'ao Ts'ao to honor Tsao Chih.

51. The south and central part of the present province of Honan.

52. These assets probably included implements and facilities that were carried along by the Yellow Turbans.

53. In A.D. 192, Ts'ao Ts'ao defeated the Ch'ing Chou branch of the Yellow Turbans. Some three hundred thousand combatants of the Yellow Turbans surrendered. They were accompanied by their dependents, amounting to over one million males and females (*SKC*, Wei-chih, 1:8a). In A.D. 196, the *t'un-t'ien* system was established to assure a source of provisions, following the proposal of Tsao Chih and Han Hao (ibid., 1:11b).

30. *T'UN-T'IEN* PEOPLE ENGAGING IN COMMERCE

Ca. A.D. 230

Later [Ssu-ma Chih] was appointed the grand minister of agriculture.[54] Formerly, the offices of commandant for farming each had its own staff and people engaging in business to pursue profit. [Ssu-ma] Chih memorialized: "Emperor Wu[55] specifically set up the *t'un-t'ien* office, which should be exclusively occupied in agriculture and sericulture. During the Chien-an reign period (A.D. 196-219), the granaries in the empire were full and people were rich. Since the Huang-ch'u reign period (A.D. 220-226), the commandants for farming have been permitted to become self-sufficient and each plans for his own subordinates. This, indeed, is not what is proper for the general principles of the state. The monarch should take the whole world as his own household. Therefore, the classic (*Lun-yü* IX) says, 'If the people do not have sufficient provisions, to whom can Your Majesty grant sufficiency!' The source of wealth is in not missing the Heavenly time and in using fully the productivity of the land. Now although the evident profit sought by trading is twice [that of farming], in the general scheme of things there has already been an immeasurable loss: it is not as good as reclaiming land and increasing the harvest from one more mu. The farmer working in the field, starting in the first month of the year, plows, plants, weeds, hoes, strips leaves from mulberry trees, burns the grass, plants wheat, harvests, and pounds the drying floor for the grain; not until the tenth month is he finished. Then he must repair the granary, tie up the well-sweep, ship grain as rent and tax, clean roads and repair bridges, and plaster his residence. Because of these activities throughout the year the farmer does not have a day in which he is not engaged in farming matters. Now the commandants of farming each report that those who remain at home devote themselves to agriculture on behalf of those who are traveling [as traders], that a calculation of the labor power shows that there is no alternative, and that, since nothing is being abandoned, there is, as before, surplus labor. Your subject[56] thinks that it is not suitable for business matters to be any longer mixed [with] agriculture. Exclusive devotion to agriculture and sericulture is beneficial to the interests of the state. Emperor Ming followed it" (i.e., his advice). [*SKC,* Wei-chih, 12:22a-b]

54. This is, in fact, the Office of Financial Affairs.
55. Ts'ao Ts'ao.
56. Ssu-ma Chih himself.

APPENDIX

HAN UNITS OF MEASURE

Equivalents, by period:

Han Unit		Western Han A. Early Period	Western Han B. Late Period	Wang Mang	Eastern Han
Linear	*ch'ih*	23.5 cm.	23.5/23.75 cm.	23.1 cm.	23/23.75/ 23.8 cm.
Volume	*sheng*	200 $cm.^3$	188/200 $cm.^3$	200 $cm.^3$	200/ 201.8 $cm.^3$
Weight	*chin*	250 gr.	250/264 gr.	240 gr.	250 gr.

A. From inception through reign of Emperor Wu, 206-87 B.C.; B. Emperor Chao and after, 87 B.C.-A.D. 9.

Data based on T'ien Shih, "Hsi Han tu-liang-heng lüeh-shou," *Wen-wu*, 1975, no. 12, pp. 79-89.

Linear	Weight
1 *li* = 300 *pu*	1 *shih* = 4 *chün*
1 *pu* = 6 *ch'ih*	1 *chün* = 30 *chin*
1 *chang* = 10 *ch'ih*	1 *chin* = 16 *liang*
1 *ch'ih* = 10 *ts'un*	1 *liang* = 24 *shu*
1 *ts'un* = 10 *fen*	

Volume	Area
1 *shih* or 1 *hu* = 10 *tou*	Pre-87 B.C.
1 *tou* = 10 *sheng*	1 *ch'ing* = 100 mu
1 *sheng* = 10 *ko*	1 mu = 100 sq. paces
	Post-87 B.C.
3 *ta-shih* = 5 *hsiao-shih*	1 mu = 240 sq. paces

(According to Michael Loewe, *hu* was substituted for *shih* as a term of volume after A.D. 9; cf. his "The Measurement of Grain during the Han Period" *T'oung Pao*, series 2, 49:75-95.)

GLOSSARY

盎	*ang*	white wine
臿	*ch'a*	spade
占緡	*chan-min*	tax declaration
丈	*chang*	measure of length, 10 *ch'ih*
長吏	*chang-li*	senior officials
常平	*ch'ang-p'ing*	"ever-normal" granary
常山	*ch'ang-shan*	fever medicine
鉊	*chao*	large scythe
朝令而暮改	*chao ling erh mu kai*	"order issued in the morning to be altered in the evening"
辰	*ch'en*	fifth of the "earthly branches" twelve-day cycle
烝	*cheng*	sacrifice, festival
丞	*ch'eng*	assistant
澄	*ch'eng*	wine

稷	*chi*	millet
穊種	*chi-chung*	thickly sow the seed
籍田	*chi-t'ien*	ceremonial field
氣	*ch'i*	breath
起居注	*ch'i-chü-chu*	daily record of imperial activities
奇民	*ch'i-min*	ne'er-do-wells and upstarts, vicious people
七大夫	*ch'i-ta-fu*	noble rank, seventh step
假	*chia*	lease of land
枷	*chia*	flail
薑	*chiang*	ginger
橿	*chiang*	handle of a hoe *(chu)*
節	*chieh*	staff of authority
箭	*chien*	species of small bamboo
建除十二神	*chien-ch'u shih-erh-shen*	"twelve spirits of *Chien* and *Ch'u*," day designations
踐更	*chien-keng*	payment of military duties
錢	*ch'ien*	[lit., money] a weeding tool
錢	*ch'ien*	cash
鈐	*ch'ien*	large plow
枳	*chih*	orange, *Citrus trifoliata*
銍	*chih*	rice sickle
稙禾	*chih-ho*	a variety of spiked millet
治粟都尉	*chih-su-tu-wei*	chief commandant for grain
尺	*ch'ih*	measurement of length, a foot
池	*ch'ih*	water reservoir
池囿	*ch'ih-yu*	imperial reservoir and lands
斤	*chin*	measure of weight, catty, 16 *liang*
井田	*ching-t'ien*	land system based on units of eight families, well-field system

卿	*ch'ing*	noble rank, above *tai-fu*
頃	*ch'ing*	unit of land measure, 100 mu
清明	*Ch'ing-ming*	"clear and bright," one of 24 divisions of year, *ca.* April 5
秋節	*ch'iu-chieh*	autumn season
州	*chou*	administrative division, province
州牧	*chou-mu*	provincial shepherd (functionally, governor)
丑	*ch'ou*	second of the "earthly branches" twelve-day cycle
諸侯子	*chu-hou-tzu*	descendants of pre-Ch'in feudal lords
除	*Ch'u*	one of the twelve-day, "twelve spirits" day designators
鉏 鋤	*ch'u*	hoe
杵	*ch'u*	pestle to husk grain
初伏	*ch'u-fu*	division of summer
處暑	*Ch'u-shu*	"limit of heat," one of 24 solar divisions of year, *ca.* Aug. 23
梟	*chü*	hoe
車前子	*chü-ch'ien-tzu*	medicinal herb, *Plantago major*
衢	*ch'ü*	four-road junction
鑺	*ch'ü*	rake with four teeth
渠	*ch'ü*	reservoir
趨	*ch'ü*	to hasten to, attend to
甽	*ch'üan*	furrow
泉	*ch'üan*	cash
钁	*chüeh*	large hoe, mattock
鈞	*chün*	measure of weight, 30 *chin*
郡縣	*chün-hsien*	commanderies and prefectures
種	*chung*	to plant

鍾	*chung*	measure of volume, 6 *hu* and 4 *tou*
中丞	*chung-ch'eng*	palace assistant
中伏	*chung-fu*	division of summer
中郎	*chung-lang*	gentlemen of the palace
伐	*fa*	to cut, a cutting tool, to fell
法	*fa*	method, the law
法家	*fa-chia*	Legalist
泛	*fan*	wine
防	*fang*	dike
分	*fen*	measure of length
分田 刼假	*fen-t'ien-chieh-chia*	forced transfer of land leased by the poor
柫	*fo*	flail
夫	*fu*	100 mu
賦	*fu*	poll tax (Han)
服官	*fu-kuan*	Office of Textile Manufacturing
脯田	*fu-t'ien*	field like jerked beef
亥	*hai*	twelfth of the "earthly branches" twelve-day cycle
禾	*ho*	spiked millet
畦	*hsi*	50 mu
西成	*hsi-ch'eng*	"west maturity," autumn
葸耳	*hsi-erh*	herb, species untranslatable
蝦蟆	*hsia-ma*	toad
小黃門	*hsiao huang-men*	lesser attendant of the Yellow Gate
小麥	*hsiao-mai*	wheat
小奴	*hsiao-nu*	minor slave
小石	*hsiao-shih*	volume measure, 60 percent of a *ta-shih*
孝弟	*hsiao-ti*	filial and brotherly loving
秈	*hsien*	a kind of rice

先農	*hsien-nung*	agricultural deity
辛	*hsin*	eighth of the "celestial stems" ten-day cycle
宿	*hsiu*	lunar mansions
戌	*hsü*	eleventh of "earthly branches" twelve-day cycle
斛	*hu*	measure of volume, ten *tou*
芣	*hua*	two-pronged hoe
黃連	*huang-lien*	medicinal herb, *Coptis chinensis*
楎	*hun*	plow with six forks
火耕水耨	*huo-keng-shui-nou*	to till the land with fire and hoe with water
鎰	*i*	unit of gold, 20 taels
乙卯	*i-mao*	date
仞	*jen*	measure of length, 8 *ch'ih*
壬寅	*jen-yin*	date
槈	*jou*	weeding implement
戎	*jung*	barbarian
戎菽	*jung-shu*	soybean
告緡	*kao-min*	report of tax cheating
膏澤	*kao-tse*	[lit., "greasy moisture"]; connotes rich benefits and sweet blessings
各	*ke*	each
秔	*keng*	nonglutinous rice
更	*keng*	military duties [lit., rotation]
更錢	*keng-ch'ien*	corvée duty fees
合	*ko*	measure of volume
耩	*kou*	spade
溝	*kou*	canal
口錢	*k'ou-ch'ien*	poll tax (minors)
穀	*ku*	tree of mulberry family
穀雨	*Ku-yü*	"grain rains," one of 24 solar divisions of the year, *ca.* April 20

瓜蓏	*kua-lo*	gourds, *Tricosanthes japonica*
關內侯	*kuan-nei-hou*	marquis within the Pass, marquis without domain
穬	*kuang*	oats
廣種	*kuang-chung*	broadly sow the seed
澮	*kuei*	large drainage trench
耒圭	*kuei*	implement to cut wheat
恭	*kung*	term of respect to an elder
公乘	*kung-ch'eng*	honorific rank, high rank
工官	*kung-kuan*	Office of Manufacturing
公大夫	*kung-ta-fu*	noble rank, seventh step
公田	*kung-t'ien*	public land
功曹	*kung-ts'ao*	merit evaluator
宮苑地	*kung-yüan-ti*	unused side lots between inner and outer palace walls
過更	*kuo-keng*	cash paid in lieu of duties
臘	*la*	[sacrifice], festival held in twelfth month
臘田	*la-t'ien*	field like dried meat
莨菪	*lang-tang*	medicinal herb, *Scopolia japonica max.*
耒	*lei*	a crooked wooden plow, digging stick
耒耜	*lei-ssu*	plow
犁	*li*	plow
里	*li*	measure of length, 1364.1 feet, 300 *pu*
厘	*li*	unit of measure
醴	*li*	sweet wine
立秋	*Li-ch'iu*	beginning of autumn, one of 24 solar divisions of the year, *ca.* Aug. 7
立春	*Li-ch'un*	beginning of spring, one of 24 solar divisions of the year, *ca.* Feb. 5
立夏	*Li-hsia*	beginning of summer, one of 24 so-

		lar divisions of the year, *ca.* May 5
力田	*li-t'ien*	diligent cultivator
粱	*liang*	a kind of millet
兩	*liang*	measure of weight, 24 *shu*
列	*lieh*	furrow, ditch
鎌	*lien*	sickle
令	*ling*	prefect
令	*ling*	to be ordered
羅枷	*lo-chia*	a type of flail
壠	*lung*	ridge
麥	*mai*	wheat, barley
漫田	*man-t'ien*	fields not arranged in furrows
芒種	*Mang-chung*	"grain in ear," one of 24 solar divisions of the year, *ca.* June 4
卯	*mao*	fourth of twelve-day "earthly branches" cycle
米	*mi*	husked rice
緡	*min*	tax on merchants and artisans
命	*ming*	to teach, to mandate
命(家)	*ming*	(as modifier to *chia*) households with titles and ranks
名田	*ming-t'ien*	registration of land
末都	*mo-tu*	soybean sauce
麰	*mou*	barley
南威	*nan wei*	"south ripening," summer
耨	*nou*	hoe
農官	*nung-kuan*	Office of Agriculture
惡	*o*	evil
堊	*o*	plaster for building walls
區	*ou*	pit
耦	*ou*	pair
耦	*ou*	two *fa*
區種	*ou-chung*	pit farming

藕荷	*ou-ho*	Indian lotus
耦耕	*ou-keng*	paired tilling
區田	*ou-t'ien*	fields of shallow pits
杷	*pa*	tool used in harvesting wheat
白露	*Pai-lu*	"white dew," mid-autumn period, one of the 24 solar divisions of the year, *ca.* Sept. 8
糒	*pei*	dried cooked rice
盆	*p'en*	bowl, basin, pre-Han volume measure
匹	*p'i*	measure for cloth, bolt
賓客	*pin-k'e*	guests
丙	*ping*	third in the "celestial stems" ten-day cycle
平準	*p'ing-chun*	Office of Price Standardization
鏺	*po*	two-edged blade with handle for cutting grass
佈	*pu*	to distribute
步	*pu*	linear measurement, 6 *ch'ih*
部	*pu*	circuit, region
布	*pu*	coin
部曲	*pu-ch'ü*	near-slave dependents, personal troops
三	*san*	three
三伏	*san-fu*	last of the three divisions of summer
三輔	*san-fu*	Three Administrations
三老	*san-lao*	thrice venerable
善者倍之	*shan-che-pei-chih*	high-yielding, furrowed land
上農夫	*shang-nung-fu*	superior farmer
上造	*shang-tsao*	noble rank, second step
少	*shao*	small, minor

少府	*shao-fu*	Privy Treasury
社	*she*	ceremony during spring and autumn, community deity
申	*shen*	ninth of "earthly branches" twelve-day cycle
參	Shen	a constellation
深耕	*shen-keng*	deep tilling
升	*sheng*	measure of volume, 200 cc.
聲色	*sheng-she*	"singers and beautiful women"
石	*shih*	measure of volume, picul, 10 *tou*
市井	*shih-ching*	market well, marketplace
施布	*shih-pu*	to distribute
守	*shou*	administrator of commandery
守宮令	*shou-kung-ling*	official in charge of writing materials and sealing clay
銖	*shu*	measure of weight
朮	*shu*	a kind of millet
黍	*shu*	glutinous millet
暑	*shu*	summer
蜀漆	*shu-ch'i*	*ch'ang-shan,* fever medicine
霜降	*Shuang-chiang*	"frost descends," one of 24 solar divisions of the year, *ca.* Oct. 23
水衡都尉	*shui-heng tu-wei*	chief commandant of waters and parks
水排	*shui-p'ai*	watermill
四	*ssu*	four
耜	*ssu*	farming implement, hoe or spade
司直	*ssu-chih*	director of justice
司市	*ssu-shih*	master of market
粟	*su*	a kind of millet

宿麥	*su-mai*	winter wheat
算	*suan*	a tax unit
算賦	*suan-fu*	poll tax levied on adults
遂	*sui*	small trench for distributing water
大蜡	*ta-cha*	ritual to thank all miscellaneous deities
大夫	*ta-fu*	noble rank, fifth step
大夫士	*ta-fu-shih*	great officers
大石	*ta-shih*	husked grain measure
大暑	*Ta-shu*	"great heat," one of 24 solar divisions of the year, *ca.* July 22
大庶長	*ta shu-chang*	noble rank, eighteenth step
大司空	*ta-ssu-k'ung*	grand minister of works
大司馬	*ta-ssu-ma*	grand minister of war
大司農	*ta-ssu-nung*	grand minister of agriculture (the state treasurer)
大司徒	*ta-ssu-t'u*	grand minister over the masses
代田	*tai-t'ien*	alternating land
太常	*t'ai-ch'ang*	grand minister of ceremonies
太官	*t'ai-kuan*	grand provisioner
太僕	*t'ai-p'u*	grand keeper of equipages
稻	*tao*	nonglutinous rice
稻人	*tao-jen*	rice man
地支	*ti-chih*	"earthly branches" cycle of days
緹	*t'i*	wine
薙氏	*t'i-shih*	field cleaner
銚	*tiao*	weeding tool
鐵	*t'ieh*	iron
鐵官	*t'ieh-kuan*	Office of Iron Production
天雄	*t'ien-hsiung*	medicinal herb, *Aconitum fischeri*
天門冬	*t'ien-men-tung*	medicinal herb, *Asparagus lucidus*

丁	*ting*	fourth of the ten-day "celestial stems" cycle of days
亭	*t'ing*	defense unit, guard station, local administration office
斗	*tou*	measure of volume, 10 *sheng*
投	*t'ou*	to throw
澤	*tse*	moisture
柞氏	*tso-shih*	agent in charge of felling trees
左庶長	*tso-shu-chang*	noble rank, tenth grade
卒更	*tsu-keng*	military service for one month each year
酣醾	*tsu-mo*	drink made from butter skimmings
酢漿	*ts'u-chiang*	sour rice drink
漕	*ts'ao*	shipping canal
草人	*ts'ao-jen*	agent in charge of soil improvement
寸	*ts'un*	measure of length, inch, 10 *fen*
緵	*tsung*	unit of measure for cloth density
宗	*tsung*	organization based on kinship
土 栝	*t'u-kua*	*Tricosanthes cucumeroides,* plant from which starch is obtained
屯田	*t'un-t'ien*	military use of soldiers for farming, garrison farms
東作	*tung-tso*	"eastern work," spring activities
同產	*t'ung-ch'an*	siblings, children of siblings
子	*tzu*	first of the "earthly branches" twelve-day cycle
耔	*tzu*	roots, to cover roots
貲算	*tzu-suan*	property tax
蛙	*wa*	toad

王不留行	*wang-pu-liu-hsing*	medicinal herb, *Vaccaria pyramidata medic.*
未	*wei*	seventh of the "earthly branches" twelve-day cycle
尉	*wei*	commandant
衞將軍	*wei-chiang-chün*	general of the guard
伍	*wu*	pentad; organizational unit of five households
戊	*wu*	fifth of the "celestial stems" ten-day cycle
屋	*wu*	measure of area, three *fu*
烏韮	*wu-chiu*	a kind of moss
五大夫	*wu-ta-fu*	noble rank, ninth grade
烏頭	*wu-t'ou*	medicinal herb, *Aconitum carmichaeli* Debx.

Soils (see p. 94, table 10)

五桀	*wu-chieh*
五壏	*wu-chien*
五殖	*wu-chih*
五𡉏	*wu-chuan*
五怚	*wu-chü*
五浮	*wu-fo*
五鳧	*wu-fu*
五瞉	*wu-hu*
五弘	*wu-hung*
五壤	*wu-jang*
五塥	*wu-ke*
五纑	*wu-lu*
五剽	*wu-p'iao*
五沙	*wu-sha*
五粟	*wu-shu*

五位	*wu-wei*	
五沃	*wu-wu*	
五隱	*wu-yin*	
五猶	*wu-yu*	
丫	*ya*	a type of flail
陽	*yang*	trees growing on a south slope
羊	*yang*	sheep
謁者	*yeh-che*	internuncio
堰沮	*yen-chu*	water reservoir for flowing water
塩官	*yen-kuan*	Office of Salt Production
寅	*yin*	third of the twelve-day "earthly branches" cycle
櫌	*yu*	implement to smooth a field surface
芋	*yü*	taro
虞	*yü*	woodsman, watchman for the forests; imperial hunting ground, garden
雨水	*Yü-shui*	"rain water," one of 24 solar divisions of the year, *ca.* Feb. 19
耘	*yün*	weed, to remove weeds
用	*yung*	to avail himself of

BIBLIOGRAPHY

WORKS IN CHINESE

Ssu-pu pei-yao 四部備要 edition has been abbreviated as SPPY; *Ssu-pu ts'ung-k'an* 四部叢刊 edition has been abbreviated as SPTK. Also abbreviated are the following works:

CCP — *Chi-chiu-p'ien.* SPTK ed.
急就篇

CHHW — *Ch'üan Hou-Han-wen.* In *Ch'uan Han-wen.* Edited by Yen K'e-chün. 1894. Taipei: World Book Co. reprint, 1961.
全後漢文 嚴可均

CKT — *Chan-kuo-ts'e.* SPPY ed.
戰國策

CLCS — *Chou-li chu-shu.* SPPY ed. Compiled by Cheng Hsüan.
周禮注疏

CS — *Chin-shu.* SPPY ed.
晉書

EY — *Erh-ya.* SPTK ed.
爾雅

HFT — *Han Fei-tzu.* SPPY ed.
韓非子

HHS	*Hou-Han-shu chi-chieh*. Edited by Wang Hsien-ch'ien. 1915 Taipei: I-wen reprint. 後漢書集解
HNT	*Huai-nan-tzu*. SPPY ed. 淮南子
HS	*Han-shu pu-chu*. Edited by Wang Hsien-ch'ien. 1900. Taipei: I-wen reprint. 漢書補注　王先謙
HT	*Hsün-tzu*. SPPY ed. 荀子
KT	*Kuan-tzu*. SPPY ed. 管子
LCCS	*Li-chi chu-shu*. SPPY ed. 禮記注疏
LS	*Li-shih*. SPTK ed. 隸釋
LSCC	*Lü-shih ch'un-ch'iu*. SPPY ed. 呂氏春秋
MSCI	*Mao-shih cheng-i*. SPPY ed. 毛詩正義
MT	*Mo-tzu*. SPPY ed. 墨子
SC	*Shih-chi hui-chu k'ao-cheng*. 1934. Taipei: I-wen reprint. 史記會注考證
SCC	Li Tao-yüan. *Shui-ching-chu*. SPPY ed. 酈道元　水經注
SKC	Chao I-ch'ing. *San-kuo-chih pu-chu*. Taipei: I-wen reprint. 趙一清　三國志 補注
SSCS	*Shang-shu chu-shu*. SPPY ed. 尙書注疏
SWCT	Hsü Shen. *Shuo-wen chieh-tzu*. SPTK ed. 許愼　說文解字
TPYL	Li Fang et al. *T'ai-p'ing-yü-lan*. Taipei: Chung-hua reprint, 1967. 李昉　太平御覽
YTL	Huan K'uan. *Yen-t'ieh-lun*. SPPY ed. 桓寬　鹽鐵論

Anhwei-sheng po-wu-kuan. "An-hui hsin-shih-ch'i shih-tai i-chih-ti tiao-ch'a." *K'ao-ku hsüeh-pao,* 1957, no. 1, pp. 21-30.
安徽省博物館　安徽新石器時代遺址的調查　考古學報

Chang Cheng-lang. "Pu-tz'u p'ou-t'ien chi ch'i hsiang-kuan chu-wen-t'i." *K'ao-ku hsüeh-pao,* 1973, no. 1, pp. 93-120.
張政烺　卜辭裒田及其相關諸問題

Ch'ang Ch'ü. *Hua-yang kuo-chih.* SPPY ed.
常璩　華陽國志

Ch'en Chih. *Liang-Han ching-chi shih-liao lun-ts'ung.* Sian: Shensi Jen-min, 1958.
陳直　兩漢經濟史料論叢

Ch'en Heng-li. "Kuan-tzu Mu-min, Ch'eng-ma, Pa-kuan teng p'ien yen-chiu." *Hsin-chien-she,* 1956, no. 5, pp. 51-59.
陳恒立　管子牧民乘馬八觀等篇研究　新建設

Ch'en Hsiao-chiang. "Kuan-yü i-mu-san-ch'üan wen-t'i-ti shang-chüeh ping-ta Yang-chün." *Shih-huo* 2 (1935), no. 1: 3-15.
陳嘯江　關於一畝三甽問題的商榷並答楊君　食貨

Ch'en Kung-jou and Hsü P'ing-fang. "Kuan-yü Chü-yen Han-chien-ti fa-hsien ho yen-chiu." *K'ao-ku,* 1960, no. 1, pp. 45-53.
陳公柔、徐苹芳　關於居延漢簡的發現和研究　考古

Ch'en Liang-tso. "Chung-kuo ku-tai nung-yeh shih-fei chih shang-ch'üeh." *Bulletin of the Institute of History and Philology* 42, pt. 4 (1971): 829-42.
陳良佐　中國古代農業施肥之商榷　歷史語言研究所集刊

——. "Wo-kuo li-tai nung-t'ien shih-yung-chih lü-fei." *Ta-lu tsa-chih* 46, no. 5:20-47.
我國歷代農田使用之綠肥　大陸雜誌

Ch'en P'an. "Yu Han-chien-chung-chih chün-li-ming-chih shuo-ch'i." *Ta-lu tsa-chih* 2, no. 8 (1951): 15, 19, 32.
陳槃　由漢簡中之軍吏名籍說起

Ch'en Tsu-kuei. "Chung-kuo wen-hsien-shang-ti shui-tao tsai-p'ei." *Nung-shih yen-chiu chi-k'an,* vol. 2. Peking: K'e-hsüeh, 1960.
陳祖槼　中國文獻上的水稻栽培　農史研究集刊

Ch'eng Shu-te. *Chiu-ch'ao lü-k'ao.* Shanghai: Chung-hua, 1926.
程樹德　九朝律考

Ch'i Hsia et al. *Ch'in-Han nung-min chan-cheng-shih.* Peking: San-lien, 1962.
漆俠　秦漢農民戰爭史

Chia-yu-kuan wen-wu hsiao-tsu. "Chia-yü-kuan Han hua-hsiang-chuan mu." *Wen-wu,* 1972, no. 12, pp. 24-41.
嘉峪關文物小組　嘉峪關漢畫象磚墓　文物

Ch'ien Mu. "Chung-kuo ku-tai pei-fang nung-tso-wu k'ao." *Hsin-ya hsüeh-pao* 1, no. 2 (1956): 1-27.
錢穆 中國古代北方農作物考 新亞學報

Ch'ien Ta-chao. *Hou-Han-shu pu-piao*. In *Erh-shih-wu-shih pu-pien*. Peking: Chunghua, 1955.
錢大炤 後漢書補表 二十五史補編

Chin Fa-ken. *Yung-chia-luan-hou pei-fang-ti hao-tsu*. Taipei, 1964.
金發根 永嘉亂後北方的豪族

Ching-tien shih-wen. SPTK ed.
經典釋文

Chiu-chang suan-shu. SPTK ed.
九章算術

Chu K'o-chen [Chu Ke-chen]. "Chung-kuo chin-wu-ch'ien-nien-lai ch'i-hou pien-hua-ti ch'u-pu yen-chiu." *K'ao-ku hsüeh-pao*, 1972, no. 1, pp. 15-38.
竺可楨 中國近五千年來氣候變化的初步研究

——. "Lun wo-kuo ch'i-hou-ti chi-ke t'e-tien chi ch'i yü liang-shih tso-wu sheng-ch'an-ti kuan-hsi." *Ti-li hsüeh-pao* 30, no. 1 (1964): 1-14.
論我國氣候的幾個特點及其與糧食作物生產的關係 地理學報

Ch'ü Wan-li. "Lun Yü-kung chu-ch'eng-ti shih-tai." *Bulletin of the Institute of History and Philology* 35 (1964): 53-86.
屈萬里 論禹貢著成的時代

Ch'üan-Han-wen. Includes *Ch'üan Ch'ien-Han-wen* and *Ch'üan Hou-Han-wen*. Compiled and edited by Yen K'e-chün. 1894. Taipei: World Book Co. reprint, 1961.
全漢文

Ch'üan-kuo chi-pen chien-she kung-ch'eng-chung ch'u-t'u wen-wu chan-lan t'u-lu. Peking, 1954.
全國基本建設工程中出土文物展覽圖錄

Chung-kuo k'e-hsüeh-yüan, K'ao-ku yen-chiu-so, comp. *Chü-yen Han-chien chia-pien*. Peking, 1959.
中國科學院考古研究所 居延漢簡甲編

Chung-kuo nung-yeh i-ch'an yen-chiu-shih, ed. *Chung-kuo nung-hsüeh-shih*. Peking: K'e-hsüeh, 1959.
中國農業遺產研究室 中國農學史

Han-shu shu-cheng. Compiled by Shen Ch'in-han. Che-chiang kuan shu-chü, 1900.
漢書疏證 沈欽韓

Ho Ch'ang-ch'ün. "Ch'in-Han-chien ke-t'i-hsiao-nung-ti hsing-ch'eng ho fa-chan." In *Han-T'ang-chien feng-chien t'u-ti-so-yu-chih hsing-shih yen-chiu*. Shanghai: Jen-min, 1964.
賀昌群 秦漢間個體小農的形成和發展 漢唐間封建土地所有制形式研究

Ho Ping-ti. *Huang-t'u yü Chung-kuo nung-yeh-ti ch'i-yüan*. Hong Kong: Chinese University, 1969.
何炳棣 黃土與中國農業的起源

Honan-sheng wen-hua-chü wen-wu kung-tso-tui. *Kung-hsien T'ieh-sheng-kou Han-tai yeh-t'ieh i-chih*. Peking, 1963.
河南省文化局文物工作隊 鞏縣鐵生溝漢代冶鐵遺址

Hou Jen-chih et al., eds. *Chung-kuo ku-tai ti-li ming-chu hsüan-tu*. Peking: K'e-hsüeh, 1959.
侯仁之 中國古代地理名著選讀

Hsia Nai. *Hui-hsien fa-chüeh pao-kao*. K'e-hsüeh, 1956.
夏鼐 輝縣發掘報告

Hsia Wei-ying. *Kuan-tzu Ti-yüan-p'ien chiao-shih*. Peking: Chung-hua, 1958.
夏緯瑛 管子地員篇校釋

——. *Lü-shih ch'un-ch'iu Shang-nung teng-ssu p'ien chiao-shih*. Shanghai, 1956.
呂氏春秋上農等四篇校釋

Hsü Cho-yun. "Han-tai chia-t'ing-ti ta-hsiao." *Ch'ing-chu Li Chi hsien-sheng ch'i-shih-sui lun-wen-chi*. Taipei, 1967.
許倬雲 漢代家庭的大小 慶祝李濟先生七十歲論文集

——. "Liang-Chou nung-tso yü chi-shu." *Bulletin of the Institute of History and Philology* 42, pt. 4 (1971): 803-27.
兩周農作與技術

——. "San-kuo Wu-ti-ti ti-fang shih-li." *Bulletin of the Institute of History and Philology*, vol. 37, pt. 1 (1967): 185-200.
三國吳地的地方勢力

Hsü Chung-shu. "Lei-ssu-k'ao." *Bulletin of the Institute of History and Philology*, vol. 2, pt. 1 (1930): 11-59.
徐仲舒 耒耜考

Hsü-Han-chih. Compiled by Ssu-ma Piao, Po-na edition.
續漢志 司馬彪

Hsü Fu-kuan. *Chou Ch'in Han cheng-chih she-hui chieh-kou-chih yen-chiu*. Hong Kong: Hsin-ya, 1972.
徐復觀　周秦漢政治社會結構之研究

Hsün Yüeh. *Shen-chien*. Han-Wei-ts'ung-shu ed.
荀悅　申鑒

Hu Hsi-wen, "Chung-kuo hsiao-mai tsai-p'ei chi-shu chien-shih." *Nung-yeh i-ch'an yen-chiu chi-k'an* 1 (1958): 51-75.
胡錫文　中國小麥栽培技術簡史　農業遺產研究集刊

Hu Tao-ching. "Shih-shu-p'ien—Shih-lun wo-kuo ku-tai nung-min tui ta-tou ken-liu-ti jen-shih." *Chung-hua wen-shih lun-ts'ung* 3:111-19. Shanghai, 1963.
胡道靜　釋菽篇一試論我國古代農民對大豆根瘤的認識
中華文史論叢

Huang Chan-yüeh. "Chin-nien ch'u-t'u-ti Chan-kuo Liang-Han t'ieh-ch'i." *K'ao-ku hsüeh-pao*, 1957, no. 3, pp. 93-108.
黃展岳　近年出土的戰國兩漢鐵器

Huang Lieh. "Shih Han-chien-chung yu-kuan Han-tai she-hui hsing-chih chu-li." *Li-shih yen-chiu*, 1957, no. 6, pp. 65-78.
黃烈　釋漢簡中有關漢代社會性質諸例　歷史研究

Huang Sheng-chang. "Chiang-ling Feng-huang-shan Han-mu chien-tu chi ch'i tsai li-shih ti-li yen-chiu-shang-ti chia-chih." *Wen-wu*, 1974, no. 6, pp. 66-67.
黃盛璋　江陵鳳凰山漢墓簡牘及其在歷史地理研究上的價值

Huang Yao-neng. "Shui-ching-chu shih-tai so-ch'u-hsien-ti Chung-kuo ku-tai ch'ü-pai fen-pu chi ch'i so-tai-piao i-i." *Yu-shih yüeh-k'an* 43, no. 5 (1976): 56-63.
黃耀能　水經注時代所出現的中國古代渠陂分佈及其所代表意義　幼獅月刊

Hung I. "Chiang-ling Feng-huang-shan shih-hao Han-mu chien-tu ch'u-tan." *Wen-wu*, 1974, no. 6, pp. 78-84.
弘一　江陵鳳凰山十號漢墓簡牘初探

Kao P'ing-tzu. *Hsüeh-li san-lun*. Taipei: Academia Sinica, 1969.
高平子　學曆散論

K'ao-ku yen-chiu-so. *Hsin Chung-kuo-ti k'ao-ku shuo-huo*. Peking, 1962.
考古研究所　新中國的考古收獲

Kiangsu-sheng wen-kuan-hui. "Chiang-su Wu-hsi Hsi-shan kung-yüan ku-i-chih ch'ing-li chien-pao." *Wen-wu ts'an-k'ao tzu-liao*, 1956,
江蘇省文管會　江蘇無錫錫山公園古遺址清理簡報

no. 1, pp. 26-32.

Kiangsu-sheng wen-wu kuan-li wei-yüan-hui, eds. *Chiang-su Hsü-chou Han-hua-hsiang-shih.* Peking: K'e-hsüeh, 1959.
江蘇省文物管理委員會編　江蘇徐州漢畫象石

——. "Chiang-su Wu-hsi Hsien-li-tun hsin-shih-ch'i shih-tai i-chih ch'ing-li chien-pao." *Wen-wu ts'an-k'ao tzu-liao,* 1955, no. 8, pp. 48-59.
江蘇無錫仙蠡墩新石器時代遺址清理簡報　文物參考資料

Ku Chieh-kang. "Yü-kung." In *Chung-kuo ku-tai ti-li ming-chu hsüan-tu.* Edited by Hou Jen-chih. Peking: K'e-hsüeh, 1959.
顧頡剛　禹貢　中國古代地理名著選讀

Ku-shih-hsüan. SPPY ed.
古詩選

Kuo Mo-jo. *Liang-Chou chin-wen-ts'u ta-hsi.* Shanghai, 1936.
郭沫若　兩周金文辭大系

——. *Nu-li-chih shih-tai.* Peking: K'e-hsüeh, 1956.
奴隸制時代

Kuo Pao-chün. "Lo-yang hsi-chiao Han-tai chü-chu i-chih." *K'ao-ku t'ung hsün,* 1956, no. 1, pp. 18-26.
郭寶鈞　洛陽西郊漢代居住遺址　考古通訊

—— et al. "Lo-yang Chien-pin ku-wen-hua i-chih chi Han-mu." *K'ao-ku hsüeh-pao,* 1956, no. 1, pp. 11-28.
洛陽澗濱古文化遺址及漢墓

Kuo Wen-t'ao. "Chung-kuo ku-tai nung-tso-chih chih-shih-ti k'ao-ch'a." *Chung-kuo nung-pao,* 1963, no. 9, pp. 41-48; no. 10, pp. 33-39.
郭文韜　中國古代農作制知識的考察　中國農報

Lao Kan. *Chü-yen Han-chien: K'ao-shih-chih-pu.* Taipei: Academia Sinica, 1960.
勞榦　居延漢簡考釋之部

——. *Chü-yen Han-chien k'ao-shih: Shih-wen-chih-pu.* Nanking: Academia Sinica, 1949.
居延漢簡考釋釋文之部

——. *Chü-yen Han-chien: Shih-wen-chih-pu.* Taipei: Academia Sinica, 1960.
居延漢簡釋文之部

——. "Han-chien-chung-ti Ho-hsi ching-chi sheng-huo." *Bulletin of the Institute of History and Philology* 11 (1943): 61-75.
漢簡中的河西經濟生活

——. "Han-tai ping-chih chi Han-chien-chung-ti ping-chih." *Bulletin of the Institute of History and Philology*, vol. 10 (1943): 23-55.
漢代兵制及漢簡中的兵制

——. "Huang-t'u yü Chung-kuo nung-yeh-ti ch'i-yüan pa." In *Huang-t'u yü Chung-kuo nung-yeh-ti ch'i-yüan*. Edited by Ho Ping-ti. Hong Kong: Chinese University, 1969.
黃土與中國農業的起源跋

——. "Liang-Han chün-kuo mien-chi-chih ku-chi chi k'ou-shu tseng-chien-chih t'ui-ts'e." *Bulletin of the Institute of History and Philology* 5, no. 2: 215-40.
兩漢郡國面積之估計及口數增減之推測

——. "Liang-Han hu-chi yü ti-li-chih kuan-hsi." *Bulletin of the Institute of History and Philology* 5, no. 2 (1935): 179-214.
兩漢戶籍與地理之關係

——. "Lun Han-tai-ti nei-ch'ao yü wai-ch'ao." *Bulletin of the Institute of History and Philology* 13 (1948): 227-67.
論漢代的內朝與外朝

——. "Ta-shih yü hsiao-shih." *Ta-lu tsa-chih* 1, no. 11: 21.
大石與小石

Li-chi Cheng-chu. SPPY ed.
禮記鄭注

Li Chien-nung. *Hsien-Ch'in Liang-Han ching-chi-shih kao*. Rev. ed. Peking: Chung-hua, 1962.
李劍農　先秦兩漢經濟史稿

Li Meng-yang and Liu Yu-chu. "Mou shou-pai-shih-chih-mi." *Wen-shih-che*. 1964, no. 2, pp. 47, 54-55.
李孟揚、劉有菊　畝收百石之謎　文史哲

Li Wen-hsin. "Ku-tai t'ieh-nung-chü." *Wen-wu ts'an-k'ao tzu-liao*, 1954, no. 9, pp. 80-86.
李文信　古代鐵農具　文物參考資料

Li Ya-nung. *Hsin-jan-chai shih-lun-chi*. Shanghai: Jen-min, 1962.
李亞農　欣然齋史論集

Li-shih yen-chiu, ed. *Chung-kuo ku-tai-shih fen-ch'i wen-t'i t'ao-lun-chi*. Peking: San-lien, 1957.
歷史研究所編輯部　中國古代史分期問題討論集

Liu Chen. *Tung-kuan Han-chi*. Shao-yeh-shang-fang, 1795.
劉珍　東觀漢記

Liu Chih-yüan. "Ch'eng-tu T'ien-hui-shan yai-mu ch'ing-li chi." *K'ao-ku hsüeh-pao,* 1958, no. 1, pp. 87-103.
劉志遠　成都天廻山崖墓清理記

Liu Hsi, comp. *Shih-ming.* SPTK ed.
劉熙　釋名

Liu Hsien-chou. *Chung-kuo ku-tai nung-yeh chi-chieh fa-ming shih.* Peking, 1963.
劉仙洲　中國古代農業機械發明史

Liu Ping-jen. "Liang-Han-ti jen-k'ou yü shih-liang cheng-ts'e." *Yü-kung* 7, nos. 1-3: 171-82.
劉秉仁　兩漢的人口與食糧政策　禹貢

Liu Tao-yüan. "Shang Yang pien-fa yü Liang-Han t'ien-fu chih-tu." *Shih-huo* 1, no. 3: 1-13.
劉道元　商鞅變法與兩漢田賦制度

Lo Chen-yü. *Liu-sha chui-chien k'ao-shih.* N.p.
羅振玉　流沙墜簡考釋

Lo-yang-chü k'ao-ku fa-chüeh-tui. *Lo-yang Shao-kou Han-mu.* Peking: K'e-hsüeh, 1959.
洛陽區考古發掘隊　洛陽燒溝漢墓

Lo-yang fa-chüeh-tui. "Lo-yang hsi-chiao Han-mu fa-chüeh pao-kao." *K'ao-ku hsüeh-pao,* 1963, no. 2, pp. 1-58.
洛陽發掘隊　洛陽西郊漢墓發掘報告

Lun-heng. SPPY ed.
論衡

Lun-yü chu-shu. SPPY ed.
論語注疏

Ma Ch'eng-feng. *Chung-kuo ching-chi-shih.* Shanghai, 1939.
馬乘風　中國經濟史

Ma Ch'eng-yüan. "Shang Yang fang-sheng ho Chan-kuo liang-chih." *Wen-wu,* 1972, no. 6, pp. 17-24.
馬承源　商鞅方升和戰國量制

Ma Fei-pai. "Ch'in-Han ching-chi-shih tzu-liao (3)." *Shih-huo* 3, no. 1: 9-33.
馬非白　秦漢經濟史資料(3)

——. "Ch'in-Han ching-chi tzu-liao (7): Tsu-shui chih-tu." *Shih-huo* 3: 9-33.
秦漢經濟資料(7)：租稅制度

Ma Kuo-han, comp. *Nung-chia i-shu chi-pen chiu-chung.* Taipei: World Book Co., reprint, 1964.
馬國翰　農家佚書輯本九種

Mao-shih chu-shu. SPPY ed.
毛詩注疏

Miao Ch'i-yü et al. "Shang-chün-shu shih-tai-ti she-hui pien-ke ho nung-yeh pien-ke." In *Chung-kuo nung-hsüeh-shih,* chap. 3.
繆啟愉　商君書時代的社會變革和農業變革

Nan-ching po-wu-yüan. "Li-kuo-i ku-tai lien-t'ieh-lu-ti tiao-ch'a chi ch'ing-li." *Wen-wu,* 1960, no. 4, pp. 46-47.
南京博物院　利國驛古代錬鐵爐的調查及清理

P'an Hung-sheng and Yang Ch'ao-po. "Chan-kuo shih-tai-ti liu-kuo nung-yeh sheng-ch'an." *Nung-shih yen-chiu chi-k'an* 2: 48-63. Peking: K'e-hsüeh, 1960.
潘鴻聲、楊超伯　戰國時代的六國農業生產

Shang-chün-shu. SPPY ed.
商君書

Shansi-sheng wen-wu kuan-li wei-yüan-hui. "Shan-hsi P'ing-lu Tsao-yüan-ts'un pi-hua Han-mu." *K'ao-ku,* 1959, no. 9, pp. 462-63.
山西省文物管理委員會　山西平陸棗園村壁畫漢墓

Shensi-sheng po-wu-kuan et al. "Mi-chih Tung-Han hua-hsiang-shih-mu fa-chüeh chien-pao." *Wen-wu,* 1972, no. 3, pp. 69-76.
陝西省博物館　米脂東漢畫象石墓發掘簡報

Shih Nien-hai. "Shih-chi Huo-chih lieh-chuan so-shuo-ti T'ao-wei-t'ien-hsia-chih-chung. Chien-lun Chan-kuo shih-tai-ti ching-chi tu-hui." *Jen-wen tsa-chih,* 1958, no. 2, pp. 77-87.
史念海　史記貨殖列傳所說的陶爲天下之中，兼論戰國時代的經濟都會　人文雜誌

Shih Sheng-han. *Ch'i-min-yao-shu chin-shih.* Peking: K'e-hsüeh, 1957.
石聲漢　齊民要術今釋

——. *Ssu-min-yüeh-ling chiao-chu.* Peking: Chung-hua, 1965.
四民月令校注

——. *Fan Sheng-chih-shu chin-shih.* Peking: K'e-hsüeh, 1959.
氾勝之書今釋

Ssu-ma Kuang. *Tzu-chih t'ung-chien.* SPTK ed.
司馬光　資治通鑑

Sun Ch'ang-hsü. *Lei-ssu-ti ch'i-yüan chi ch'i fa-chan.* Shanghai, 1964.
孫常敍　耒耜的起源及其發展

Sun Hsing-yen. *Shang-shu chin-ku-wen chu-shu.* SPPY ed.
孫星衍　尚書今古文注疏

Szechwan-sheng po-wu-kuan. "Szu-ch'uan sheng Mu-ma-shan Kuan-kai-ch'ü ku-mu ch'ing-li chien-pao." *K'ao-ku,* 1959, no. 8, pp. 419-29.
四川省博物館　四川省牧馬山灌溉渠古墓清理簡報

Szechwan-sheng po-wu-kuan wen-wu kung-tso-tui. "Szu-ch'uan Hsin-chin-hsien Pao-tzu-shan yai-mu ch'ing-li chien-pao." *K'ao-ku t'ung-hsün,* 1958, no. 8, pp. 31-37.
四川省博物館文物工作隊　四川新津縣抱子山崖墓清理簡報

T'ang Ch'ang-ju. *Wei-Chin Nan-pei-ch'ao shih lun-ts'ung.* Peking: San-lien, 1955.
唐長孺　魏晉南北朝史論叢

T'ao Hsi-sheng. *Hsi-Han ching-chi-shih.* Shanghai: Commercial Press, 1935.
陶希聖　西漢經濟史

T'ien Ch'ang-wu. "Han-Wu-ti." *Jen-min Chung-kuo,* 1960, no. 8, pp. 5-8.
田昌五　漢武帝　人民中國

T'ien Shih. "Hsi-Han tu-liang-heng lüeh-shou." *Wen-wu,* 1975, no. 12, pp. 79-89.
天石　西漢度量衡略説

Ting Ying. "Chiang-Han p'ing-yüan hsin-shih-ch'i shih-tai hung-shao-t'u-chung-ti tao-ku-k'o k'ao-ch'a." *K'ao-ku hsüeh-pao,* 1959, no. 4, pp. 31-34.
丁穎　江漢平原新石器時代紅燒土中的稻穀殼考察

——. "Chung-kuo tsai-p'ei-tao-chung-ti ch'i-yüan chi ch'i yen-pien." In *Tao-tso k'e-hsüeh lun-wen hsüan-chi.* Edited by Chung-kuo nung-yeh k'e-hsüeh-yüan. Peking, 1959.
中國栽培稻種的起源及其演變　稻作科學論文選集　中國農業科學院

Ts'ao Lung-kung. "Chung-kuo nung-shih wen-hsien-shang su-ti tsai-p'ei." *Nung-shih yen-chiu chi-k'an* 2:94-108.
曹隆恭　中國農史文獻上粟的栽培

Tseng Jung. "Han-tai-ti t'ieh-chih kung-chü." *Wen-wu,* 1959, no. 1, pp. 16-19.
曾庸　漢代的鐵製工具

Tseng Yung-ch'ien. *Shang-shu cheng-tu*. Peking: Chung-hua, 1964.
曾永乾 尙書正讀

Tsou Pao-chün. "Chung-kuo wen-hua ch'i-yüan ti." *Tsing Hua Journal of Chinese Studies,* n.s., 6:22-34.
鄒豹君 中國文化起源地

Tsou Shu-wen. "Shih-ching shih-tai su-chi-pien." *Nung-shih yen-chiu chi-k'an* 2:18-34.
鄒樹文 詩經時代黍稷辨

Tung-pei po-wu-kuan. "Liao-yang San-tao-hao Hsi-Han ts'un-lo i-chih." *K'ao-ku hsüeh-pao,* 1957, no. 1. pp. 119-26.
東北博物館 遼陽三道壕西漢村落遺址

T'ung Shu-yeh. "Ts'ung Chung-kuo k'ai-shih yung-t'ieh-ti shih-tai wen-t'i p'ing Hu Shih-p'ai-ti chih-hsüeh fang-fa." *Wen-shih-che,* 1955, no. 2, pp. 30-33.
童書業 從中國開始用鐵的時代問題評胡適派的治學方法

Wan Kuo-ting. "Lü-shih ch'un-ch'iu-chung-ti keng-tso yüan-li." In *Chung-kuo nung-hsüeh-shih*. Peking: K'e-hsüeh, 1959.
萬國鼎 呂氏春秋中的耕作原理

——. "Lü-shih chün-ch'iu-ti hsing-chih chi ch'i tsai nung-hsüeh-shih-shang-ti chia-chih." *Nung-shih yen-chiu chi-k'an* 2:175-85.
呂氏春秋的性質及其在農學史上的價值

——. *Wu-ku shih-hua.* Chung-hua, n. p., n. d.
五谷史話

——, et al. *Chung-kuo nung-hsüeh-shih*. Chung-kuo nung-yeh i-ch'an yen-chiu shih ed. Peking: K'e-hsüeh, 1959.
中國農學史

Wang Chung-lo. "Kuan-yü Chung-kuo nu-li she-hui-ti wa-chieh chi feng-chien kuan-hsi-ti hsing-ch'eng wen-t'i." *Chung-kuo ku-tai-shih fen-ch'i wen-t'i t'ao-lun-chi.* Edited by Li-shih yen-chiu. Shanghai: San-lien, 1957.
王仲犖 關於中國奴隸社會的瓦解及封建關係的形成問題
中國古代史分期問題討論集 歷史研究

Wang Chung-shu. "Han-tai wu-chih wen-hua lüeh-shuo." *K'ao-ku t'ung-hsün,* 1956, no. 1, pp. 57-76.
王仲殊 漢代物質文化略說

Wang Shu-min. *Lü-shih ch'un-ch'iu chiao-pu*. Taipei: Academia Sinica, n.d.
王叔珉 呂氏春秋校補

Wang Yü-hu. *Ou-chung shih-chung.* Shanghai, 1955.
王玉瑚 區種十種

Wu Shan-ch'ing. "Lüeh-lun Ch'ing-lien-kang wen-hua." *Wen-wu,* 1973, no. 6, pp. 45-61.
吳山菁 略論青蓮崗文化

Yang Chung-li. "Tsai-lun i-mou-san-ch'üan sui-tai-ch'u." *Shih-huo* 2, no. 4:29-40.
楊仲禮 再論一畝三甽歲代處

Yang K'uan. *Chung-kuo ku-tai yeh-t'ieh chi-shu-ti fa-ming ho fa-chan.* Shanghai, 1956.
楊寬 中國古代冶鐵技術的發明和發展

Yang Lien-sheng. "Ts'ung Ssu-min-yüeh-ling so-chien-tao-ti Han-tai chia-tsu-ti sheng-ch'an." *Shih-huo* 1, no. 6:8-11.
楊聯陞 從四民月令所見到的漢代家族的生產

——. "Tung-Han-ti hao-tsu." *Tsing-hua hsüeh-pao* 11, no. 4 (1936): 1007-63.
東漢的豪族 清華學報

Yen Keng-wang. *Chung-kuo ti-fang hsing-cheng chih-tu-shih.* Taipei: Academia Sinica, 1961.
嚴耕望 中國地方行政制度史

Yin Chung-jung. *Lü-shih ch'un-ch'iu chiao-shih.* Taipei: Chung-hua ts'ung-shu, 1958.
尹仲容 呂氏春秋校釋

Yin Ti-fei. "An-hui-sheng Shou-hsien An-feng-t'ang fa-hsien Han-tai cha-pa kung-ch'eng i-chih." *Wen-wu,* 1960, no. 1, pp. 61-62.
殷滌非 安徽省壽縣安豐塘發現漢代閘壩工程遺址

Yu Hsiu-ling. "Tui Ho-mu-tu i-chih ti-ssu-wen-hua-ts'eng ch'u-t'u tao-ku ho ku-ssu-ti chi-tien k'an-fa." *Wen-wu,* 1976, no. 8, pp. 20-23.
游修齡 對河母渡遺址第四文化層出土稻穀和骨耜的幾點看法

Yü Ching-jang. "Shu-chi-su-liang yü kao-liang." *Ta-lu tsa-chih* 13, no. 43:1-10; 13, no. 4: 15-20.
于景讓 黍稷粟粱與高粱

——. *Tsai-p'ei chih-wu-k'ao,* vol. 2. Taipei: I-wen, 1972.
栽培植物考

Yü Ying-shih. "Tung-Han cheng-ch'üan-chih chien-li yü shih-tsu ta-hsing-chih kuan-hsi." *Hsin-ya hsüeh-pao* 1, no. 2 (1956): 209-80.
余英時 東漢政權之建立與世族大姓之關係

Yu Yü. "Kuan-tzu Ti-yüan-p'ien yen-chiu." *Nung-shih yen-chiu chi-k'an* 1: 17-36.
友于　管子地員篇研究

——. "Kuan-tzu Tu-ti-p'ien t'an-wei." *Nung-shih yen-chiu chi-k'an* 1: 1-16.
管子度地篇探微

——. "Yu Hsi-Chou tao Ch'ien-Han-ti keng-tso chih-tu yen-ke." *Nung-shih yen-chiu chi-k'an* 2: 1-17.
由西周到前漢的耕作制度沿革

Works in Japanese

Amano Motonosuke. "Chūgoku bosei kō." *Tōa keizei kenkyū hō*. n.s., no. 3.
天野元之助　中國畝制考　東亞經濟研究報

——. "Chūgoku kodaishika no shosetsu o hyōsu." *Rekishigaku kenkyū* 180 (1955): 29-37.
中國古代史家の諸說を評す　歷史學研究

——. "Chūgoku ni okeru suki no hattatsu." *Tōhōgakuhō* (Kyoto) 26 (1956): 104-172.
中國における犂の發達　東方學報

——. *Chūgoku nōgyōshi kenkyū*. Tokyo: Ochanomizu shobō, 1962.
中國農業史研究

——. "Daiden to kōden–Kandai nōgyō gijutsu kō." *Matsuyama shōdai ronshū,* 1950, pp. 146-69.
代田と區田—漢代農業技術考　松山商大論集

——. " 'Kakō-suidō' no ben: Chūgoku kodai Kōnan suidō gijutsu kō." *Shigaku zasshi* 61, no. 4 (1952): 346-49.
「火耕水耨」の辯—中國古代江南水稲技術考　史學雜誌

——. "Kandai gōzoku no daitochi keiei shiron." In *Takikawa hakushi kanreki kinen ronbun shū* 1: 1-17. Tokyo: Takikawa hakushi kanreki kinen ronbun shū kankō iinkai, 1957.
漢代豪族の大土地經營試論　瀧川博士還曆記念論文集

Goi Naohiro. "Kandai kōden ni okeru kasaku ni tsuite." *Rekishigaku kenkyū* 220, no. 6 (1958): 11-20.
五井直弘　漢代公田における假作について

Hamaguchi Shigekuni. "Chūgoku shijō no kodai shakai mondai ni kansuru oboegaki." *Yamanashi daigaku gakugeibu kenkyū hōkoku,* no. 4, pp. 55-70.
濱口重國　中國史上の古代社會問題に關する覺書　山梨大學學藝部研究報告

——. "Senkō to kakō: Jojun-setsu no hihan." *Tōyōgakuhō* 19, no. 3 (1931): 370-93, and 20, no. 2 (1932): 292-93.
踐更と過更—如淳說の批判　東洋學報

——. *Shin Kan Zui Tō shi no kenkyū.* Tokyo: Tōkyō daigaku shuppankai, 1966.
秦漢隋唐史の研究

Hiranaka Reiji. *Chūgoku kodai no densei to zeihō.* Tōyōshi kenkyū sōkan, no. 16. Kyoto: Tōyōshi kenkyūkai, 1967.
平中苓次　中國古代の田制と稅法　東洋史研究叢刊

——. "Kandai kōden no 'ka': Entetsuron Enchi-hen no kisai ni tsuite." In *Wada hakushi koki kinen Tōyōshi ronsō.* Tokyo: Kōdansha, 1961. pp. 781-92.
漢代公田の「假」—鹽鐵論園池篇の記載について 和田博士古稀記念東洋史論叢

——. "Kyoen Kan-kan to Kandai no zaisanzei." *Ritsumeikan daigaku jinbun kagaku kenkyūjo kiyō,* no. 1, 1953, pp. 166-82.
居延漢簡と漢代の財產稅　立命館大學人文科學研究所紀要

——. "Shindai no jujitsuden ni tsuite." *Ritsumeikan bungaku* 180 (1960): 325-41.
秦代の自實田について　立命館文學

——. "Shindai tochi seido no ichi kōsatsu: 'Myoden-taku' ni tsuite." *Ritsumeikan bungaku* 70 (1961): 23-39.
秦代土地制度の一考察—「名田宅」について

Itō Norio. "Daidenhō no ichi kōsatsu." *Shigaku zasshi* 69, no. 11 (1960): 61-78.
伊藤德男　代田法の一考察　史學雜誌

Kageyama Takeshi. "Kandai no keizaikan ni tsuite." In *Wada hakushi koki kinen Tōyōshi ronsō,* pp. 291-301. Tokyo: Kōdansha, 1960.
影山剛　漢代の經濟觀について

Kato Shigeshi. *Shina keizaishi kōshō.* vol. 1. Tokyo: Tōyō bunko, 1965.
加藤繁　支那經濟史考證

Kimura Masao. *Chūgoku kodai teikoku no keisei: Tokuni sono seiritsu no kiso jōken.* Tokyo: Fumai-do shōten, 1965.
木村正雄　中國古代帝國の形成—特にその成立の基礎條件

——. "Shin Kan jidai no denso to sono seikaku." *Rekishigaku kenkyū* 232 (1958): 1-10.
秦漢時代の田租とその性格　歴史學研究

——. "Zengo-Kan kōtaiki no nōmin hanran: Sono tenkai katei." *Tōkyō kyōiku daigaku bungaku kiyō* 16 (1967): 1-50.
前後漢交替期の農民叛亂–その展開過程　東京教育大學文學紀要

Makino Tatsumi. "Kandai ni okeru raikohō shinpo no imisurumono." *Tōzai gakujutsu kenkyūjō ronsō* 10 (1953): 1-16.
牧野巽　漢代における犂耕法進歩の意味するもの　東西學術研究所論叢

Masubuchi Tatsuo. *Chūgoku kodai no shakai to kokka: Shin Kan teikoku seiritsu katei no shakaiteki kenkyū.* Tokyo: Kōbundō, 1960.
增淵龍夫　中國古代の社會と國家—秦漢帝國成立過程の社會的研究

——. "Kandai gunkensei no chiikibetsuteki kōsatsu ni tsuite." In *Chūgoku kodaishi kenkyū,* pp. 291-325. Edited by Chūgoku kodaishi kenkyū-kai. Tokyo: Yoshikawa kōbun-kan, 1960.
漢代郡縣制の地域別的考察について　中國古代史研究

——. "Sen-Shin jidai no sanrin sōtaku to Shin no kōden." *Chūgoku kodai no shakai to kokka,* pp. 239-306. Tokyo: Kōbun-dō, 1960.
先秦時代の山林藪澤と秦の公田　中國古代の社會と國家

Moriya Mitsuo. *Chūgoku kodai no kazoku to kokka.* Kyoto: Kyōto daigaku bungakubunai Tōyōshi kenkyūkai, 1968.
守屋美都雄　中國古代の家族と國家

——. *Chūgoku kosaijiki no kenkyū: Shiryō fukugen o chūshin to shite.* Tokyo: Teikoku shoin, 1963.
中國古歲時記の研究—資料復元を中心として

Morohashi Tetsuji, ed. *Dai Kanwa jiten.* Tokyo: Taishukan, 1955-60.
諸橋轍次　大漢和辭典

Nagata Hidemasa. "Kandai nintōzei no hōkai katei: Tokuni sanpu o chūshin to shite." *Tōyōshi kenkyū* 18, no. 4 (1960): 50-71.
永田英正　漢代人頭稅の崩壞過程—特に算賦を中心として　東洋史研究

Nishijima Sadao, *Chūgoku keizaishi kenkyū*. Tōkyō daigaku bungakubu kenkyū hōkoku II. Tokyo: Tōkyō daigaku bungakubu, 1966.
西嶋定生　中國經濟史研究　東京大學文學部研究報告

——. *Chūgoku kodai teikoku no keisei to kōzō*. Tokyo: Tōkyō daigaku shuppan-kai, 1961.
中國古代帝國の形成と構造

——. "Kandai no tochi shoyūsei: tokuni myōden to senden ni tsuite." *Shigaku zasshi* 58, no. 1 (1949): 23-60.
漢代の土地所有制—特に名田と占田について　史學雜誌

——. *Shin Kan teikoku. Chūgoku no rekishi,* vol. 2. Tokyo: Kōdansha, 1974.
秦漢帝國　中國の歷史

Okazaki Fumio. "Shina kodai no tōbei inasaku kō." In *Nanbokuchō ni okeru shakai keizai seido,* pp. 18-57. Tokyo: Kōbun-dō shobō, 1935.
岡崎文夫　支那古代の稻米稻作考　南北朝における社會經濟制度

Okazaki Satoshi. "Kandai meiki deishō ni arawareta suiden suichi ni tsuite: Shisenshō shutsudohin o chūshin to shite." *Kōkogaku zasshi* 44, no. 2 (1958): 65-80.
岡崎敬　漢代明器泥象にあらわれた水田水池について—四川省出土品を中心として　考古學雜誌

Oshima Riichi. "Tonden to daiden." *Tōyōshi kenkyū* 14, nos. 1, 2 (1955): 1-22.
大島利一　屯田と代田

——. "Ryo-shi shunjū nōgyō shihen ni mieru nōgyō gijutsu ni tsuite." *Shirin* 49, no. 1 (1966): 57-73.
呂氏春秋農業四篇に見える農業技術について　史林

Sekino Takeshi. "Shin Rai-shi kō." *Tōyōbunka kenkyūjō kiyō* 19 (1959): 1-77.
關野雄　新耒耜考　東洋文化研究所紀要

Shinoda Sumeru. "Gokoku no kigen." *Shizen to bunka,* no. 2 (1951), pp. 37-69.
篠田統　五穀の起源　自然と文化

Shirakawa Shizuka. "Shikyō ni mieru nōjishi (jō)." *Ritsumeikan bungaku* 1, no. 138 (1956): 18-45.
白川靜　詩經に見える農事詩(上)

Tada Kensuke. "Kō Kei no ran ˋshi." *Tōyōshi kenkyū* 6, no. 4 (1968): 160-83.
多田狷介　侯景の亂前史

Tokunaga Shigeyasu and Naora Nobuo. *Manshū teikoku Kichirin-shō Kōgōton dai-ikkai hakkutsu hōkoku,* 1934.
德永重康，直良信夫　滿州帝國吉林省顧鄉屯第一回發掘報告

Utsunomiya Kiyoyoshi. "Kandai gōzoku ron." *Tōhōgakuhō* 23 (1962): 6-23.
宇都宮清吉　漢代豪族論　東方學報

——. *Kandai shakai keizaishi kenkyū.* Rev. ed. Tokyo: Kōbun-dō, 1967.
漢代社會經濟史研究

Yoneda Kenjirō. "Kandai denso sateihō kanken." *Shiga daigaku kyōiku gakubu kiyō,* vol. 17 (1967).
米田賢次郎　漢代田租查定法管見　滋賀大學教育學部紀要

——. "Kandai yōeki nissū ni kansuru ichi shiron." *Tōhōgakuhō* 27 (1967): 189-230.
漢代徭役日數に關する一試論

——. "O Shō 'Kakō-suidō' chū yori mitaru Gokan Kō-Wai no suitō-saku gijutsu ni tsuite." *Shirin* 38, no. 5 (1955): 349-66.
應劭「火耕水耨」注より見たる後漢江淮の水稻作技術について

——. "Seimin yōjutsu to ninen sanmō-saku." *Tōyōshi kenkyū* 17, no. 4 (1959): 407-430.
齊民要術と二年三毛作

Yoshida Torao. *Ryōkan sozei no kenkyū.* 2nd ed. Tokyo: Daian, 1966.
吉田虎雄　兩漢租稅の研究

Works in Western Languages

Almond, Gabriel A. "A Functional Approach to Comparative Politics." In *The Politics of Developing Areas,* edited by G. A. Almond and J. S. Coleman. Princeton, N.J.: Princeton University Press, 1960.

Balazs, Etienne. *Chinese Civilization and Bureaucracy,* translated by H. M. Wright. New Haven: Yale University Press, 1964.

Bielenstein, Hans. "The Restoration of the Han Dynasty." *Bulletin of the Museum of Far Eastern Antiquities* (Stockholm), no. 26 (1954), pp. 1-209.

Blau, Peter. "The Implication of Weber's Construct." In *Bureaucracy in Historical Perspective,* edited by Michael T. Dalby and Michael S. Werthman. Glenview, Ill.: Scott, Foresman & Co., 1971.

Bodde, Derk. *China's First Unifier: A Study of the Ch'in Dynasty as Seen in the Life of Li Ssu.* Sinica Leidensia, vol. 3. Leiden: E. J. Brill, 1938.

Boserup, Ester. *The Conditions of Agricultural Growth.* Chicago: Aldine Atherton, 1965.

Bunting, Brian T. *The Geography of Soil.* Chicago: Aldine, 1967.

Chang, Chun-shu. "The Colonization of the Ho-hsi Region—A Study of the Han Frontier System." Ph.D. dissertation, Harvard University, 1963.

——. "The Han Colonists and Their Settlements on the Chü-yen Frontier." *Tsing Hua Journal of Chinese Studies,* n.s. 5, no. 2 (1966): 154-269.

Chi, Ch'ao-ting. *Key Economic Areas in Chinese History.* 2nd ed. New York: Paragon, 1963.

Ch'ü, T'ung-tsu. *Han Social Structure.* Han Dynasty China, vol. 1. Seattle: University of Washington Press, 1972.

de Crespigny, Rafe, trans. *The Last of the Han.* Canberra: University of Australia, 1970.

Dubs, Homer H., trans. *History of the Former Han Dynasty, by Pan Ku.* 3 vols. Baltimore: Waverley Press, 1938, 1944, 1955.

——. trans. *The Works of Hsüntze.* London: Arthur Probsthain, 1928.

——. "Wang Mang and His Economic Reforms." *T'oung Pao* 35 (1946): 205-19.

Duyvendak, J. J. L., trans. *The Book of Lord Shang: A Classic of the Chinese School of Law.* London: Arthur Probsthain, 1928.

Eberhard, Wolfram. "Bemerkungen zu statistischen Angaben der Han-Zeit." *T'oung Pao* 36 (1940): 1-25.

——. *Conquerors and Rulers.* 2nd ed. Leiden: E. J. Brill, 1968.

Ebrey, Patricia B. "Estate and Family Management in the Later Han as Seen in the Monthly Instructions for the Four Peoples." *Journal of the Economic and Social History of the Orient* 17 (1974): 173-205.

Eisenstadt, S. N. *The Political Systems of Empires: The Rise and Fall of the Historical Bureaucratic Societies.* New York: The Free Press of Glencoe, 1963.

Gale, Esson M., trans. *Discourses on Salt and Iron: A Debate on State Control of Commerce and Industry in Ancient China.* Leiden: E. J. Brill, 1931.

Harrison, John A. *The Chinese Empire.* New York: Harcourt Brace Jovanovich, 1972.

Ho, Ping-ti. "The Loess and the Origin of Chinese Agriculture." *American Historical Review* 75, no. 1 (1969): 1-36.

——. *The Cradle of the East.* Chicago: University of Chicago Press, 1975.

Hoselitz, Bert F. "Economic Policy and Economic Development." In *The State and Economic Growth,* edited by Hugh G. J. Aitken, pp. 325-52. New York: Social Science Research Council, 1959.

——. "Levels of Economic Performance and Bureaucratic Structures." In *Bureaucracy and Political Development,* edited by Joseph La Palombara, pp. 168-98. Princeton, N.J.: Princeton University Press, 1963.

Hsu, Cho-yun. *Ancient China in Transition.* Stanford, Calif.: Stanford University Press, 1965; paperback, 1968.

——. "The Changing Relationship between Local Society and the Central Political Power in Former Han 206 B.C. - 8 A.D.," *Comparative Studies in Society and History,* vol. 7, no. 4 (1965).

Hulsewe, A. F. P. *Remnants of Han Law.* Vol. 1. Leiden: E. J. Brill, 1955.

Hymer, Stephen, and Stephen Resinick. "A Model of an Agrarian Economy with Nonagricultural Activities." *American Economic Review,* vol. 59, no. 4, pt. 1 (Sept. 1969).

Karlgren, Bernhard, trans. "The Book of Documents." *Bulletin of the Museum of Far Eastern Antiquities* (Stockholm), no. 22 (1950), pp. 1-81.

Legge, James, trans. *The Chinese Classics.* 5 vols. 1893-95. Hong Kong: Hong Kong University Press, 1970.

Levy, Howard. "Yellow Turban Religion and Rebellion at the End of Han." *Journal of the American Oriental Society* 76 (1956): 214-27.

Loewe, Michael. *Records of Han Administration.* London: Cambridge University Press, 1967.

——. "Measurement of Grain during the Han Period." *T'oung Pao,* series 2, 49:75-95.

Malthus, T. R. *First Essay on Population.* 1798. London: Royal Economic Society, 1926.

Rand McNally. *Illustrated Atlas of China.* Chicago: Rand McNally, 1972.

Mei, Y. P., trans. *The Ethical and Political Works of Motse.* London: Arthur Probsthain, 1929.

Needham, Joseph. *Science and Civilization in China,* vol. 3. 5 vols. Cambridge: Cambridge University Press, 1959.

Pokora, Timoteus, trans. *Hsin-lun (New Treatise) and other Writings by Huan T'an (43 B.C. - 28 A.D.).* Ann Arbor, Mich.: Center for Chinese Studies, University of Michigan, 1975.

Shen, T. H. *Agricultural Resources of China.* Ithaca: Cornell University Press, 1951.

Shih, Sheng-han. *On Fan-Sheng-chih-shu, an Agricultural Book of China Written by Fan Sheng-chih in the First Century B.C.* Peking: Science Press, 1959.

Skinner, G. William. "Chinese Peasants and the Closed Community: An Open and Shut Case." *Comparative Studies in Society and History,* vol. 12, no. 3 (1971).

——. "Marketing and Social Structure in Rural China," Part I. *Journal of Asian Studies* 24, no. 1 (1964): 3-43.

Smith, T. L., and P. E. Zopf. *Demography: Principles and Methods.* Philadelphia: Davis, 1970.

Swann, Nancy, trans. *Food and Money in Ancient China: The Earliest Economic History of China to A.D. 25.* Princeton, N.J.: Princeton University Press, 1950.

Vale, Joshua. "Irrigation of the Ch'eng Tu Plain." *Journal of the North China Branch of the Royal Asiatic Society,* vol. 33 (1899-1900).

Wang, Yü-ch'üan. "An Outline of the Central Government of the Former Han Dynasty." *Harvard Journal of Asiatic Studies* 12 (1949): 134-87.

Watson, Burton, trans. *Records of the Grand Historian of China: From the Shih Chi of Ssu-ma Ch'ien.* 2 vols. New York: Columbia University Press, 1961.

Whitney, Milton. *Soil and Civilization.* New York: Van Nostrand, 1925.

Wilbur, C. Martin. *Slavery in China during the Former Han Dynasty, 206 B.C. - A.D. 25.* 1943. New York: Russell and Russell, 1967.

Wrigley, E. A. *Population and History.* 1969. New York: McGraw-Hill, 1971.

Yang, Lien-sheng. *Studies in Chinese Institutional History.* Cambridge, Mass.: Harvard University Press, 1961.

Yü, Ying-shih. *Trade and Expansion in Han China.* Berkeley: University of California Press, 1967.

INDEX

Chung-shan commandery, 143
Chung-t'iao Mountains, 265

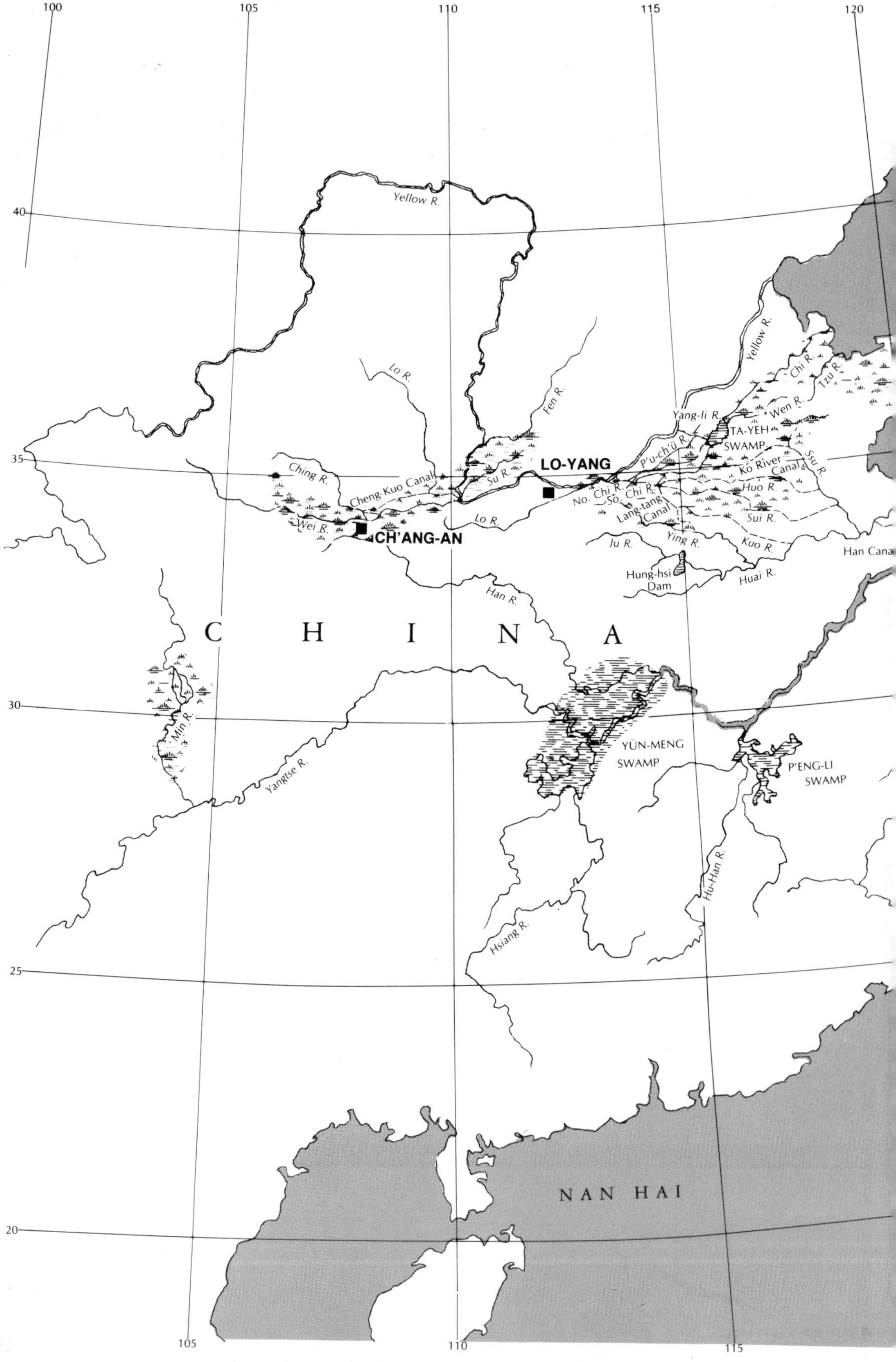

100
105
110
115
120
40
35
30
25
20
Yellow R.
Lo R.
Fen R.
Ching R.
Cheng-Kuo Canal
Su R.
LO-YANG
Wei R.
CH'ANG-AN
Lo R.
No. Chi R.
So. Chi R.
Lang-tang Canal
Yang-li R.
P'u-ch'ü R.
TA-YEH SWAMP
Yellow R.
Chi R.
Tzu R.
Wen R.
Ko River Canal
Ssu R.
Huo R.
Sui R.
Ying R.
Ju R.
Kuo R.
Hung-hsi Dam
Huai R.
Han Canal
Han R.
C H I N A
Min R.
Yangtse R.
YÜN-MENG SWAMP
P'ENG-LI SWAMP
Hu-Han R.
Hsiang R.
NAN HAI